International Workshop on Positron Studies of Defects 2008

(PSD 08)

Journal of Physics: Conference Series Volume 265

Prague, Czech Republic
1 – 5 September 2008

Editors:

Jan Kuriplach
Ivan Prochazka

ISBN: 978-1-61782-268-1
ISSN: 1742-6588

Printed from e-media with permission by:

Curran Associates, Inc.
57 Morehouse Lane
Red Hook, NY 12571

Printed by Curran Associates, Inc. (2011)

Institute of Physics
Dirac House, Temple Back
Bristol BS1 6BE UK

Phone: 44 1 17 929 7481
Fax: 44 1 17 920 0979

techtracking@iop.org

Additional copies of this publication are available from:

Curran Associates, Inc.
57 Morehouse Lane
Red Hook, NY 12571 USA
Phone: 845-758-0400
Fax: 845-758-2634
Email: curran@proceedings.com
Web: www.proceedings.com

TABLE OF CONTENTS

The 6th International Workshop on Positron Studies of Defects (PSD) took place in Prague, Czech Republic, from September 1 to 5, 2008. It was hosted by the Faculty of Mathematics and Physics of the Charles University in Prague. The PSD Workshop brought together positron scientists interested in studying various defects in mainly crystalline materials, and provided an opportunity to report on new results and achievements as well as on novel experimental and theoretical methods in this field. The workshop topics can be characterized as follows:

- Defect formation, migration, agglomeration and annealing
- Momentum distribution studies of defects: Doppler broadening, angular correlation of annihilation radiation (ACAR)
- Slow positron beam studies of defects at surface and near surface regions
- Theoretical calculations and simulations of momentum distributions, positron lifetimes and other characteristics for defects
- Defects in unusual materials: quasicrystals, nanostructures
- Advances in positron experimental methods applicable to defect studies
- Complementary experimental methods in defect studies (TEM, XRD, AP, SANS, DLTS, PL and others)
- Industrial applications of positron defect studies

The first PSD workshop was organized in Wernigerode, Germany in 1987. The next four workshops were held in Halle, Germany (1994), Hamilton, ON, Canada (1999), Sendai, Japan (2002) and Pullman, WA, USA (2004) under the name Positron Studies of Semiconductor Defects (PSSD). As studying defects in metals is once again gaining importance – as is also documented in these Proceedings – the name of the Workshop in Prague was changed to the original version PSD. The PSD workshops will be organized every three years and the next one will be held in Delft, The Netherlands at the turn of August and September, 2011.

We would like to express our gratitude to all of the workshop participants for their presentations and contributions to discussions, which made the PSD Workshop a successful scientific event. In total 65 scientists and students from 15 countries took part in the PSD Workshop. The workshop programme comprised 24 invited lectures, 19 contributed talks and 22 posters. Thirty contributions are included in these Proceedings, covering various aspects of positron defect studies. In particular, proceedings papers are divided into five categories: defects in semiconductors, defects in metals, nanostructures, larger free volumes and experimental apparatus including data evaluation. For the previous PSD/PSSD workshops proceedings were not published on a regular basis and, hopefully, these Proceedings will be the start of a new tradition.

Jan Kuriplach
Ivan Procházka
Editors

Journal of Physics: Conference Series **265** (2011) 012001 doi:10.1088/1742-6596/265/1/012001

Activation energies for vacancy migration, clustering and annealing in silicon

P G Coleman

Department of Physics, University of Bath, Bath BA2 7AY, UK

E-mail: p.g.coleman@bath.ac.uk

Abstract. A series of measurements have been performed at the University of Bath to study the evolution of vacancy-type structures in silicon. Isothermal annealing performed during positron beam-based Doppler broadening measurements have yielded activation energies for vacancy cluster formation and evaporation in silicon of approximately 2.5 and 3.7 eV, respectively. The clusters, which could predominantly be the stable hexavacancy, appear to form between 400-500°C, and anneal at ~ 600°C. A similar technique applied to low-temperature *in situ* measurements have yielded the migration energies for the silicon monovacancy and interstitial (of ~ 0.5 and 0.08 eV, respectively). Interesting observations of positronium formation at the surface of the samples studied during isothermal annealing are presented.

1. Introduction

For many years researchers – including many using positron beam spectroscopy - have studied the thermal evolution of vacancy-type defects in ion-implanted silicon [1-3]. There appears to be a consensus (a) that the defects left in room-temperature silicon are predominantly divacancies (V_2), (b) that at 200-300°C V_2 become mobile and migrate to sinks or – if their concentration is high enough – agglomerate to form small clusters, (c) that at 400-500°C even larger clusters are formed, and (d) that at ≥ 600°C these large clusters are annealed. There have been observed differences between these behaviours in different types of silicon (e.g., float zone (FZ), Czochralski (Cz), highly-doped [4,5]), and some debate concerning the various stages of evolution, especially stage (b) – where positron annihilation spectroscopies (both lifetime and Doppler broadening) have in some studies implied that the V_2-like response has survived to temperatures well above that accepted for V_2 migration, while other techniques applied in parallel have indicated their disappearance [6].

There is also the question of possible differences between the evolution of V_2 in the bulk and near the surface of silicon, and the dependence on V_2 concentration. Although lifetime spectroscopy is held to yield greater insights into the sizes and relative concentrations of open-volume point defects in silicon, we shall consider in this paper the application of single-detector Doppler broadening spectroscopy (formally called variable-energy positron annihilation spectroscopy, VEPAS, and sometimes variable-energy Doppler broadening of annihilation radiation, VEDBAR) to identify various small vacancy clusters and, as it continues to provide the fastest method for following *changes* in defect structure during, for example, annealing, to deduce activation energies for various processes involved in vacancy evolution.

2. Experimental details

All the measurements described here were performed on the magnetic-transport positron beam at the University of Bath, full details of which can be found elsewhere [7]. Positrons from a ^{22}Na source are moderated by annealed tungsten meshes and transported magnetically to the sample, mounted on 0.1 mm-diameter tungsten wires, via an ***E***×***B*** velocity filter through which unmoderated positrons are not transmitted and an accelerator section which determines the final incident positron energy. Because the electrostatic accelerating field (potential difference V) and guiding magnetic field are not perfectly aligned the positron beam moves laterally as V is changed; the beam is brought back to the same position at the sample target by two orthogonal trim coils. The beam is positioned using a small washer-shaped dummy sample and observing the image created by those positrons which pass by the target and proceed along a beam dump line to a dual electron multiplier array mounted in front of a phosphor screen. The energy spectra of annihilation gamma photons from the sample are measured by a Ge detector, and the Doppler-broadened linewidth parameters S and W [8] are measured for each incident positron energy. Any positron which misses the sample – for example, if the sample is slightly smaller than the 8mm-diameter beam - decay on at the end of the beam dump, out of sight of the gamma detector. The probability of detection of annihilation photons from backscattered positrons is minimized by the use of a large sample chamber. Beam energy and position, together with all aspects of data collection, are computer-controlled. Individual run times are chosen so that ~ 10^6 events are recorded in each annihilation line, and are ~10^3 s.

All Si samples (except those discussed in section 4.2) were ion-implanted at room temperature at the Surrey Ion Beam Centre (Nodus Laboratory, under the direction of Professor Russell Gwilliam). Typical sample sizes were ~ 10×10mm, although sizes ranging from 5×5 to 20×20mm have been used.

3. Isochronal annealing studies

Almost all previous work using positrons has been in this category. With the exception of He-implanted Si, where the ramp annealing rate is important in void growth mechanisms [9], samples are usually held at the chosen temperature for a set time and the ramp rate is not considered important. In the first example we consider FZ Si samples implanted with 50 keV Si^+ ions at a dose of 5×10^{13} cm^{-2} and annealed for 30s at temperatures between 250 and 850°C. The raw data (S parameter vs incident positron energy E) are shown in Fig. 1 [10], normalised to the value for bulk Si. The peak in $S(E)$ represents the non-saturated VEPAS response to divacancies (which have a characteristic S value of ~1.04). The data show that there is no measurable change in VEPAS response for samples annealed to 450°C, with a rapid change at 550°C and essentially complete annealing at 650°C and above. This result was confirmed by in-situ annealing experiments by measuring the fraction of implanted positrons able to diffuse to the surface as the temperature was raised (via measurement of the amount of positronium formed at the surface). The result is surprising at first sight because V_2 should be

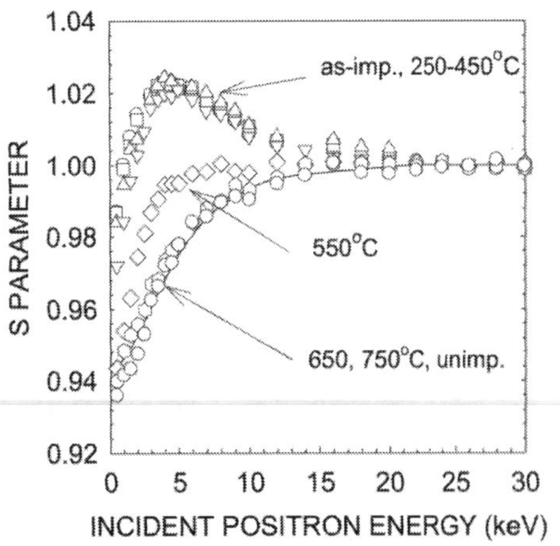

Figure 1. $S(E)$ for FZ Si implanted with 50 keV Si^+ ions at 5×10^{13}cm^{-2} and annealed at temperatures from 250 to 750°C [10].

mobile at ~ 300°C and annealing should therefore progress at this temperature. As for clustering (agglomeration), one might still expect to see a change in $S(E)$ as this progresses, for the following reasons.

The conventional model describing VEPAS response to clustering states that (a) the S value characteristic of V_n (a cluster of n vacancies) increases with n, and the specific positron trapping rate ν_n for V_n is $\sim n\nu_1$, where $\nu_1 \approx 3.4 \times 10^{14}\ s^{-1}$. Let us consider here the dependence of S on n for Si. Here we attempt to apply the dependence on n of the positron lifetime results of Staab *et al* [11], assuming an asymptotic value of 499 ps (the spin-averaged lifetime of positronium) with the calculations of S by Hakala *et al* [12] convoluted with our experimental detector resolution. This is reasonably straightforward for low n, but extrapolation to large clusters is difficult and the fit in figure 2 should be taken with caution. The asymptotic value of S at large n here is suggested to be 1.13, close to the largest values measured directly (see, for example, ref [8]); however, a value somewhat higher than this may be possible.

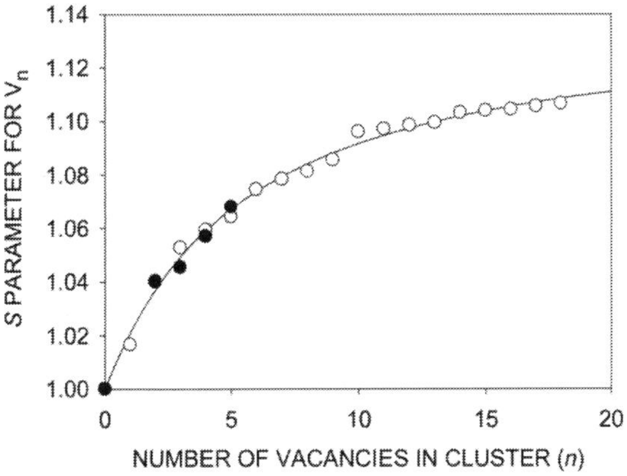

Figure 2. S parameter characteristic of vacancy clusters V_n vs n. Black circles: convoluted calculations of Hakala *et al* [12]. Open circles: normalised positron lifetime results of Staab *et al* [11]. Asymptotic S value (large n) = 1.13.

The total positron trapping rate is νC, where C is the defect concentration per atom. Therefore, if *all* the V_2 agglomerate into clusters V_n, the trapping rate is $\nu_n C_n = (n/2)\nu_2\ (2/n)C_2 = \nu_2 C_2$ – i.e., it remains unchanged; however, S_n is larger than S_2 and so the measured S value should increase. This is the case even if the simple proportionality between ν_n and n is too crude an approximation.

Let us consider four possible models which might explain why this is not the case. (a) Only a fraction of the V_2 cluster and the rest diffuse to sinks (particularly the surface). For the measured S value to remain unchanged then the increase in S_n resulting from clustering would have to be balanced exactly by the decrease in C. This is possible but an exact balance is nevertheless unlikely. (b) The V_2 are pinned by impurities and become mobile at higher temperatures than the 300°C expected. This model is also not very convincing because we are using FZ Si and the impurity concentrations are expected to be too low to pin all the V_2. (c) The V_2 do cluster, say to V_4, but the S value does not change measurably; this is the scenario invoke by Poirier et al. [6] to explain a similar observation after measuring no significant change in positron lifetime on annealing. This is contrary to the $S(n)$ dependence suggested by figure 2; if it were to be the case then the clusters would have to have a structure different from that assumed in the calculations shown in the figure – as suggested by Poirier et al. - for example, a linear structure whose S value was similar to that for V_2. Another possibility, earlier proposed by Poirier *et al* [13] is that V_2 for 'virtual' clusters – i.e., they form loose connections without close bonding, so that they become less mobile but positrons are still trapped in a defect of the size of V_2 and thus the annihilation parameters remain unchanged. (d) The initial defects are not V_2 but are already an array of small clusters which are not mobile below 500°C. While possible, this is unlikely as there have been a large number of positron measurements indicating that the defects remaining in ion-implanted Si at room temperature are overwhelmingly V_2.

In order to remove the possible complications associated with the close proximity of the surface, measurements on 4 MeV Si-implanted FZ Si have been performed [14]. Here the peak of the damage created by the ion implantation is close to 2.5 μm below the surface. Figures 3 and 4 show *S*(*E*) for implant doses between 10^{12} and 10^{14} cm^{-2} before an after 30 mins one-shot annealing to 600°C, respectively. Comparison of the two figures leads to the conclusion that the V_2 formed at depths from the surface to ~2 μm are annealed below 600°C, but that between 2-3 μm a layer of small clusters is formed which survives to higher temperatures. By fitting the data shown in figure 3, and those obtained by employing an etch-and-measure technique to bring the damaged layer progressively towards the surface, with the standard code VEPFIT [15], the average size of the defects in this buried layer is suggested to be 3.5 – i.e., most likely dominated by tetravacancies. It was also found that progressive annealing up to 600°C removed all VEPAS response to vacancy-type damage, confirming that cluster formation is critically dependent upon annealing history.

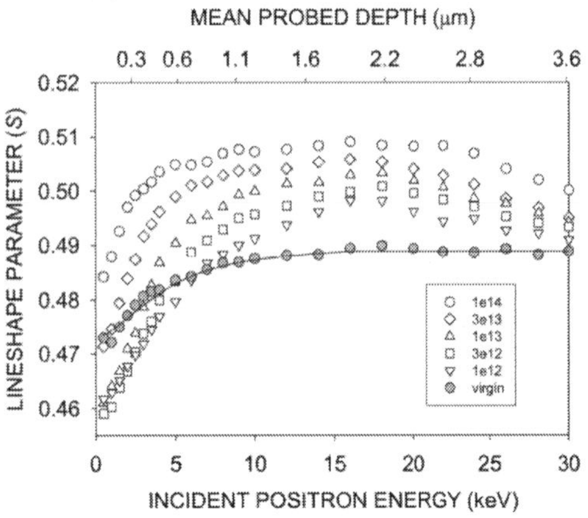

Figure 3. *S*(*E*) for 4MeV Si ions in Si at doses from 10^{12}-10^{14} cm^{-2} (1e12-1e14) [14].

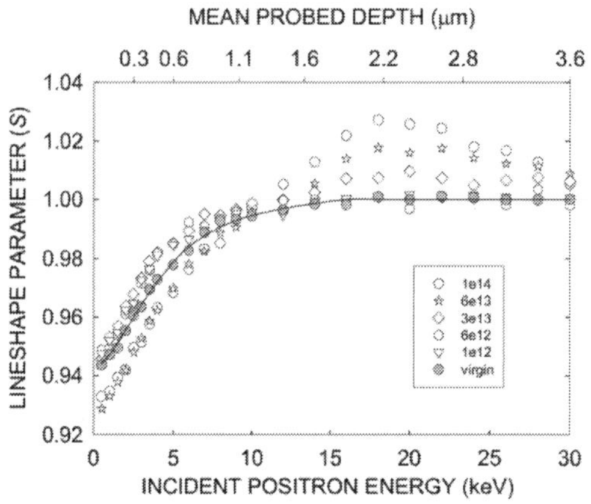

Figure 4. As figure 3, after one-shot annealing for 30 mins at 600°C [14].

The explanation for the formation of small clusters in the 2-3 μm region is most likely linked to the initial distribution of vacancies following implantation. A simulation for 4 MeV Si^+ in Si using the code SRIM [16] is shown in figure 5. The concentration of vacancies in the 2-3 μm region is almost an order of magnitude higher than in the tail stretching to the surface; it is therefore considerably more likely that V_2 in the former region find each other on migrating at high temperatures than in the latter –

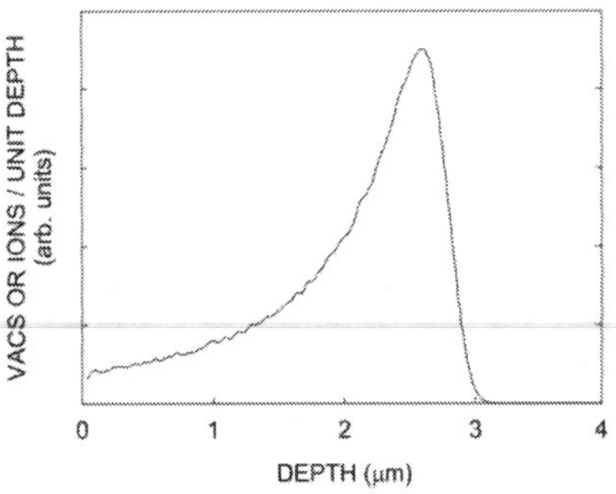

Figure 5. SRIM simulation of vacancy damage in Si implanted with 4MeV Si^+ ions.

which, additionally, is closer to the surface sink. On extracting a defect concentration in the buried layer from data fitting, one obtains a super-linear dependence on ion dose which is consistent with the probability of agglomeration depending on the relative magnitudes of the mean divacancy separation and the diffusion length of the migrating V_2.

4. Isothermal annealing studies

4.1. Divacancy evolution

The first isothermal annealing studies in the author's laboratory aimed at extracting more information on the thermal evolution of vacancy defects in Si were performed on 2 MeV Si-implanted Cz Si [17]. A high dose of 10^{15} cm^{-2} ensured that the V_2 concentration was high enough to lead to agglomeration, and the 2 MeV ion energy meant that the damage extended to depths where positron (and V_2) diffusion to the surface might not present significant interpretational problems when analyzing the data.

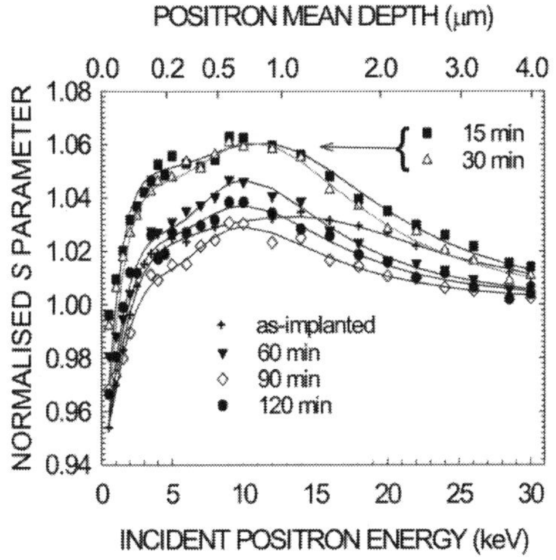

Figure 6. Normalized *S*(*E*) energy for Cz Si implanted with 2 MeV, 10^{15} Si^+ cm^{-2} after annealing at 600°C for different time intervals. The solid lines are fits using VEPFIT. Adapted from [17].

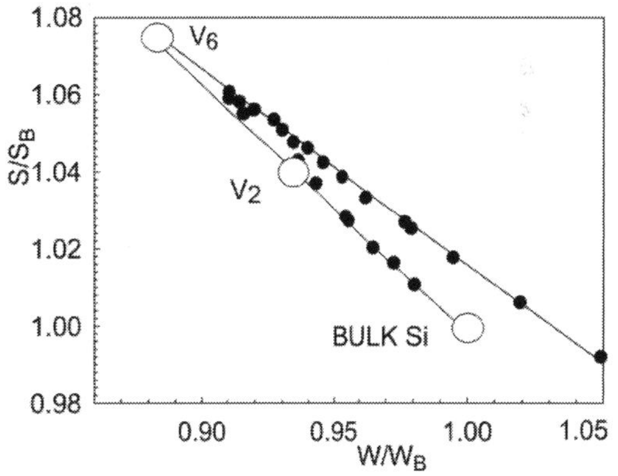

Figure 7. *S*-*W* plots for samples implanted with 2 MeV Si ions at 10^{15} cm^{-2} dose after annealing at 600°C for 30 min. The circles are the dominant annihilation sites for positrons. Adapted from ref [17].

Examples of the data taken are shown in figure 6. *S*(*E*) for the as-implanted sample (crosses in figure 6) show a broad peak with a maximum response of ~ 1.03 at *E* ~ 15 keV (equivalent to a mean positron implantation depth of ~ 1.3 μm). On annealing at 600°C for times up to 30 mins *S* increases remarkably at all *E*, pointing to the formation of clusters, particularly at depths ~ 750 nm, where there is a broad peak in the response. On annealing for longer times the response decreases and the shape of *S*(*E*) becomes progressively more peaked at ~ 10 keV (~ 750 nm), a response which survives annealing for 2h.

Data such as those in figure 6 are routinely fitted with the code VEPFIT, which gives values for *S* and for the positron diffusion length (*L*) in each layer (the *S* value would only be measured if at some energy *E* all the positrons were annihilated in the layer; this is not generally the case). A graphical equivalent of VEPFIT is the parameter-parameter map – illustrated by the *S*-*W* plot in figure 7. Here the parameters *S* and *W* for every value of *E* are plotted against each other. On such a graph can be plotted points which are characteristic of each layer – and additionally of each 'pure' annihilation state (sample surface, bulk Si, V_2, etc). In figure 7 the surface state is in the bottom right-hand corner of the graph, and the S,W points for bulk Si and annihilations in V_2 are represented by circles. If the linear trends suggested by the data points are extrapolated as shown in figure 7, then they meet at the S,W point characteristic of a defected layer, even if experimentally positrons do not all suffer annihilation in the layer. In this case the straight lines meet at $S \sim 1.075$, which is the value predicted in figure 2 for the hexavacancy V_6. The data are thus consistent with saturation trapping of those positrons which are annihilated in the layer centred at ~ 750 nm in defects of the size of V_6.

Let us now turn to the time dependence of the evolution seen in figure 6. In these experiments the concentration of defects (C_D) was first deduced by remembering that the S value for a layer in which defect trapping sites exist with a characteristic $S = S_D$ is a linear combination of S_D and S_B, where S_B is the S value for bulk Si, which is normalised to unity here: $S = fS_D + (1-f)$. f, the fraction of positrons annihilated in the layer from the trapped state, is linked to C_D via $f = \nu C_D/(\lambda+\nu C_D)$, where ν is the specific trapping rate for the defects in question and λ is the positron annihilation rate in bulk defect-free Si. By inserting this expression for f into that for S one arrives at

$$C_D = \lambda(S - 1)/[\nu(S_D - S)] \quad . \tag{1}$$

At a given temperature T one can now measure the time $t_{1/2}$ taken for C_D to decrease to 50% of its room-temperature value. If we assume that at a temperature T C_D decreases as

$$C_D = C_0 \exp(-E_A/k_BT) \quad , \tag{2}$$

where C_0 is the defect concentration at time zero and E_A is the activation energy for the annealing process, then the slope of an Arrhenius plot of $\ln(t_{1/2})$ vs $(1/T)$ should yield E_A. In the case of the data shown in figure 6 ($T = 873$K), $t_{1/2} \approx 39$ min (see figure 8).

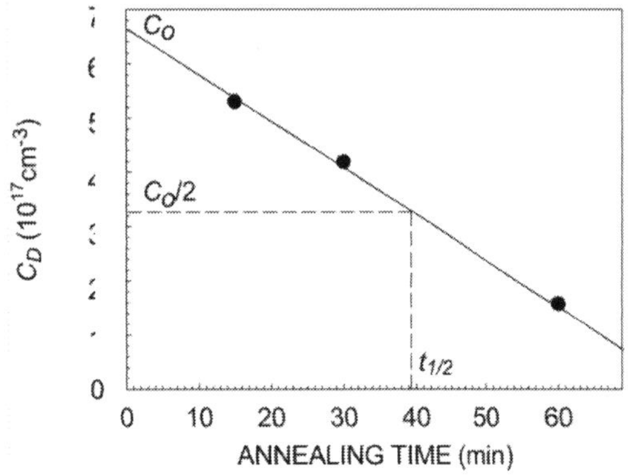

Figure 8. Defect concentration C_D (cm^{-3}) deduced from the data in figure 6 vs annealing time at 600°C. The vacancies are assumed to be all V_6 [17].

Unfortunately, there were not enough data points to enable the unambiguous extraction of E_A for the annealing of the vacancy clusters (assumed to be V_6) formed in this experiment. An upper limit of 7.5 min could be assigned to annealing at 700°C, and the annealing at 650°C was complicated by the possible presence of a mixture of V_6 and smaller clusters, so that the analysis could only suggest a minimum value for E_A of 1.2 eV. This is considerable lower than the figure of ~ 3.5 eV postulated for the binding of V to V_6 [11,18-20] and of the same order as E_A for V_2 migration [21]. A further series of experiments was therefore performed to endeavour to resolve these problems.

A set of samples of Cz and epitaxially-grown Si were implanted with 50 keV Si^+ ions at a dose of 5 × 10^{13} cm^{-2}. The samples were then isothermally annealed *in situ* (i.e., on the sample holder at the target position in the positron beam system) at temperatures between 300 and 640°C for times of up to ~ 100 h, and the value of S at a positron energy equivalent to a probed depth similar to that predicted by SRIM to be the depth at which the vacancy-type damage was a maximum (i.e. 3.5 keV, equivalent to 120 nm) recorded every 600s (reduced to 300s at the highest temperatures) [22]. The *S*(*E*) plots shown in figure 9 were taken after *S*(3.5 keV) had stopped changing in the epi-Si samples held at 470 and 580°C; the samples were cooled to room temperature before the data were recorded. Note that the samples were in this case not HF-etched to remove surface oxide, to increase the absolute size of the change in *S*; as a test, other measurements were taken after etching and no effect on the final results was noted.

Examples of *S*(3.5 keV) vs time plots are shown in figure 10. Data were also taken at $E = 1$ and 24

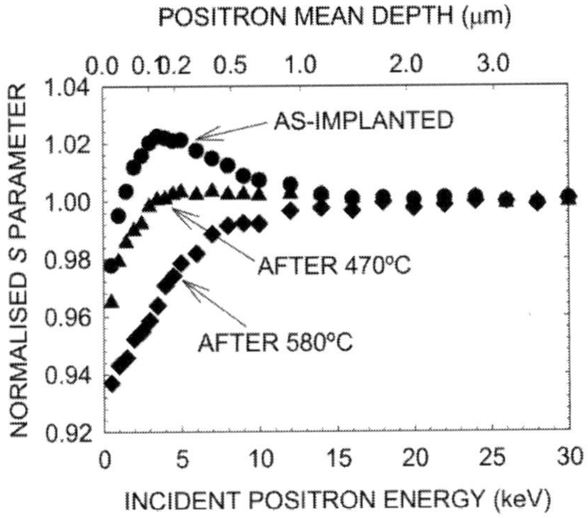

Figure 9. *S*(*E*) for epi-Si implanted with 50keV Si^+ ions at a dose of 5×10^{13} cm^{-2}, as-implanted and after annealing at 470 and 580°C. Adapted from [23].

keV to check for physical changes near the surface and systematic changes (e.g. in the detector/amplifier performance). The three plots in figure 10 illustrate the three temperature regions of interest. At temperatures below ~ 400°C essentially no change is seen in the value of *S*, in line with earlier observations. Between ~ 400 and 500 °C *S* falls to an intermediate level (here coincidentally ~1 – see figure 9). Between ~ 500-600 °C the defects are annealed away and the positron response falls to that for unimplanted Si (at 3.5 keV this is ~ 0.97 here).

The fall in *S* at higher temperatures reflects the change in the defect structure; at 470°C in figure 10,

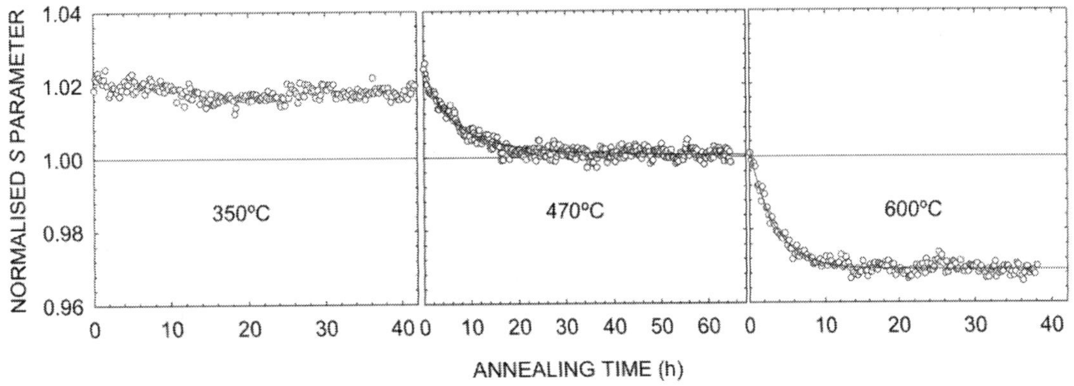

Figure 10. *S*(3.5 keV) vs annealing time for implanted epi Si at 350, 470 and 600°C. The solid lines through the points in are exponential fits. *S* values are normalised to that for bulk Si (i.e., unity – as shown in the graphs). Adapted from [23].

for example, the peak *S* falls to an equilibrium value characteristic of an intermediate state. The fact that *S* falls (i.e., as opposed to rises) indicates that any changes have involved a decrease in the trapping rate, or a sublinear increase in defect *S* value with defect size, or the migration of some V_n to sinks, or a combination of these processes. In principle it does not matter whether *S* increases or decreases; it is the ability to extract a time rate of change of *S* which is important here. Thus, if plots of *S*(*t*) such as those shown in figure 10 are fitted to an exponential $S_0\exp(-\lambda t)$, the decay constant λ represents the change which is occurring, and can be used in Arrhenius plots to evaluate the activation energy for the process. Figure 11 shows such plots for the low and high temperature regions (400-500°C and 560-640°C) for both Cz Si and epi-Si.

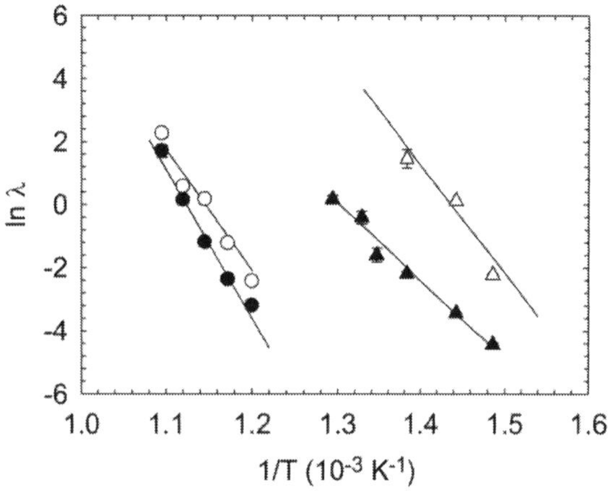

Figure 11. Arrhenius plots for epi and Cz-Si samples. The slopes of the fitted lines give activation energies for two vacancy evolution processes (lower temperatures, triangles and higher temperatures, circles). Solid symbols – epi-Si: open symbols – Cz Si. From [23].

The activation energies E_A obtained the gradients of the lines fitted to the data in figure 11 are 2.1 (2) and 2.7 (7) eV for the low-temperature process, and 3.9 (3) and 3.6 (3) eV for the high-temperature process, the first values being for epi-Si and the second for Cz Si.

4.2. Monovacancy evolution

Studying monovacancies (V_1) in Si with VEPAS presents the practical problem that V_1 are mobile below room temperature, which is the reason why room-temperature samples contain predominantly V_2 (with the possibility of small clusters for samples implanted with ions at high doses). In order to observe the formation and annealing of V_1 the Si sample was mounted in the Bath positron beam on a copper cold finger attached to the head of a closed-cycle helium cryostat [23]. The 25mm-diameter copper rod was broken by a 0.3mm-thick sapphire disk, affording good thermal and poor electrical conductivity. The Si sample was cooled to below 20K and was held at -20kV while low-density He gas was admitted to the sample chamber; He ions were accelerated into the sample from the ensuing discharge. A measure of *S*(*E*) after implantation showed a peak between 2 and 4 keV (mean depth ~ 130 nm, consistent with 20 keV He implantation), close to the saturation value of 1.027 (assigned to V_1).

Two types of Si – low-impurity FZ Si and highly As-doped (n^+) Si – were studied in these experiments [24]. *S*(3.5 keV) was, as in the V_2 experiments in subsection 4.1, monitored as a function of time for periods of up to 6 days at a range of constant temperatures from 200K to over 300K. Figure 12 shows examples of raw data for FZ and n^+ Si. In n^+ Si two processes in different temperature ranges were observed directly, whereas in FZ Si both processes could proceed at the same temperature.

The Arrhenius plots obtained by plotted the natural logarithm of the decay constant representing the time rate of change in *S* (shown to be proportional to that for the change in defect concentration) are shown in figure 13, therefore consist of two straight lines for n^+ Si and a nonlinear plot for FZ which tends to one slope at low temperatures and a higher slope at higher temperatures, as the relative probability of the two processes changes. Note that the uncertainties attached to the measurements of the first stage of annealing of defects in n^+ Si were too high for data to be included in figure 13. The

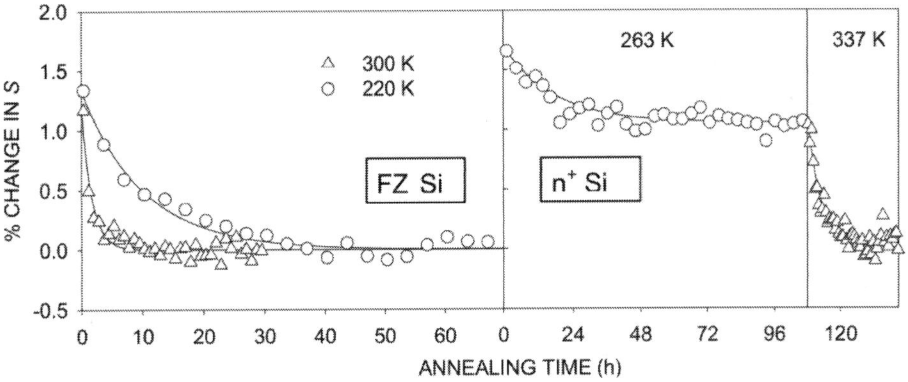

Figure 12. Percentage difference between S parameter at E = 3.5 keV for low-temperature He-implanted FZ and n^+ Si and its asymptotic value as a function of time at four temperatures. Adapted from [24].

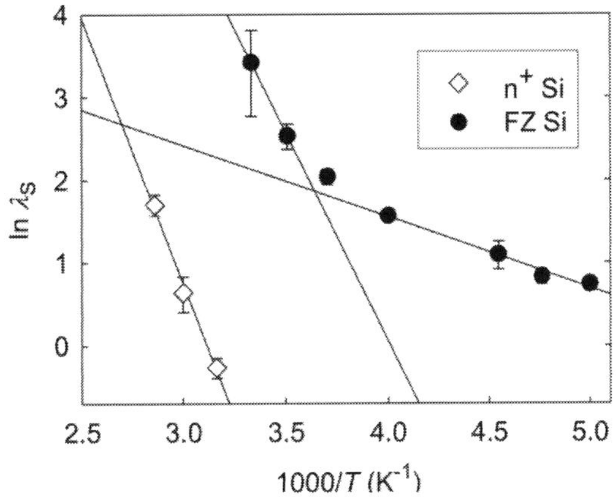

Figure 13. Arrhenius plots of lnλ_S vs 1000/T for low-temperature He-implanted FZ and n^+ Si, where λ_S is the decay constant from exponential fits of data such as those in figure 12, and T is the absolute temperature in K [24].

activation energies derived from the slopes of the fit lines in figure 13 were 0.078(7) and 0.46 (28) eV for FZ Si, and 0.59 (6) eV for n^+ Si. Let us now proceed to discuss the activation energy results in the following section.

5. Activation energies

The Arrhenius plots in figures 11 and 13 yield activation energies for processes involving the evolution of vacancy-type defect structures in different types of silicon and over different temperature ranges. The challenge of identifying the most likely candidates for these processes is not simple, and we can only at this stage offer what appears to be the most likely. The most straightforward interpretation is that for the high-temperature process in figure 11. The measured activation energy of about 3.75 eV for both epi and Cz Si describes the disappearance of the VEPAS response to defects in the samples, and lies in the range of published values for the binding energy of vacancies to small clusters [11,18-20], particularly the 3.76 eV calculated by Hastings et al for the hexavacancy. This high-temperature process is therefore likely to be the evaporation of clusters – the hexavacancy is the most probably candidate here – via the removal of successive vacancies which are then able readily to migrate to sinks.

The lower-temperature process seen in both Cz and epi-Si, with activation energies of ~ 2.5 eV, is more difficult to interpret. The transition from V_2 to the clusters which evaporate at higher temperatures is the most likely explanation, but this does not easily sit with the accepted knowledge that V_2 migrate at or below 300°C, or that the migration energy for V_2 is 1.3 eV [21]. One therefore has to introduce the possibility that V_2 may first undergo some kind of transition into small clusters - say V_4 – which may exists in an unstable chain-like form as proposed by Poirier et al. [6], with a negligible change in the VEPAS response. At the temperatures studied this agglomeration may happen very rapidly. 2.5 eV may then be a measure of the migration energy of V_4. The agglomeration occurring between 400-500°C proceeds measurably more quickly in Cz Si than in epi-Si at the same temperatures; this could be due to the action of impurities as seeds for cluster formation.

The plots in figure 13 also suggest the existence of two steps in the annealing process for (neutral) V_1 in Si, with activation energies of about 0.08 and 0.5 eV. The fact that the two processes appear to dominate in different temperature ranges suggests that the former involves a higher number of steps than the second. Comparison with well-accepted values [25] suggests that the high-temperature process is associated with neutral V_1 migration. This energy has been measured previously by EPR and other spectroscopies. However, more importantly, the other process (with the 0.08 eV activation energy) is attributed to the migration of Si interstitials. This energy is very difficult to measure and has been the subject of much debate over recent decades; the VEPAS value agrees within uncertainties with that measured by Hallén et al. [26] and may go some way towards ending the debate in this area.

The survival of a fraction of vacancies formed in highly As-doped n^+ Si (figure 12) is consistent with the scenario in which a significant fraction of the V_1 form complexes with the dopant impurities [4] and thus consequently become considerably less mobile.

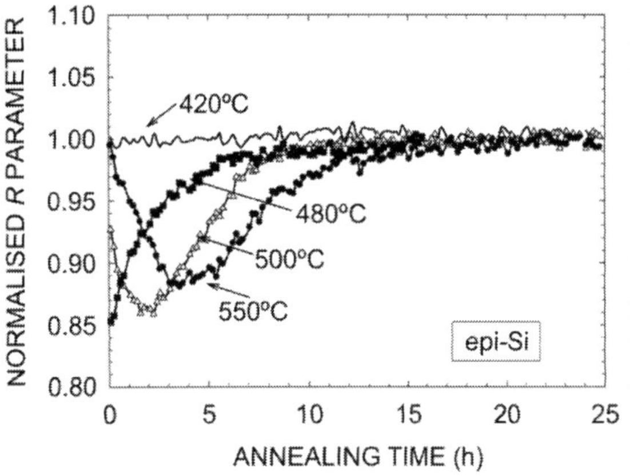

Figure 14. Normalised R parameter as a function of annealing time for epi-Si samples at four temperatures. The sample at 420°C was unetched prior to the measurements.

6. Loose ends

A number of unexpected and interesting observations were made during the isothermal annealing studies described in section 4. Perhaps the most intriguing was the behaviour of the peak-to-valley ratio, defined as the ratio of the total counts in the annihilation line (Ge detector photopeak centred at 511 keV) divided by those in the ('valley') region immediately to the left of the photopeak. This ratio has long been used as a measure of the amount of positronium (Ps) formation at the surface of a metal or semiconductor sample, enhanced at elevated temperatures by the desorption of positrons trapped in the surface as thermal Ps [27]. The peak-to-valley ratio, which we shall here call R, was measured automatically along with the *S* and *W* lineshape parameters; as the system was set up for measurement of these parameters the extent of the valley region was smaller than would normally be used – however, *R* could be measured with relatively high precision as a function of time at each temperature.

Figure 14 shows an example of the behaviour of *R* in four samples of implanted Si during isothermal annealing. The values are normalised to that for bulk Si (measured directly by implanting room-temperature samples with 25 keV positrons, so that a negligible fraction of the implanted positrons are able to diffuse back to the surface to form Ps). The fraction of positrons forming Ps at the surface of the samples is clearly varying significantly with time, increasing as *R* decreases and tending to zero for each sample at long times.; the exception is the sample annealed at 420°C, which had not been etched using HF prior to the measurements. It is not unexpected that the Ps fraction greatly depends on the presence or otherwise of native oxide on the surface of the Si; what is not currently understood by the author is the cause of the variation of Ps formation with time at the various temperatures. One notes that the variation does not show a clear trend with annealing temperature, and so must critically depend on the nature of the individual surfaces of each sample. The presence of native oxide seems to prevent, or at least greatly reduce, the variation seen. Oxidation may hold the key, but it is yet to be confirmed that oxide growth, or the state of the oxide-Si interface, may vary in such a way to explain the observed data. It is also possible that heating is creating changes in formation potential for Ps at the surface. These observations may have a prosaic or an exotic explanation; in either case they warrant further investigation.

7. Conclusions

This paper has tried to summarise the current situation with regard to the study of vacancy-type defect evolution at the University of Bath. There are many pitfalls and potential problems with making measurements at high temperatures, both practical and interpretational, and the conclusions summarised above must be taken in the spirit of work in progress. However, the extraction of activation energies has not been widely performed with beam-based positron spectroscopies and it is to be hoped that this work will encourage more researchers to consider applying this method to gain information on defect evolution not only in semiconductors but also in metals, alloys and other solid materials. Caution must always be invoked in these studies; for example, positron detrapping from defects could affect the results (in a recent paper Makkonen and Puska [28] predict an extremely small binding energy – perhaps a few tens of meV – for positrons in neutral Si monovacancies). There may be much more scope for working at low temperatures, perhaps by incorporating a positron beam into an ion implanter and performing measurements with the samples in situ; and the unambiguous identification of vacancy cluster size may require the application of beam-based positron lifetime spectroscopy (VEPALS). It is clear that there is still much interesting work to be done in this area.

Acknowledgements

The author is grateful to the many people who have contributed in various important ways to the research reported in this paper. Drs Dana Abdulmalik, Paul Burrows, Ruth Mason, Ruth Harding and Ilham Al-Qaradawi, and Prof. Gordon Davies, were directly involved in taking and analysing data and in the many discussions which followed, and Nathan Potter worked on void formation in He-implanted Si. I am especially grateful to Dr Andy Knights for many stimulating discussions and ideas. Most of the ion implantion of samples was performed at the Surrey Ion Beam Centre. Work at the positron laboratory of the University of Bath was supported financially by EPSRC, UK throughout the time of the work reported herein, for which the author is grateful.

References

[1] Simpson P J, Vos M, Mitchell I V, Wu C and Schultz P J 1991 *Phys. Rev.* B **44** 12180
[2] Fujinami M, Tsuge A and Tanake K 1996 *J. Appl. Phys.* **79** 9017
[3] Nielsen B, Holland O W, Leung T C and Lynn K G 1993 *J. Appl. Phys.* **74** 1636
[4] Ranki V, Nissilä J and Saarinen K 2002 *Phys. Rev. Lett.* **88** 105506

[5] Dannefaer S, Avalos V, Kerr D, Poirier R, Shmarovoz V and Zhang S H 2006 *Phys. Rev.* B **73** 115202
[6] Poirier R, Avalos V, Dannefaer S, Schiettekatte F and Roorda S 2003 *Physica* B **340-2** 609
[7] Chilton N B and Coleman P G 1995 *Meas. Sci. Technol.* **6** 53
[8] van Veen A, Schut H and Mijnarends P E 2000 *Positron Beams and their Applications* ed P G Coleman (World Scientific: Singapore) p 191
[9] Ruffell S, Simpson P J, Knights A P 2007 *J.Phys.: Condens. Matter* **19** 466202
[10] Coleman P G, Malik F and Knights A P 2002 *J.Phys.: Condens. Matter* **14** 681
[11] Staab T E M, Sieck A, Haugk M, Puska M J, Frauenheim Th and Leipner H S 2002 *Phys. Rev.* B **65** 115210
[12] Hakala M, Puska M J and Nieminen R M 1998 *Phys. Rev.* B **57** 7621
[13] Poirier R, Avalos V, Dannefaer S, Schiettekatte F and Roorda S 2003 *Nucl. Instrum. Methods* B **206** 85
[14] Coleman PG, Harding R E, Davies G, Tan J and Wong-Leung J 2007 *J. Mater. Sci. Electron.* **18** 695
[15] van Veen A, Schut H, De Vris J, Hakvoort R A and Ijpma M R 1990 *AIP Conf. Series* **218** 171
[16] Ziegler J 2005 www.srim.org
[17] Abdulmalik D A, Coleman P G and Al-Qaradawi I Y 2006 *Appl. Surf. Sci.* **252** 3209
[18] Hastings J L, Estreicher S K and Pedders P A 197 *Phys. Rev.* B **56** 10215
[19] Makhov D V and Lewis L J 2004 *Phys. Rev. Lett.* **92** 255504
[20] Kalyanaraman R, Haynes T E, Holland O W, Gossman H-J L and afferty C S 2001 *Nucl. Instrum. Methods* B **175-7** 182
[21] Watkins G D and Corbett J W 1965 *Phys. Rev.* **138** A543
[22] Abdulmalik D A and Coleman P G 2008 *Phys. Rev. Lett.* **100** 095503
[23] Mason R E and Coleman P G 2006 *Appl. Surf. Sci.* **252** 3228
[24] Coleman P G and Burrows C P 2007 *Phys. Rev. Lett.* **98** 265502
[25] Watkins G D 2000 *Mater. Sci. Semicond. Process.* **3** 227
[26] Hallén A, Keskilato N, Josyula L and Svensson BG 1999 *J. Appl. Phys.* **86** 214
[27] Mills A P Jr 1995 *Positron Spectroscopy of Solids* ed A Dupasquier and A P Mills Jr (IOS: Amsterdam) p 209
[28] Makkonen I and Puska MJ 2007 *Phys. Rev.* B **76** 054119

Journal of Physics: Conference Series **265** (2011) 012002 doi:10.1088/1742-6596/265/1/012002

Activities towards *p*-type doping of ZnO

G Brauer[1], J Kuriplach[2], C C Ling[3] and A B Djurišić[3]
[1] Institut für Ionenstrahlphysik und Materialforschung, Forschungszentrum Dresden-Rossendorf, Postfach 510119, D-01314 Dresden, Germany
[2] Department of Low Temperature Physics, Charles University, V Holešovičkách 2, CZ-18000 Prague, Czech Republic
[3] Department of Physics, University of Hong Kong, Pokfulam Road, Hong Kong, P.R. China

E-mail: g.brauer@fzd.de

Abstract. Zinc oxide (ZnO) is an interesting and promising semiconductor material for many potential applications, e.g. in opto-electronics and for sensor devices. However, its *p*-type doping represents a challenging problem, and the physical reasons of its mostly *n*-type conductivity are not perfectly clear at present. Efforts to achieve *p*-type conductivity by ion implantation are reviewed, and ways to achieve *p*-type ZnO nanorods and thin films through various growth conditions are summarized. Then, issues associated with the preparation of Schottky contacts is discussed in some detail as this is a requirement of the device formation process. Finally, the possible incorporation of hydrogen and nitrogen into structural defects, which can act as trapping sites for positrons, is discussed in the context of experimental and theoretical positron results and the estimated H and N content in a variety of ZnO materials.

1. Introduction

Zinc oxide (ZnO) is an interesting and promising semiconductor material for many potential applications, e.g. in opto-electronics and for sensor devices [1-4]. The interplay of dopants, impurities, and intrinsic defects is essential for the electrical and optical properties of ZnO, especially if one would like to achieve *p*-type conductivity.

In a recent approach for *p*-type doping by repeated temperature modulation growth [5] it has been shown exemplarily that high quality crystal growth and doping with nitrogen have to be done in rather different temperature regimes. On the other hand, ion implantation with nitrogen into commercially available ZnO single crystals and subsequent thermal annealing has also been demonstrated to result in the creation of a *p*-*n* junction and electroluminescence [6, 7]. However, in any case defect creation and dopant incorporation remain to be studied systematically, if possible with methods rendering structural information on an atomistic level.

Positron Annihilation Spectroscopy (PAS) [8, 9], especially in the form of Slow Positron Annihilation Spectroscopy (SPIS) using mono-energetic positrons [10], is now a well-established tool for the study of electronic and defect properties of bulk solids and thin films. In particular, it can primarily reveal information about open-volume defects and their chemical surrounding provided they are not positively charged. Temperature-dependent measurements may also reveal some information about the presence of shallow positron traps, like interstitial impurity ions, but the interpretation of experimental data is not straightforward and more difficult.

A consistent theoretical modeling of bulk and defect positron properties of ZnO with and without lattice relaxation has been published for the first time only recently [11] including a summary of all types of PAS studies published so far. All positron lifetime data available in the literature have been collected [12] and related to the values calculated for bulk and defect configurations within the same scheme presented in Ref. [11]. From this comparison, the idea of the existence of V_{Zn} – hydrogen complexes in ZnO sampled by positrons was born.

A very systematic study of various, nominally undoped ZnO single crystals, either hydrothermally grown (HTG) or melt grown (MG), has been performed [13]. The crystal quality has been assessed by X-ray diffraction (XRD), and a comprehensive estimation of the detailed impurity and hydrogen contents by Inductively Coupled Plasma Mass Spectrometry (ICP-MS) and Nuclear Reaction Analysis (NRA), respectively, has also been made. High precision positron lifetime experiments then showed that a single positron lifetime is observed in all crystals investigated, clustering at 180-182 ps and 165-167 ps for HTG and MG crystals, respectively. Furthermore, hydrogen has been detected in all crystals in a bound state with a high concentration (at least 0.3 at.-%), whereas the concentrations of other impurities have been found to be comparatively very small. From ab initio calculations it is suggested that the existence of Zn vacancy-hydrogen complexes is the most natural explanation for the given experimental facts at present. All experimental findings imply that the role of H in ZnO is yet far from being completely understood, and that various types of Zn vacancy-hydrogen complexes need to be taken into account in future studies.

Nitrogen substituting oxygen (N_O) and N^{2-} molecules have been identified in ZnO by Electron Paramagnetic Resonance (EPR) [14, 15]. Complex centres containing nitrogen and Zn interstitial (Zn_i-N_O) atoms [16] and "pure" intrinsic defects created by low temperature, high-energy electron irradiation [17] have been observed by optically detected magnetic resonance.

The influence of nitrogen on the lattice dynamics of ZnO has also been investigated in a series of samples grown by Chemical Vapour Deposition (CVD) containing different nitrogen concentrations, as determined by Secondary Ion Mass Spectroscopy (SIMS) [18]. The Raman spectra revealed several local vibrational modes (LVM's) in addition to the host phonons of ZnO, and their intensity has been found to correlate linearly with the nitrogen concentration. This fact has been used as a quantitative measure of nitrogen in ZnO. Furthermore, SIMS has shown that a correlation also exists between the concentration of incorporated N and unintentional H, similar to the incorporation of the *p*-dopant Mg and H in GaN during metal-organic CVD (MOCVD) [19].

Experimental investigations on CVD grown, N-doped ZnO [20-23] revealed that these samples contain not only a large amount of H, but also C, as well as an excess of N in unwanted forms. These works also provide some insight into the difficulties of creating *p*-type ZnO through the incorporation of nitrogen, and suggest that these defects may play important roles in compensating or passivating the N acceptors. Moreover, infra-red (IR) absorption measurements on N-doped ZnO showed several peaks in the range of 1800–2000 cm^{-1}, which are surprisingly close to the IR frequencies of free diatomic molecules but significantly higher than the phonon frequencies of the ZnO host.

In a theoretical work [24], real-space multi-grid electronic structure calculations within the local-density-functional approximation have been exploited to find out a mechanism for the compensation of N acceptors in ZnO which could explain the difficulty in achieving low-resistivity *p*-type ZnO. There, it has been found that at low N doping levels using a normal N_2 source, O vacancies are the main compensating donors for N acceptors, while N acceptors are compensated via the formation of defect complexes with Zn anti-sites at high doping levels. When an active plasma N_2 gas is used to increase the N solubility, N acceptors are still greatly compensated by N_2 molecules at oxygen sites and N-acceptor-N_2 complexes.

First-principles calculations [25] have shown that *AB* defects substituting on an O site in ZnO, where *A*, *B* = N, O, or C, are an important class of defects whose physical properties cannot be described by the usual split interstitials but rather by substitutional diatomic molecules. The molecular nature of the $(AB)_O$ defects are reflected in their vibrational frequencies, which are red-shifted from those of the corresponding free molecules – but only by about 10%. These calculated results agree with the frequency range observed by IR measurement on N-doped ZnO [20-23]. Moreover, most $(AB)_O$ defects are donors in *p*-type samples. The $(NC)_O$ and $(N_2)_O$ defects have been found to have sufficiently low energies to convert substitutional N_O acceptors into donors, thereby hindering the efforts of doping ZnO *p*-type.

Engineering of the optical band gap of ZnO has already been demonstrated, e.g. by shifting to higher or lower energy by alloying with MgO [26-28], or CdO [29, 30], respectively. In a more recent work [31], Raman measurements have indicated an increased solubility of N in MgZnO compared with that in ZnO. Using first-principles band structure methods, the formation energy and transition energy level for the nitrogen acceptor in ZnO and a randomly chosen $Mg_xZn_{1-x}O$ alloy (with $x = 0.25$) have been calculated. The results support the experimental findings and point out that the acceptor energy level deeper in the alloy can be ascribed to the downward shift of the valence-band maximum edge in the presence of magnesium.

It is certain that all types of applications rely on a full understanding of the role of lattice defects in ZnO, and provision of high-quality bulk material, films, and also nanostructures. However, its *p*-type doping represents a challenging problem, and the physical reasons of its mostly *n*-type conductivity are not perfectly clear at present.

In the present work, three different activities will be summarized. First, efforts to achieve *p*-type conductivity by ion implantation are reviewed, and the creation of a *p-n* junction by N^+ ion implantation and thermal annealing is discussed. Second, ways to achieve p-type ZnO of thin films and nanorods by various growth conditions are presented. Third, the preparation of Schottky contacts on ZnO is discussed in some detail as this is a serious requirement of the device formation process. Finally, the possible incorporation of hydrogen and nitrogen into structural defects, which can act as trapping sites for positrons, is considered. This latter issue will be discussed in the context of experimental and theoretical positron results and the estimated H and N contents in a selected variety of ZnO materials.

2. Activities to achieve *p*-type ZnO

2.1. Ion implantation

The creation of semiconductor devices is mainly based on ion implantation as a primary step to create *p-n* junctions because this method allows tailoring their depth in the material and geometrical structure in accordance with industrial production needs. However, the implantation is unavoidably connected with the creation of point defects (extrinsic defects) which might be removed by post-implantation annealing. But the material of choice for doping by ion implantation may already possess intrinsic defects which are mainly determined from the growth conditions, like the impurity content and lattice defects. On the other hand, electrical activation of dopant atoms after implantation also requires thermal annealing at higher temperatures. The time spent by the implanted material at the annealing temperature will allow diffusion of the defects and dopant atoms as well. Thus, it is necessary to characterize any material regarding its defects, and to investigate their annealing in order to optimize the annealing conditions for a given material.

Possible ways for achieving *p*-type conductivity of ZnO are shown in figure 1 as a schematic illustration. First, doping by ion implantation will be considered.

Based on first-principles calculations, a method for fabricating *p*-type ZnO with group-I elements such as Li and Na has been suggested [32]. It has been found that with group-I dopants alone, substitutional acceptors are mostly self-compensated by interstitial donors, whereas in ZnO co-doped with H impurities, the formation of compensating interstitials is severely suppressed, and thus the acceptor solubility is greatly enhanced by forming H-acceptor complexes. Then, H atoms could be easily dissociated from these defect complexes at relatively low annealing temperatures, and thus low resistivity *p*-type ZnO would be achievable with dopants different from group-V elements.

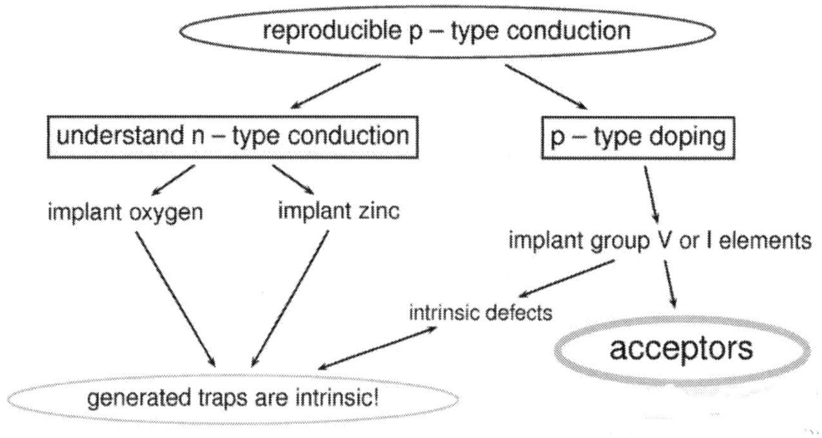

Figure 1. Cartoon showing how to achieve *p*-type conductivity of ZnO, by courtesy of Matthias Schmidt (U Leipzig).

Other authors present their first-principles calculations of doping effects in ZnO with group-IB elements, such as Cu, Ag, and Au [33]. The calculated transition energies for substitutional Cu, Ag, and Au have been estimated to be very low for these group-IB elements on the substitutional sites, but rather high at the interstitial sites under oxygen-rich growth conditions. Under these conditions, the formation of major hole-killer defects, such as oxygen vacancies (V_O) and Zn interstitials (Zn_i), will be suppressed. Thus, Ag has been suggested to may be a good candidate for producing *p*-type ZnO.

Based on first-principles calculations, a model for large-size-mismatched group-V dopants in ZnO has also been proposed [34]. The dopants have been found to not occupying the O sites as is widely perceived, but rather the Zn sites: each forms a complex with two spontaneously induced Zn vacancies (V_{Zn}) in a process that involves fivefold As coordination. Moreover, an As_{Zn}-$2V_{Zn}$ complex has been estimated to have maybe lower formation energy than any of the parent defects. The model agrees with the recent observations that both As and Sb have low acceptor-ionization energies and that to obtain *p*-type ZnO requires O-rich growth or annealing conditions. Interestingly, *p*-type conductivity in ZnO with exceptionally large-size-mismatched dopants such as P [35], As [36, 37], and Sb [38] has already been reported.

Ion implantation with nitrogen, having an ionic radius comparable to oxygen, into commercially available ZnO single crystals and subsequent thermal annealing has already been shown to result in the creation of a *p-n* junction [6, 7]. The rectifying property of the *p-n* junction was found to persist after the annealing temperature up to 1200 °C [7]. With the fabrication of a 50 nm indium-tin oxide (ITO) n^+-film on top of the *p*-layer, the diode under a forward bias emitted light at room temperature [7]. The luminescence intensity increased with increasing forward bias voltage and the electroluminescence spectra showed peaks at ~530 nm(~2.34 eV) and ~740 nm (~1.68 eV). A Deep Level Transient Spectroscopy (DLTS) study on the N^+-implanted ZnO diode showed that the ion implantation had the effects of introducing a deep level at E_C–0.95 eV (D1) and enhancing the intensity of another deep level E_C –0.30 eV (E3), which originally existed in the as-grown sample [7]. D1 and E3 disappeared after the annealing at 750 °C and 900 °C, respectively [7]. After annealing at 750 °C, another trap having the activation energy of 0.17 eV (D2) was formed [7].

Second, for an understanding of *n*-type conductivity of ZnO, the role of intrinsic defects – i.e. native defects which are a result of the growth conditions and impurity atoms of the given ZnO material – would be invaluable in the design of a treatment to achieve tailored *p*-type conductivity. However, in spite of extensive research for many years the role of these intrinsic defects is still not perfectly clear at present.

Using first-principles methods based on density functional theory and pseudopotentials, a detailed study of native point defects in ZnO has been performed recently [39]. Contrary to the conventional wisdom, it has been estimated that native point defects are unlikely to be the cause of the frequently observed unintentional *n*-type conductivity. V_O defects, which have most often been invoked as shallow donors, have been found to possess high formation energies in *n*-type ZnO, and thus are actually deep donors with very high ionization energy. Zn_i defects have been calculated to be shallow donors, but to have high formation energies in *n*-type ZnO; in addition, they are fast diffusers, and thus unlikely to be stable in *n*-type ZnO. Zn antisites (Zn_O) have been found to be also shallow donors, and to have even higher formation energies than Zn_i defects. However, they may play a role under non-equilibrium conditions, such as in irradiation experiments. V_{Zn} defects have been estimated to be deep acceptors which may act as compensating centres in *n*-type ZnO. Oxygen interstitials (O_i) have been calculated to be stable in the form of electrically inactive split interstitials as well as deep acceptors at the octahedral interstitial site under *n*-type conditions. These results [39] provide a guide to more in-depth experimental studies of point defects in ZnO and their influence on the control of doping.

A comprehensive discussion of results from PAS studies aimed at the identification of structural defects, both intrinsic and extrinsic, can be found in Refs. [11-13].

Investigations of intrinsic defects in ZnO, generated by implantation of oxygen and zinc ions into ZnO thin films, have now been started by us [40]. The pulsed laser deposition (PLD) technique on a-plane sapphire substrates [41] was employed for the growth of c-plane oriented ZnO films of ~ 1 μm thickness. Ion implantation was carried out with oxygen at four different energies (30, 70, 120, and 200 keV) to obtain a ~ 400 nm box-like implanted region with an additional oxygen concentration of $\sim 1 \times 10^{17}$ cm^{-3} (as calculated by SRIM [42]). The zinc implantation was carried out with a single energy of 250 keV to a fluence of 5×10^{11} cm^{-2}.

Thermal annealing has been employed to reduce the density of implantation-induced defects in the film samples. In preparation for capacitance spectroscopic measurements Schottky contacts, realised by thermally evaporating Pd, have been deposited onto the samples. To obtain the energetical depth and concentration of electrically chargeable traps present in the film samples, DLTS and Thermal Admittance Spectroscopy (TAS) measurements at the implanted as well as reference samples have been performed.

These experiments are suitable for the investigation of electronic states having thermal activation energies in the range from several meV up to 1 eV below the conduction band edge. Furthermore, they allow a distinction between defects that preferentially form under either oxygen or zinc rich conditions. Such measurements finally revealed that oxygen implantation generated a ~170 meV deep trap (yet unlabeled), and the well-known E4 [43] trap. After the Zn implantation and thermal annealing of the film sample, no new defect species could be detected but the concentration of a (yet unlabeled) ~210 meV deep trap was observed to increase.

2.2. Variation of growth conditions

2.2.1. ZnO nanostructures. Compared to thin film samples, *p*-type doping of ZnO nanostructures has been less comprehensively studied. Doping of ZnO nanostructures with acceptor impurities during growth [44-46], as well as by post-growth treatment [47-49] has been demonstrated. However, the doped nanostructures are usually characterized mainly in terms of their structural and optical properties, while characterization of electronic properties and confirmation of *p*-type conductivity is rarely performed, mainly due to experimental difficulties in electronic characterization of single nanostructures.

Examples of incorporation of acceptor impurities by post-fabrication treatment include nitrogen doping by calcination of ZnO nanoparticles in ammonia [47], as well as annealing of nanorods grown on a GaAs substrate, which results in diffusion of As acceptors into the nanorods [48]. *p-n* homo-junctions have been fabricated by achieving *p*-type doping of HTG ZnO nanorods by exposure to a NH_3 plasma [49].

Concerning doping during growth, one of the major difficulties is the change of the morphology of ZnO nanostructures with addition of the dopant [46]. Nevertheless, there have been several examples of successful doping of ZnO nanowires with an acceptor dopant introduced during growth [44, 45]. Phosphorus doped ZnO nanowire arrays have been fabricated by the CVD method using P_2O_5 as a dopant source [45]. As-grown nanowires exhibited *n*-type conductivity, while *p*-type conductivity was obtained after annealing in nitrogen at 850 °C [45]. P_2O_5 was also used as a dopant for the fabrication of P:ZnO nanowires using PLD [44].

Doped ZnO nanostructures can also be prepared by chemical methods, where the dopant is introduced into the solution during synthesis of the nanostructures [50]. *P*-type and *n*-type ZnO nanoparticles have been prepared using this approach, and a rectifying behavior was observed in the junctions consisting of a *p*-ZnO nanoparticle layer and an *n*-ZnO nanoparticle layer [50].

In addition to the *p*-type behavior observed in samples with intentionally incorporated acceptor impurities, *p*-type behavior in undoped ZnO nanorods grown by a HTG method [51] has recently been observed. A scanning electron microscopy (SEM) image of the undoped *p*-type ZnO nanorods is shown in figure 2. The *p*-type behavior was attributed to an increased concentration of V_{Zn} and a

Figure 2. SEM image of *p*-type HTG ZnO nanorods [51].

decreased incorporation of donor impurities [51]. The type of conductivity was strongly dependent on the type of seed layer used for the nanorod growth [51]. In addition, the growth of nanorods on ITO substrates was found to be detrimental for the achievement of *p*-type conductivity due to diffusion of indium during seed/nanorod fabrication [51]. This illustrates the importance of native defects in the achievement of *p*-type conductivity in ZnO.

Defects in HTG nanorods are typically abundant due to a very low growth temperature (90 °C). However, a large variation in sample properties can be achieved by changing the fabrication conditions and/or post-fabrication processing. An ambipolar behavior of field effect transistors (FET's) based on HTG ZnO nanorods has also been attributed to native defects [52]. Measured electron and hole mobilities in these samples were comparable (3.2 and 2.1 $cm^2V^{-1}s^{-1}$, respectively) [52].

2.2.2. ZnO thin films. *p*-type behavior has been observed in ZnO films prepared by different methods [5, 53-57]. Nitrogen doped ZnO grown by laser molecular beam epitaxy (MBE) has been reported and a homojunction light emitting diode (LED) has been demonstrated [5]. The crucial part in obtaining these devices was a careful optimization of the growth conditions [5]. MOCVD grown homo-junction ZnO LED's have also been demonstrated, although in this case for the achievement of *p*-type conductivity annealing in a N_2O plasma was necessary in addition to the introduction of a N dopant during growth from an ammonia source [53]. Homojunction devices based on other dopants have also been demonstrated, such as *p*-ZnO:Sb/*n*-ZnO:GaN junction grown by MBE [54].

In addition to doping with a single acceptor, co-doping for achieving *p*-type ZnO has also been investigated [55-57]. Co-doping with both acceptor and donor impurities [55, 56] and with two acceptor impurities [57] has been demonstrated. (Al, As):ZnO [55] and (Cu, Ga):ZnO [56] films were fabricated by radio frequency (rf) magnetron sputtering, while (Li, N):ZnO films were prepared by PLD [57].

Thus, the doped ZnO films can be prepared by a variety of experimental techniques. Key issues in achieving *p*-type doping are an optimization of the growth conditions and a good control over native defects to avoid compensation effects. In addition, for some acceptor impurities *p*-type conductivity can be attributed to the formation of complexes with native defects. For example, for the As impurity it has been proposed that the actual acceptor is an As_{Zn}-2V_{Zn} complex, rather than an arsenic antisite (As_O) [34, 55].

Only recently, As-doped ZnO films were grown by the rf magnetron sputtering method [58]. As the substrate temperature during growth was raised above ~ 400 °C, the films changed from *n*-type to *p*-type. A hole concentration and mobility of ~ 6×10^{17} cm^{-3} and ~ 6 cm^2 V^{-1} s^{-1}, respectively, were achieved. The ZnO films were studied by SIMS, X-ray photoelectron spectroscopy (XPS), low temperature photoluminescence (PL), and PAS. The results were found to be consistent with the As_{Zn} – 2 V_{Zn} shallow acceptor model proposed by Limpijumnong et al. [34]. The results of XPS, PL, PAS, and thermal studies finally allowed suggestion of a comprehensive picture of the As-related shallow acceptor formation.

The importance of native defects in the achievement of *p*-type conductivity in ZnO is also confirmed by recent reports of *p*-type conductivity in undoped ZnO films, which was attributed to V_{Zn} acceptors [59] or complexes consisting of V_{Zn} and another acceptor [60]. Thus, while *p*-type behavior can be achieved by varying the growth conditions of ZnO for a certain number of dopants, and even in the absence of dopants, further work is still needed to achieve a full understanding of the relevant phenomena and stable and reproducible *p*-type doping.

2.3. *Issue of Schottky contacts*

Developing the techniques for fabricating good quality Ohmic and Schottky contacts on ZnO is an essential step for device fabrication. Ohmic contact fabrication is relatively easy, but the relevant knowledge in fabricating a Schottky contact on ZnO is not thoroughly understood. According to the Schottky-Mott model, the barrier height of an ideal Schottky contact ϕ_b is given by the difference between the metal work function ϕ_m and the semiconductor electron affinity χ_s, i.e., $\phi_b = \phi_m - \chi_s$. Thus, any metal having a work function larger than the electron affinity of ZnO (4.35 eV [61]) would be a candidate for forming a Schottky contact. The Schottky barrier height increases with the value of the metal work function. This implies that metals like Pt (ϕ = 5.65 eV), Pd (ϕ = 5.12 eV) and Au (ϕ = 5.10 eV) [62] would form a Schottky contact with the calculated Schottky barrier equal to 1.30 eV, 0.77 eV, and 0.75 eV, respectively. However in practice, depositing these metals onto an untreated ZnO surface usually results in a contact which is Ohmic in nature.

In practice, the electrical property of the fabricated Schottky contact also depends on other factors such as surface contamination, Fermi level pinning, chemical reaction at the interface, surface morphology, etc. [3, 20].

There is some literature reporting the fabrication of good quality Schottky contacts using the metals Au [63-70], Ag [67, 71], Pt [72-74] and Pd [63, 64, 75, 76]. Most of these successful cases involved surface pre-treatments, like wet etching (H_2O_2 [73, 77-79] and sulphide [74]), ozone treatment [72] and plasma cleaning [65, 70].

With the aim to explore the mechanisms that determine the contact electrical properties, efforts have also been devoted to study the correlations between the contact's electrical properties and the surface conditions (like contamination, interfacial defects, interfacial layer morphology, etc.) with different pre-treatments. However, the problem was not unambiguously understood. The improvement of the contact rectifying property has been associated by different authors with a good surface morphology, the removal of deep level defects [73, 80], the formation of a low conductivity layer [69, 73, 76-78], and the removal of OH and C contaminations [70, 72, 77-79].

2.4. Incorporation of H and N into structural defects

A full chemical analysis of each material is essential for two reasons: (i) to determine the detailed content of any impurities and their corresponding concentrations, and (ii) to compare the values with the sensitivity limits of PAS. In a previous study [13] of various, nominally undoped, HTG and MG ZnO single crystals chemical compositions have generally been estimated by ICP-MS using a Perkin-Elmer ELAN-9000 spectrometer.

To account for the possible influence of hydrogen on the properties of ZnO – as there is an ongoing debate in the literature – an experimental estimation of the real H content of each crystal has been performed by NRA [13] also using 6.64 MeV ^{15}N ions, as successfully demonstrated earlier for HTG ZnO nanorods [81]. NRA has a depth resolution of ~ 5 nm, and the hydrogen detection limit is ~200 ppm. The depth chosen for the analysis is estimated by SRIM calculations [42].

From SRIM it is also found that the analyzing ^{15}N ions lose their kinetic energy E with increasing penetration depth x mainly by electronic stopping ($(dE/dx)_{electronic}$ = 2.52 keV nm^{-1}), whereas nuclear stopping is about three orders of magnitude smaller ($(dE/dx)_{nuclear}$= 0.006 keV nm^{-1}). This energy transfer to the crystal during analysis can be sufficient to release a weekly bound hydrogen atom from its bonding site, so that it will be able to start diffusion. It is generally assumed that single H atoms would be trapped again immediately so that this diffusion should most probably take place in the form of H_2 molecules. As this diffusing hydrogen is then no longer available at the analysis position, this is seen as a drop in concentration with increasing ^{15}N fluence. Throughout this paper we will call this "unbound H (H-u)". All H atoms which are not moving during analysis because the energy transfer is not sufficient to release them from their bonding site will be called "bound H (H-b)". However, it has to be clearly stated also that from NRA it is impossible to draw any conclusion on the kind of bonding of H atoms in the crystal.

It has been found that H-b exists in all crystals with a high concentration (at least 0.3 at.-%). On the other hand, saturated positron trapping has generally been observed in the crystals known to occur usually for vacancy-like defect concentrations > 0.01 at.-% [9]. However, all non-hydrogen impurities detected were found to have their concentrations apparently below this limit and thus can hardly influence the positron lifetime data even if they would form a complex with vacancies. NRA results further have led to the conclusion that H-u has no influence upon the positron lifetimes observed. From the comparison of theoretical calculations with the experimental results it has been concluded that most probably a (V_{Zn} + 1H) complex could account for the positron lifetime observed in HTG crystals. However, a (V_{Zn} + 2H) complex has been estimated to be unable to trap a positron, and thus the explanation of the positron lifetime generally observed in MG crystals (165-167 ps) remained unexplained [13].

To account for the possible influence of nitrogen on measurable positron properties, its concentration had to be determined. To estimate and compare the total N content of an HTG (MT-08) and an MG (B12) ZnO single crystal, a "multi N/C 2100S" device from Analytik Jena AG has been used.

The corresponding sample of a given weight is dissolved in 1ml of 30% HCl (suprapure quality) and diluted to 10 ml with Milli-Q-water®, then an amount of 500 μl is injected into an oven, where by use of a Pt catalyst in a stream of oxygen gas the dissolved sample is converted into NO_X. The amount of NO_X is determined by a well calibrated chemo-luminescence detector. Routinely this is done for samples having a nitrogen content of the order mg/l. The detection limit of the commercial device is given to be 100 μg/l. Our HTG sample (MT-08) had a total weight of 0.241 g, whereas the MG sample (B12) had a total weight of 0.0839 g only. No nitrogen has been detected in both samples, i.e. the nitrogen content should be <4 μg/g for sample MT-08 and <12 μg/g for sample B12. These values of an upper limit of the nitrogen content can be expressed also as 11.6 ppm (9.6×10^{17} cm^{-3}) for the HTG (MT-08) sample and 34.9 ppm (29.3×10^{17} cm^{-3}) for the MG sample (B12), respectively.

On the other hand, these estimated upper limits of the nitrogen content should be related with the experimentally determined concentration of H-b = (0.14 ± 0.03) at.-% (1400 ppm) and (0.40 ± 0.03) at.-% (4000 ppm) for the HTG and MG samples, respectively, where on the other hand no indication of the presence of H-u could be seen from NRA in both crystals. Thus, a ratio of H:N = 121 for the HTG sample (MT-08) and H:N = 115 for the MG sample (B12) can be estimated. These ratios are almost equal for both investigated samples, which had to be expected because for their evaluation the estimated upper limits of the given method for both crystals have been used. Thus, a ratio H:N ~ 120 should represent a lower limit for the HTG and MG ZnO crystals investigated here because the real nitrogen concentration in both types of samples should be less (and absolute concentrations remain to be determined by a more sensitive method in the future).

Nevertheless, the given results allow one to draw two major conclusions so far: (1) H again turns out to be the impurity with the highest concentration in both types of crystals, compared to other detected impurities and to nitrogen, which has not yet been really detected [13], and (2) taking the estimated upper limits of the nitrogen concentration to represent the real content would mean that N-related open volume defects could play a role in the variety of conceivable structural defects which are able to trap positrons and thus influence the measurable positron lifetime spectrum. The latter conclusion challenges and justifies the calculation of positron properties of such defects, and the results of first calculations are presented and discussed in the next chapter.

3. Theory

3.1. Brief description of calculations

In order to understand the structure of defects related to N and H impurities, we have performed ab initio studies of such defects aiming mainly at the defects' structure and their positron characteristics. For this purpose we employ the Vienna ab initio package (VASP) [82] considering projected augmented wave potentials [83] in order to get relaxed defect geometries. Positron calculations have been performed using a real space method as implemented in the atomic superposition method (ATSUP) [84]. More details regarding calculations can be found in our recent work [13].

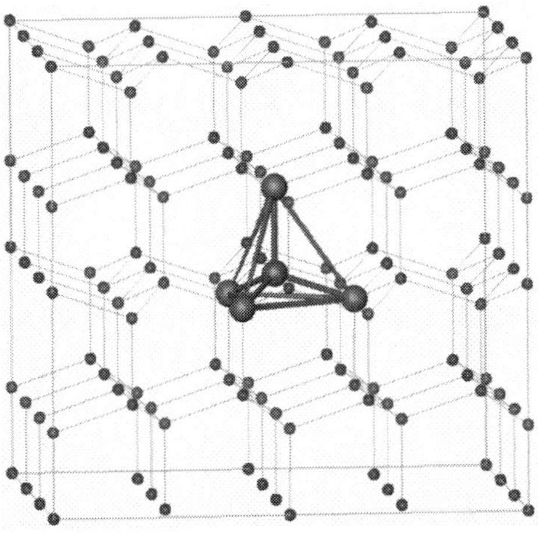

Figure 3. Perfect ZnO lattice (hexagonal, wurtzite) together with a ZnO_4 tetrahedron. Red (blue) spheres mark O (Zn) atoms.

3.2. Summary of results for H-related defects

Hydrogen can be incorporated into the ZnO lattice in several ways [85]. First, these are 'bond-center' positions in which H atoms are located approximately in the middle of the original Zn-O bonds (inside the corresponding ZnO_4 tetrahedron; see figure 3), which causes a substantial lattice distortion at the position of defects. Second, there also exist 'antibonding' positions where the H bond is located outside ZnO_4 tetrahedra. Furthermore, hydrogen atoms may occupy vacancies on both Zn and O sublattices as suggested in [39, 86]. In addition, weakly bound H atoms (in the form of H_2 molecules) may reside in 'channels' (along the hexagonal axis) in the ZnO structure [85]. Hydrogen atoms can also be bound to other impurities.

As for hydrogen incorporated in single vacancies, it was found that the H atom resides in the center of an oxygen vacancy. Bonding properties of H in such a defect is a matter of ongoing discussion [39, 87], but positrons do not trap in such a defect [11, 87]. Concerning the Zn vacancy V_{Zn}, incorporation of one, two and three H atoms has been studied [13]. Like interstitial H in the ZnO lattice [85], H atoms make a bond inside V_{Zn} with a neighboring oxygen atom with a bonding length of about 1 Å. The hydrogen molecule is not stable inside V_{Zn}. The details of configurations are given in Ref. [13]. Here we point out that with the increasing number of H atoms inside the vacancy the open volume available for positron trapping (and the depth of the positron potential well) is decreasing and already the V_{Zn} + 2H complex does not trap positrons. In the case of the V_{Zn} + 1H complex, positrons can be trapped and, depending on the configuration, the positron lifetime ranges from 177 to 179 ps (considering positron induced forces). Such lifetimes are very close to those measured for HTG samples and it was therefore suggested that the V_{Zn} + 1H complex is responsible for saturated trapping in HTG samples, though other defect configurations need to be yet investigated to prove this suggestion. In particular, the investigation of the zinc plus oxygen (Zn+O) di-vacancy containing a varying number of hydrogen atoms is in progress.

3.3. First calculations for N-vacancy complexes

We present here the first results of investigations of charge neutral V_{Zn} + 1N and V_{Zn} + 2N complexes that could possibly trap positrons and thereby would become detectable by PAS. The structure of the studied defects is shown in figure 4. In positron calculations we restricted ourselves to the superposition of atomic densities and potentials, neglecting thus charge transfer effects and positron induced forces.

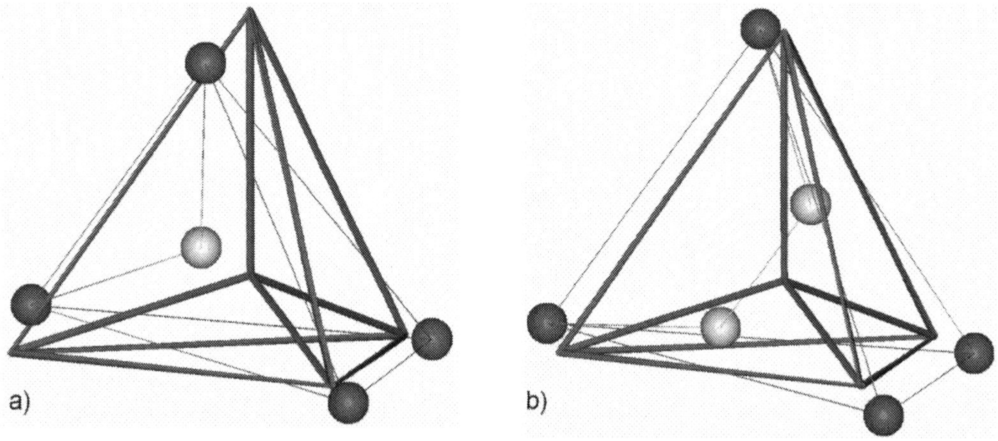

Figure 4. Atomic configurations of a) V_{Zn} + 1N and b) V_{Zn} + 2N defects studied. The color designation is the same as in figure 3 and yellow spheres mark nitrogen atoms. Thick lines indicate the original (undistorted) VO_4 tetrahedron.

The V_{Zn} + 1N defect (figure 4a) was obtained using the starting configuration in which the N atom was placed into the same position where the hydrogen atom resides. After relaxation the N atom moves from the starting position and is bound to two neighboring O atoms (bond length ~1.4 Å, O-N-O bond angle ~111°). Thereby the original VO_4 tetrahedron becomes substantially distorted. The positron lifetime calculated for such a defect amounts to 166 ps, which indicates weak positron trapping. This could be eventually changed into a non-trapping defect after considering charge transfer effects and positron induced forces, but such calculations are not straightforward and still need to be performed.

As for the V_{Zn} + 2N defect (figure 4b), there is an O-N-N-O chain of atoms with the O-N (N-N) bond length about ~1.45 Å (~1.25 Å); N-N-O bond angle is about 127°. Again the original VO_4 tetrahedron is found to be largely distorted. The corresponding positron lifetime is 159 ps and indicates no positron trapping as this value has also been obtained for the defect-free bulk of ZnO [11]. We note that N-O (N-N) bond lengths given above are somewhat longer (by about 10%) than those for NO/NO_2 (N_2) free molecules, which contrasts with the O-H bond length that is always very close to 1 Å, as in the H_2O free molecule. We may finally state that if V_{Zn}-nitrogen complexes do represent positron traps then they will be very shallow, though other possible configurations need to be further investigated.

4. Conclusions

Although the intrinsic properties of ZnO provide significant advantages for a variety of device applications, the reliable formation of *p*-type conductivity still remains a challenging goal. In particular, in this work the potential role of PAS to help identification of open volume defects that might be involved is mentioned.

First, efforts to achieve *p*-type conductivity of ZnO by ion implantation are reviewed, and the creation of a *p-n* junction by N^+ ion implantation and thermal annealing is discussed.

Second, ways to achieve *p*-type ZnO of thin films and nanorods by various growth conditions are summarized. A very recent example of achieving finally *p*-type ZnO:As by radio frequency magnetron sputtering at elevated temperature is given. This result was found to be consistent with the As_{Zn} – 2 V_{Zn} shallow acceptor model proposed theoretically by Limpijumnong et al. [34].

Third, the preparation of Schottky contacts on ZnO is mentioned in some detail as this also makes serious demands of any device formation process.

Finally, the possible incorporation of hydrogen and nitrogen into structural defects, which can act as trapping sites for positrons, is considered. The V_{Zn} + 1 H complex has been found to trap a positron, whereas the V_{Zn} + 2 H complex does no longer constitute such a trap [13]. Correspondingly, first calculations by ATSUP reveal that the V_{Zn} + 1 N complex is just a weak positron trap, whereas the V_{Zn} + 2 N complex does not constitute such trap at all. As a ratio H:N ~ 120 is determined to be a lower limit for the HTG and MG ZnO single crystals investigated here, there is a marginal influence expected of the N-related complexes considered so far on measurable positron lifetime spectra. The probability to detect the V_{Zn} + 1 N complex experimentally by PAS is very low in any case.

Acknowledgement

Financial support from German Academic Exchange Service (DAAD: D/08/01769) and the Research Grant Council of Hong Kong (HKU7037/06P, HKU7031/08P and G_HK026/07) for cooperation visits between Forschungszentrum Dresden-Rossendorf and University of Hong Kong in 2008/2009 is gratefully acknowledged. This work is also part of the research plan MS 0021620834 and the project COST OC 165 that are financed by the Ministry of Education of the Czech Republic.

References

[1] Look D C 2001 *Materials Science and Engineering* **B 80** 383
[2] *Transparent Conducting Oxides* ed D S Ginley and C Bright 2000 *MRS Bull* **25** 15

[3] Özgür Ü, Alivov Ya I, Lin C, Teke A, Reshchikov M A, Dogan S, Avrutin V, Cho S-J, Morkoc H 2005 *J Appl Phys* **98** 041301
[4] Look D C 2006 *J Electron Mater* **35** 1299
[5] Tsukazaki A *et al.* 2005 *Nature Mater* **4** 42
[6] von Wenckstern H, Pickenhain R, Schmidt H, Brandt M, Biehne G, Lorenz M, Grundmann M and Brauer G 2006 *Appl Phys Lett* **89** 092122
[7] Gu Q L, Ling C C, Brauer G, Anwand W, Skorupa W, Hsu Y F, Djurisic A B, Zhu C Y, Fung S and Lu L W 2008 *Appl Phys Lett* **92** 222109
[8] *Positron Spectroscopy of Solids* 1995 ed A Dupasquier and A P Mills Jr (Amsterdam: IOS)
[9] Krause-Rehberg R and Leipner H S 1999 *Positron Annihilation in Semiconductors – Defect Studies* (Berlin: Springer)
[10] *Slow Positron Beam Techniques for Solids and Surfaces* 2002 ed Brauer G and Anwand W *Appl Surf Sci* **194**
[11] Brauer G, Anwand W, Skorupa W, Kuriplach J, Melikhova O, Moisson C, von Wenckstern H, Schmidt H, Lorenz M and Grundmann M 2006 *Phys Rev* **B 74** 045208
[12] Brauer G, Kuriplach J, Cizek J, Anwand W, Melikhova O, Prochazka I and W. Skorupa 2007 *Vacuum* **81** 1314
[13] Brauer G, Anwand W, Grambole D, Grenzer J, Skorupa W, Cizek J, Kuriplach J, Prochazka I, Ling C C, So C K, Schulz D and Klimm D 2009 *Phys Rev* **B 79** 115212
[14] Garces N Y, Wang L, Giles N C, Halliburton L E, Cantwell G and Eaton D B 2003 *J Appl Phys* **94** 519
[15] Carlos W E, Glaser E R and Look D C 2001 *Physica* **B 308-310** 976
[16] Aliev G N, Bingham S J, Wolverson D, Davie J J, Makino H, Ko H J and Yao T 2004 *Phys Rev* **B 70** 115206
[17] Vlasenko L S and Watkins G D 2005 *Phys Rev* **B 72** 035203
[18] Kaschner A *et al.* 2002 *Appl Phys Lett* **80** 1909
[19] Kaschner A, Siegle H, Kaczmarczyk G, Strassburg M, Hoffmann A, Thomsen C, Birkle U, Einfeldt S and Hommel D 1999 *Appl Phys Lett* **74** 3281
[20] Pearton S J, Norton D P, Ip K, Heo Y W and Steiner T 2004 *J Vac Sci Technol* **B 22** 932
[21] Keyes B M, Gedvilas L M, Li X and Coutts T J *J Cryst Growth* **281** 297
[22] Li X, Keyes B M, Asher S E, Zhang S B, Wei S-H, Coutts T J, Limpijumnong S, Van de Walle C G 2005 *Appl Phys Lett* **86** 122107
[23] Perkins C L, Lee S-H, Li X, Asher S E and Coutts T J 2005 *J Appl Phys* **97** 034907
[24] Lee E-C, Kim Y-S, Jin Y-G and Chang K J 2001 *Phys Rev* **B 64** 085120
[25] Limpijumnong S, Li X, Wei S-H and Zhang S B 2005 *Appl Phys Lett* **86** 211910
[26] Ohtomo A, Kawasaki M, Koida T, Masubuchi K, Koinuma H, Sakurai Y, Yoshida Y, Yasuda T and Segawa Y 1998 *Appl Phys Lett* **72** 2466
[27] Bundesmann C, Schubert M, Spemann D, Butz T, Lorenz M, Kaidashev E M and Grundmann M 2002 *Appl Phys Lett* **81** 2376
[28] Schmidt R *et al.* 2003 *Appl Phys Lett* **82** 2260
[29] Makino T, Segawa Y, Kawasaki M, Ohtomo A, Shiroki R, Tamura K, Yasuda T and Koinuma H 2001 *Appl Phys Lett* **78** 1237
[30] Lorenz M *et al.* 2003 *Solid-State Electronics* **47** 2205
[31] Gai Y Q *et al.*2008 *Appl Phys Lett* **92** 062110
[32] Lee E-C and Chang K J 2004 *Phys Rev* **B 70** 115210
[33] Yan Y, Al-Jassim M M and Wei S-H 2006 *Appl Phys Lett* **89** 181912
[34] Limpijumnong S, Zhang S B, Wei S-H and Park C H 2004 *Phys Rev Lett* **92** 155504
[35] Kim K-K, Kim H-S, Hwang D-K, Lim J-H and Park S-J 2003 *Appl Phys Lett* **83** 63
[36] Morhain C, Teisseire M, Vezian S, Vigue F, Raymond F, Lorenzini P, Guion J, Neu G and Faurie J-P 2002 *Phys Stat Sol* **(b) 229** 881
[37] Ryu Y R, Lee T S and White H W 2003 *Appl Phys Lett* **83** 87

[38] Aoki T, Shimizu Y, Miyake A, Nakamura A, Nakanishi Y and Hatanaka Y 2002 *Phys Stat Sol* **(b) 229** 911
[39] Janotti A and Van de Walle C G 2006 *J Cryst Growth* **287** 58
[40] Schmidt M, Ellguth M, Czekalla C, von Wenckstern H, Pickenhain R, Grundmann M, Brauer G, Skorupa W, Helm M, Gu Q and Ling C C 2009 *J Vac Sci Technol* **B 27** 1597
[41] von Wenckstern H, Biehne G, Abdel Rahman R, Hochmuth H, Lorenz M and Grundmann M 2006 *Appl Phys Lett* **88** 092102
[42] www.srim.org
[43] Frank T et al. 2007 *Appl Phys* **A 88** 141
[44] Cao B Q, Lorenz M, Rahm A, von Wenckstern H, Czekalla C, Lenzner J, Benndorf G and Grundmann M 2007 *Nanotechnology* **18** 455707
[45] Xiang B, Wang P W, Zhang X Z, Dayeh S A, Aplin D P R, Soci C, Yu D P and Wang D L 2007 *Nano Lett* **7** 323
[46] Zeng D W, Xie C S, Zhu B L, Jiang R, Chen X, Song W L, Wang J B and Shi J 2004 *J Crystal Growth* **266** 511
[47] Wei H Y, Wu Y S, Wu L L and Hu C X 2005 *Mater Lett* **59** 271
[48] Lee W, Jeong M C, Joo S W and Myoung J M 2005 *Nanotechnology* **16** 764
[49] Lin C-C, Chen H-P and Chen S-Y 2005 *Chem Phys Lett* **404** 30
[50] Mohanta K, Batabyal S K and Pal A J 2007 *Chem Mater* **18** 3662
[51] Hsu Y F *et al.* 2008 *Adv Funct Mater* **18** 1020
[52] Wong C-Y, Lai L-M, Leung S-L, Roy V A L and Pun E Y-B 2008 *Appl Phys Lett* **93** 023502
[53] Sun J C, Liang H W, Zhao J Z, Bian J M, Feng Q J, Hu L Z, Zhang H Q, Liang X P, Luo Y M and Du G T 2008 *Chem Phys Lett* **460** 548
[54] Chu S, Lim J H, Mandalapu L J, Yang Z, Li L and Liu J L 2008 *Appl Phys Lett* **92** 152103
[55] Yun E-J, Park H-S, Lee K H, Nam H G and Jung M 2008 *J Appl Phys* **103** 073507
[56] Shet S, Ahn K-S, Yan Y, Deutsch T, Chrustowski K M, Turner J, Al-Jassim M and Ravindra N 2008 *J Appl Phys* **103** 073504
[57] Zhang Y Z, Lu J G, Ye Z Z, He H P, Zhao B H and Wang L 2008 *Appl Surf Sci* **254** 1993
[58] Fan J C, Zhu C Y, Fung S, Zhong Y C, Wong K S, Xie Z, Brauer G, Anwand W, Skorupa W, To C K, Yang B, Beling C D and Ling C C 2009 *J Appl Phys* **106** 073709
[59] Ma Y, Du G T, Yang S R, Li Z T, Zhao B J, Yang X T, Yang T P, Zhang Y T and Liu D L 2004 *J Appl Phys* **95** 6268
[60] Zeng Y J, Ye Z Z, Xu W Z, Lu J G, Zhu H P, Zhao B H, Che Y and Zhang S B 2006 *Appl Phys Lett* **88** 262103
[61] Aranovich J A, Golmayo D G, Fahrenbruch A L and Bube R H 1980 *J Appl Phys* **51** 4260
[62] Rhoderick E H and Williams R H 1988 *Metal-Semiconductor Contacts* (Oxford: Clarendon) p 48
[63] Mead C A 1965 *Phys Lett* **18** 218
[64] Neville R C and Mead C A 1970 *J Appl Phys* **41** 3795
[65] Coppa B J, Davis R F and Nemanich R J 2003 *Appl Phys Lett* **82** 400
[66] Auret F D, Goodman S A, Hayes M, Legodi M J, van Laarhoven H A and Look D C 2001 *Appl Phys Lett* **79** 3074
[67] Polyakov A Y, Smirnov N B, Kozhukhova E A, Vdovin V I, Ip K, Heo Y W, Norton D P and Pearton S J 2003 *Appl Phys Lett* **83** 1575
[68] Yuan G, Ye Z, Zhu L, Huang J, Qian Q and Zhao B 2004 *J Crystal Growth* **268** 169
[69] Oh D C, Kim J J, Makino H, Hanada T, Cho M W, Yao T and Ko H J 2005 *Appl Phys Lett* **86** 042110
[70] Mosbacker H L, Strzhemechny Y M, White B D, Smith P E, Look D C, Reynolds D C, Litton C W and Brillson L J 2005 *Appl Phys Lett* **87** 012102
[71] Sheng H, Muthukumar S, Emanetoglu N W and Lu Y 2002 *Appl Phys Lett* **80** 2132
[72] Ip K *et al.* 2004 *Appl Phys Lett* **84** 5133

[73] Kim S-H, Kim H-K and Seong T-Y 2005 *Appl Phys Lett* **86** 112101
[74] Kim S-H, Kim H-K and Seong T-Y 2005 *Appl Phys Lett* **86** 022101
[75] von Wenckstern H, Kaidashev E M, Lorenz M, Hochmuth H, Biehne G, Lenzner J, Gottschalch V, Pickenhain R and Grundmann M 2004 *Appl Phys Lett* **84** 79
[76] Weichsel C, Pagni O and Leitch A W R 2005 *Semicond Sci Technol* **20** 840
[77] Gu Q L *et al.* 2007 *Appl Phys Lett* **90** 122101
[78] Gu Q L *et al.* 2008 *J Appl Phys* **103** 093706
[79] Schifano R, Monakhov E V, Grossner U and Svensson B G 2007 Appl Phys Lett **91** 193507
[80] Brillson L J, Mosbacker H L, Hetzer M J, Strzhemechny Y, Jessen G H, Look D C, Cantwell G, Zhang J and Song J J 2007 Appl Phys Lett **90** 102116
[81] Brauer G, Anwand W, Grambole D, Skorupa W, Hou Y, Andreev A, Teichert C, Tam K H and Djurisic A B 2007 *Nanotechnology* **18** 195301
[82] Kresse G and Hafner J 1993 *Phys Rev B* **47** 558; Kresse G and Hafner J 1994 *Phys Rev B* **49** 14251; Kresse G and Furthmüller J 1996 *Comput Mater Sci* **6** 15; Kresse G and Furthmüller J 1996 *Phys Rev B* **54** 11169
[83] Kresse G and Hafner J 1994 *J Phys: Condens Matter* **6** 8245; Kresse G and Joubert D 1999 *Phys Rev B* **59** 1758
[84] Puska M J and Nieminen R M 1983 *J Phys F: Metal Phys* **13** 333; Seitsonen A P, Puska M J and Nieminen R M 1995 *Phys Rev B* **51** 14057
[85] Van de Walle C G 2000 *Phys Rev Lett* **85** 1012
[86] Wardle M G, Goss J P and Briddon P R 2006 *Phys Rev Lett* **96** 205504
[87] Takenaka H and Singh D J 2007 *Phys Rev B* **75** 241102(R)

Journal of Physics: Conference Series **265** (2011) 012003 doi:10.1088/1742-6596/265/1/012003

Vacancy defects in III-nitrides: what does positron annihilation spectroscopy reveal?

Filip Tuomisto
Department of Applied Physics, Aalto University, P.O. Box 11100, FI-00076 Aalto, Finland

E-mail: filip.tuomisto@tkk.fi

Abstract. The purpose of this paper is to present a short review and comparison of the results obtained with positron annihilation spectroscopy studies of vacancy defects in AlN, GaN and InN. The combination of positron lifetime and Doppler broadening techniques with theoretical calculations has provided the means to deduce both the identities and the concentrations of the vacancies in these materials, while performing measurements as a function of temperature has given information on the charge states of the detected defects. The III-sublattice vacancies are common defects in all the III-nitrides, and they compensate donors either by forming vacancy-impurity complexes or by providing deep states for electrons. In some cases, N vacancies have also been observed.

1. Introduction

The material family of the III-nitrides is a class of important semiconductor materials for applications in optoelectronics at blue and ultraviolet wavelengths and electronic devices operating at high temperatures and high voltage. The importance of intrinsic point defects in nitrides is due to their role in determining the electrical and optical properties of the material, as well as their influence on the material behavior in various processing steps such as ion implantation and thermal annealing. Point defects are created in semiconductor materials during growth, where their formation is governed by thermodynamics and growth kinetics. They can be introduced in concentrations much larger than those given by the thermodynamic equilibrium by means of irradiation of the material, e.g., by electrons. The study of the formation of point defects in different conditions hence yields important information on the basic physical properties of the semiconductor materials in question.

Positron annihilation spectroscopy is an effective method for the investigation of vacancy-type defects in semiconductors. Positrons implanted into a sample can get trapped in and localize at neutral and negative vacancies due to the missing positive ion core. This results in observable changes in the measurable annihilation characteristics, i.e. the positron lifetime and the momentum distribution of the annihilating positron-electron pair (Doppler broadening). The annihilation data can be used to determine the vacancy concentration as well as to distinguish between different vacancy types and their chemical environment.

Positron studies have now been performed in nitride semiconductors for about 10 years after the identification of the Ga vacancy in bulk GaN crystals [1]. In this paper, a short review of the knowledge obtained by positron annihilation spectroscopy on the vacancy defects in nitrides is given, concentrating on the recent results on group-III vacancies and related complexes. The results obtained in GaN, InN and AlN are compared and the similarities and differences are discussed.

2. Experimental details

2.1. Measurements in bulk crystals

The positron lifetimes in bulk crystals were measured with a conventional fast-fast coincidence spectrometer with a time resolution of 250 ps [2]. Two identical sample pieces were sandwiched with a 20 μCi positron source (^{22}Na deposited on 1.5 μm Al foil). Typically 2 × 10^6 annihilation events were collected in each positron lifetime spectrum. The lifetime spectrum $n(t)= \Sigma_i I_i \exp(-t/\tau_i)$ was analyzed as the sum of exponential decay components convoluted with the Gaussian resolution function of the spectrometer after subtracting the constant background and annihilations in the source material (typically a few percent, for details see *e.g.* Ref. [3]) The positron in state i annihilates with a lifetime τ_i and an intensity I_i. The state in question can be the delocalized state in the lattice or the localized state at a vacancy defect. The increase of the average lifetime $\tau_{ave} = \Sigma_i I_i\, \tau_i$ above the bulk lattice lifetime τ_B shows that vacancy defects are present in the material. This parameter that coincides with the center of mass of the spectrum is insensitive to the decomposition procedure, and even a change as small as 1 ps in its value can be reliably measured. In the case of one type of vacancy defect with specific lifetime τ_V, the decomposition of the lifetime spectrum into two components τ_1 and τ_2 is straightforward to interpret. The second lifetime component $\tau_2 = \tau_V$ gives directly the vacancy specific lifetime and the first lifetime component is $\tau_1 = (\tau_B^{-1} + \kappa_V)^{-1} < \tau_B$, where τ_B is the positron lifetime in the delocalized state in the lattice and κ_V the positron trapping rate into the vacancy defects.

The temperature dependence of the average positron lifetime τ_{ave} is analyzed with the model of trapping and escape rates of positrons, explained in detail in earlier works [2, 4, 5, 6]. In this model, the trapping coefficient μ_V to a neutral vacancy is independent of temperature and to a negatively charged vacancy it varies as $T^{-0.5}$. The trapping rate of positrons into the vacancies (concentration c_V) is $\kappa_V = \mu_V c_V$ ($\mu_V = 3 \times 10^{15}$ cm^{3}s^{-1} for Ga vacancies in GaN at 300 K). Positrons can get trapped also at hydrogen-like Rydberg states surrounding negative-ion-type defects (shallow traps for positrons). The positron trapping rate at the Rydberg state μ_R varies also as $T^{-0.5}$, which is the result predicted by theory for the transition from a free state to a bound state in a Coulomb potential [6]. The thermal escape rate from the Rydberg state can be written as

$$\delta_{st} = \mu_R \left(\frac{m_+ k_B T}{2\pi\hbar^2} \right)^{3/2} \exp\left(-\frac{E_{b,st}}{k_B T} \right) \tag{1}$$

where μ_R is the positron trapping coefficient to the lowest hydrogen-like Rydberg state, $E_{b,st}$ is the positron binding energy of the lowest Rydberg state (typically < 0.1 eV), and $m_+ \cong m_0$ is the effective mass of the positron. In principle, positrons can also escape from the Rydberg states around negatively charged vacancies, but we assume that the transition from the Rydberg state to the ground state in the vacancy is fast enough so that this effect can be neglected. An effective trapping rate of the shallow traps can thus be defined as

$$\kappa_{st}^{eff} = \frac{\kappa_{st}}{1 + \delta_{st} / \lambda_{st}} \tag{2}$$

where $\lambda_{st} \cong \lambda_B$ is the annihilation rate of positrons trapped at the Rydberg state, which coincides with the annihilation rate λ_B from the delocalized state in the bulk lattice, and $\kappa_{st} = \mu_R\, c_{st}$ is directly related to the concentration of the negative ions.

Finally, the decomposition of the lifetime spectra into several lifetime components gives the possibility to determine experimentally the fractions of positrons annihilating in various states. The average lifetime can be written as [2]

$$\tau_{ave} = \eta_B \tau_B + \sum_j \eta_{D,j} \tau_{D,j}, \tag{3}$$

where η_B and τ_B are the annihilation fraction and positron lifetime in the free state in the lattice, and $\eta_{D,j}$ and $\tau_{D,j}$ are the corresponding values in bound states at the defect D_j. The annihilation fractions are related to the trapping rates through

$$\eta_B = \frac{\lambda_B}{\lambda_B + \sum_j \kappa_{D,j}^{eff}}, \quad \eta_{D,j} = \frac{\kappa_{D,j}^{eff}}{\lambda_B + \sum_j \kappa_{D,j}^{eff}}. \tag{4}$$

2.2. Measurements in epitaxial thin films

The positron annihilation characteristics were measured in thin epitaxial films by using a mono-energetic continuous positron beam. As lifetime experiments are not possible with a continuous beam, the Doppler broadening of the annihilation radiation was monitored as a function of the beam energy, providing a depth profile of the signal. The motion of the annihilating electron-positron pair causes a Doppler shift in the annihilation radiation $\Delta E = cp_L / 2$, where p_L is the longitudinal momentum component of the pair in the direction of the annihilation photon emission. This causes the broadening of the 511 keV annihilation line. The shape of the 511 keV peak gives thus the one-dimensional momentum distribution $\rho(p_L)$ of the annihilating electron-positron pairs. A Doppler shift of 1 keV corresponds to a momentum value of $p_L = 0.54$ a.u ($\approx$.3.91 $\times$ 10^{-3} m_0c).

The Doppler broadening can be experimentally measured using a Ge gamma ray detector with a good energy resolution. The typical resolution of a detector is around 1–1.5 keV at 500 keV. This is considerable compared to the total width of 2–3 keV of the annihilation peak meaning that the experimental lineshape is strongly influenced by the detector resolution. Therefore, various shape parameters are used to characterize the 511 keV line. The low electron-momentum parameter S is defined as the ratio of the counts in the central region (typically $p_L < 0.4$ a.u.) of the annihilation line to the total number of the counts in the line. In the same way, the high electron-momentum parameter W is the fraction of the counts in the wing regions of the line (typically $p_L > 1.5$ a.u.). Due to their low momenta, mainly valence electrons contribute to the region of the S parameter. On the other hand, only core electrons have momentum values high enough to contribute to the W parameter. Therefore, S and W are sometimes called the valence and core annihilation parameters, respectively. The S and W parameters are related to the annihilation fractions in a similar manner as the average positron lifetime in Eq. (3).

The high-momentum part of the Doppler broadening spectrum arises from annihilations with core electrons that contain information on the chemical identity of the atoms. Thus the detailed investigation of core electron annihilation can reveal the nature of the atoms in the regions where positrons annihilate. In order to study the high-momentum part in detail, the experimental background needs to be reduced. A second gamma ray detector can be placed opposite to the Ge detector and the only events that are accepted are those for which both 511 keV photons are detected [7, 8]. Depending on the type of the second detector, electron momenta even up to $p \approx 8$ a.u. ($\approx 60 \times 10^{-3}$ m_0c) can be measured with the coincidence detection of the Doppler broadening.

3. Vacancy defects in GaN

Since the identification of native Ga vacancies by positron annihilation spectroscopy in high nitrogen pressure (HNP) grown bulk GaN crystals [1], more than a hundred studies of GaN with positrons have been published [9]. In the following, some of the main findings about the nature of vacancy defects in GaN are presented. Figure 1 shows the average positron lifetime measured in a variety of thick bulk or quasi-bulk GaN samples. The average positron lifetime (τ_{ave}) in most of the measured GaN samples is

Journal of Physics: Conference Series **265** (2011) 012003 doi:10.1088/1742-6596/265/1/012003

above the bulk lifetime of 160 ps [10] ($\tau_{ave} > \tau_B = 160$ ps) in some temperature range, indicating that positrons annihilate as trapped at vacancy defects. These vacancies have been identified as Ga vacancies by the second lifetime component of $\tau_2 = 235 \pm 10$ ps (figure 1b) separated from the lifetime spectra [1, 10–12]. The temperature dependence of the average positron lifetime in O-doped and undoped GaN samples grown by hydride vapor phase epitaxy (HVPE) shows enhanced positron trapping at Ga vacancies at low temperatures, indicating that the Ga vacancies are negatively charged. On the other hand, the temperature dependence of τ_{ave} in the GaN samples grown by the high nitrogen pressure (HNP) method indicates that negative non-open volume defects that act as shallow traps for positrons compete in positron trapping with the Ga vacancies at low temperatures. These negative centers have been identified as Mg_{Ga}^{-} acceptors in the HNP GaN samples by comparing to data obtained with secondary ion mass spectrometry [1, 10]. The lack of these shallow traps in the undoped and O-doped HVPE GaN samples in figure 1 and in the data in Refs. [11–13] implies that Ga vacancies are the dominant intrinsic acceptor defects in *n*-type GaN when impurities acting as acceptors are not present [14]. This conclusion has also been confirmed by the very good match of vacancy and total acceptor densities, obtained by combining positron annihilation and temperature-dependent Hall measurements [13]. The experiments have also shown the universal role of the Ga vacancy as the most important compensating acceptor over four orders of magnitude of intentional oxygen doping [11].

By combining positron lifetime and coincidence Doppler broadening experiments with state-of-the-art theoretical calculations, the in-grown Ga vacancy defects in GaN with sufficient O content have been identified as V_{Ga}–O_N pairs [11, 15]. In contrast, in electron irradiation experiments [12, 16], isolated Ga vacancies are produced. Figure 1 shows also the average positron lifetime measured in an as-irradiated (2-MeV electrons at room temperature) undoped HVPE-GaN sample and a sample subsequently annealed at 775 K. It can be seen from figure 1b that the second lifetime components extracted from the lifetime spectra are the same in the case of in-grown O-complexed and isolated Ga vacancies, in good agreement with theory [11]. The average positron lifetime in the as-irradiated sample shows the coexistence of a high concentration of Ga vacancies and irradiation-induced (and hence intrinsic) negative ions. Interestingly, after the annealing at 775 K, the temperature threshold for positron escape from these negative ions changes indicating a change in the binding energy of positrons to these defects. This has been interpreted as a change in the charge state (from 2^- to 1^-) of the negative ions.

The electron irradiation experiments [12, 16] have shown that the migration barrier of the isolated Ga vacancies is $E_M = 1.8 \pm 0.2$ eV. Annealing experiments in as-grown HVPE GaN samples at high temperatures and high nitrogen pressure, where a redistribution of the V_{Ga}–O_N pairs [15, 17, 18] was observed, have given the possibility to determine the binding energy of the pair as $E_B = 1.6 \pm 0.2$ eV. This is in excellent agreement with theoretical results [19, 20]. Also the theoretical prediction about the formation energy of the isolated Ga vacancy was confirmed in those experiments. Studies of the differences in Ga vacancy distributions in GaN grown in the Ga- and N-polar directions as well as in non-polar directions, both in homo- and hetero-epitaxial HVPE samples, have given further information on the formation of the V_{Ga}–O_N pairs [3, 10, 21, 22]. These results demonstrate that Ga vacancies are created thermally as isolated at the high growth temperatures in both HVPE and HNP GaN, but their ability to form V_{Ga}–O_N complexes determines the fraction of vacancy defects surviving the cooling down. In addition, those results show that the O incorporation from the growth ambient and the subsequent vacancy formation depends on the growth polarity due to the different sticking coefficients of O for different growth surfaces. It is important to note that Si doping does not have the same effect, as the binding between the Si donors and Ga vacancies is much less efficient due to the larger distance between the individual acceptor (Ga vacancy) and donor (substitutional Si) defects [23].

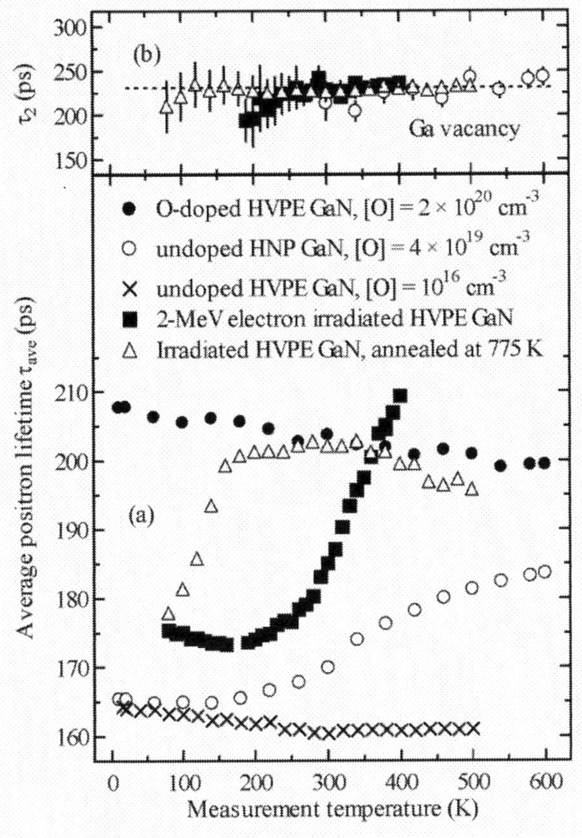

Figure 1. (a) The average positron lifetime measured as a function of temperature in different thick bulk or quasi-bulk GaN samples. Vacancies are present when the average positron lifetime is above the bulk lifetime of $\tau_B \approx 160$ ps. The temperature dependence of the average positron lifetime demonstrates the different effects of negative charge of both the vacancies and acceptor-type impurities. (b) The second lifetime component extracted from the lifetime spectra is the same, $\tau_2 \approx 235$ ps, in all the measured samples. This component is attributed to Ga vacancies, either isolated or complexed with O impurities.

The above presented results show that Ga vacancies act as dominant compensating centers in *n*-type GaN. On the other hand, in *p*-type GaN, where the formation of Ga vacancies is energetically unfavourable due their acceptor nature, the natural question is whether N vacancies could compensate the doping. The detection of vacancy defects on the N sublattice with positrons is not evident due to the small open volume generated by the missing N atom. Nevertheless evidence of the existence of N vacancies complexed with Mg (V_N-Mg_{Ga}) has been obtained with positrons in Mg-doped (*p*-type) GaN grown by metal-organic chemical vapour deposition (MOCVD) [24,25]. It is worth noting that even though the vacancy concentrations are similar relative to the doping densities (a few percent at most) in both *n*- and *p*-type GaN, the vacancies are dominant compensating centers only in *n*-type GaN, while in *p*-type GaN other defects and impurities such as hydrogen play the most important role. Isolated neutral N vacancies have also been provisionally identified in electron-irradiated (0.5 MeV energy) samples recently [12].

The effects of varying growth conditions (such as temperature, stoichiometry and growth polarity) have been studied with a slow positron beam in a variety of thin film samples grown by either molecular beam epitaxy (MBE) [26] or MOCVD [23, 27]. These studies show that the growth stoichiometry has a very strong impact on the formation of Ga vacancies: the more N rich the growth, the more there are Ga vacancies in the material. Also a higher growth temperature results in a higher Ga vacancy concentration. Interestingly, these findings hold also when comparing the thin-film techniques to bulk or quasi-bulk growth techniques, suggesting that the formation of Ga vacancies is governed by thermodynamical effects irrespective of the growth method. This conclusion is supported by the results obtained in samples grown with various polarities [3, 10, 21, 22], where it has been observed that the sticking of O on different growth surfaces and the following differences in O content in the material dictate the formation of Ga vacancies (in addition to the growth stoichiometry).

4. Vacancy defects in InN and AlN

The other two nitrides, *i.e.*, InN and AlN, have been far less studied with positrons than GaN [9]. The reason for this has been the lack of suitable samples of sufficiently high quality, and especially in the case of InN, the inexistence of (thick) bulk samples. However, the material quality has improved significantly over the past couple of years, and extensive studies of InN and AlN with positrons should be anticipated.

4.1. Vacancy defects in InN

Figure 2 shows the S parameter measured at room temperature as a function of positron implantation energy in MBE-grown InN samples of different thicknesses and dates of growth (2002 vs. 2006) [28, 29]. When positrons are implanted close to the sample surface with E = 0 – 1 keV, the same S parameter of S = 0.48 is recorded in all the InN samples. These values characterize the defects and chemical nature of the near-surface region of the samples at the depth 0 – 5 nm. The region of constant S is different from sample to sample due to the different thicknesses of the layers, and extends from a few keV up to 30 keV in the thickest sample. The data recorded at the energies where S is constant can be taken as characteristic of the layer. The decrease of the S parameter at higher implantation energies is due to the increasing fraction of positrons annihilating in the sapphire substrate, for which the characteristic S parameter is roughly S = 0.41. It is clearly seen that the S parameter of InN decreases with increasing layer thickness (when the thicknesses are below roughly 1 micron) and also with the general improvement of the material quality over the years.

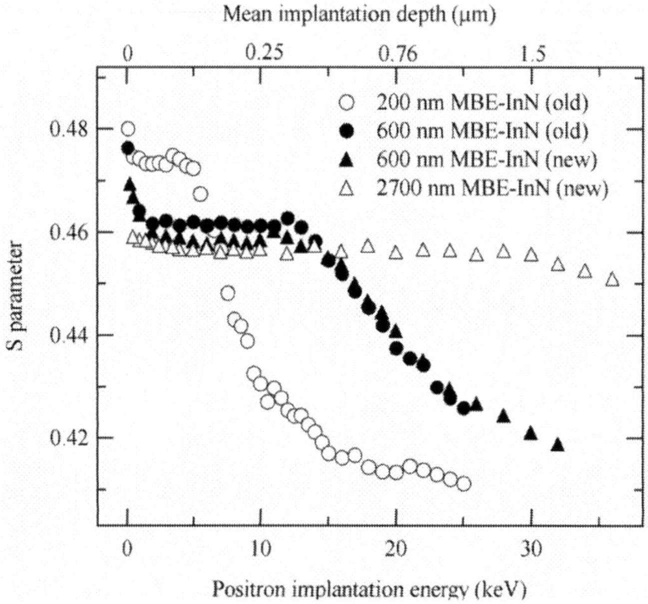

Figure 2. The S parameter measured as a function of positron implantation energy in selected MBE-grown InN layers.

The thickest (2700 nm) InN layer is essentially defect-free from the point of view of positrons [29]. The increased S parameter in the other layers indicates that the positron-electron momentum distribution is narrower than in the defect-free reference sample. The narrowing is due to positrons annihilating as trapped at vacancy defects, where the electron density is lower and the probability of annihilation with high-momentum core electrons is reduced compared to that of delocalized positrons in the lattice. Hence the increased S parameter is a clear sign of vacancy defects present in the measured layers. When the S and W parameters of the InN layers are plotted as function of each other (not shown, see Ref. [28]) they form a line indicating that positrons annihilate from two different states. These have been identified as the delocalized state in the InN lattice and the In vacancy by combining Doppler broadening and positron lifetime measurements [28], the latter performed with a pulsed positron beam.

The concentration of V_{In} can be estimated from the S parameter data with the positron trapping model. When the cation vacancies are the only defects trapping positrons, their concentration can be determined with the simple formula

$$[V_{In}] = \frac{N_{at}(S - S_B)}{\mu_V \tau_B (S_V - S)}, \quad (5)$$

where τ_B = 185 ps is the positron lifetime in the InN lattice [28], $\mu_V = 3 \times 10^{15}$ s^{-1} is the positron trapping coefficient [1, 28] and $N_{at} = 6.367 \times 10^{22}$ cm^{-3} is the atomic density of InN. For the S parameter at the In vacancies we take S_V/S_B = 1.050 [28, 29]. From these values, the In vacancy concentrations can be extracted to be in the 10^{16} – 10^{19} cm^{-3} range in the different layers. The In vacancy concentration has been shown to decrease dramatically with the layer thickness in the first micron, and to correlate negatively with the electron mobility suggesting that the In vacancies act as important scattering centers in InN, although their concentration is typically not sufficient for efficient compensation of the *n*-type conductivity [28]. Interestingly, the formation of In vacancies seems not to be influenced by the growth stoichiometry in MOCVD growth [30], and recent results suggest this holds also in the case of MBE growth [31].

As in the case of GaN, irradiation experiments have been performed in InN in order to study the introduction of In vacancies and their recovery in subsequent thermal treatments [29]. It has been shown that light particle irradiation produces In vacancies, but with an introduction rate more than one order of magnitude lower than that of Ga vacancies in GaN. Instead, intrinsic negative ion type defects are produced at a similar rate as in GaN, indicating that they are the dominant compensating defects in irradiated n-type InN. In fact, InN becomes more n-type in the irradiation experiments, and it has been suggested that this is due to the production of N vacancies that would act as donors [32]. The results obtained with positrons both before and after [31] the thermal treatments support the existence of a high concentration of N vacancies in the irradiated material, even if the N vacancies have not been observed directly.

4.2. Vacancy defects in AlN

Figure 3 presents the average positron lifetime measured as a function of temperature in bulk AlN crystal [33] grown by physical vapor transport (PVT) [34] together with the data from a bulk HNP-GaN sample [10]. The temperature dependence of the average lifetime τ_{ave} measured in AlN is qualitatively similar to that of GaN. It is constant, about 158 ps, at temperatures below 300 K, above which it starts increasing with increasing temperature reaching about 170 ps at 600 K. The lifetime spectra recorded at 450 – 600 K can be decomposed into two components, the longer of which (not shown in the figure) is constant, $\tau_2 = 210 \pm 15$ ps, as a function of temperature. By comparison to GaN, the behavior as a function of temperature is interpreted as the competition of negative ion type defects (shallow traps with no open volume) and vacancy defects in trapping of positrons. The longer lifetime component has been preliminarily attributed to Al vacancies [33], and the positron lifetime of the AlN lattice has been estimated to be about 155 ps.

The concentrations of both the Al vacancies and the negative ions can be estimated from the data using Eqs. (1) – (4) and assuming a trapping coefficient of $\mu_V = 3 \times 10^{15}$ cm^3s^{-1} for the Al vacancies at 300 K. The Al vacancy concentration of about 1×10^{17} cm^{-3} is extracted from the data measured at 600 K, taking into account the temperature dependence of the trapping coefficient, and gives a lower limit estimate as the lifetime still increases at this temperature as seen in figure 3. However, as the increase of τ_{ave} with temperature seems to slow down when approaching 600 K, the actual vacancy concentrations are not likely to be more than twice the estimated values. The concentration of the negative ions can be estimated from the data below 300 K, where no thermal escape from the hydrogenic states is observed, giving $c_{st} \approx 4 \times 10^{18}$ cm^{-3}. It is important to note that the concentration of the negative ions is an order of magnitude higher than that of the Al vacancies, hence making them the dominant negative (acceptor) centers in AlN, while in GaN with a similar O content the concentrations of Ga vacancies and negative ions are similar. Surprisingly, the comparison of the concentrations of impurities and negative ions suggests that the latter could be related to oxygen rather than to carbon impurities. It should be noted, though, that in contrast to the situation in GaN, O in AlN has been predicted to exhibit negative charge states as well [33].

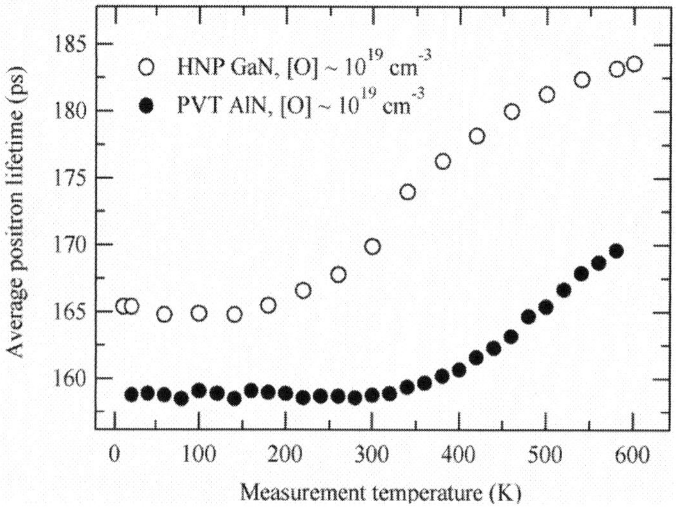

Figure 3. The average positron lifetime measured as a function of temperature in bulk GaN and AlN samples.

In addition to bulk crystal growth, both MBE and MOCVD have been used to grow AlN thin films on different substrates. Some positron studies have been performed on both kinds of materials [35, 36]. However, as proper identification of the vacancy and lattice parameters combining Doppler broadening and positron lifetime experiments is still lacking, the interpretations need to be considered with reasonable care. The results are comparable to those obtained in other nitrides (and semiconductors in general) in the sense that higher lattice mismatch between the substrate and the AlN layer seems to lead to a higher S parameter (and hence to a higher vacancy content). In addition, the dependence of the vacancy content on the growth stoichiometry seems to be rather weak for low V/III ratios, but effects become visible at very high V/III ratios. Finally, as in the case of GaN and InN, a higher growth temperature causes a higher vacancy content.

5. Discussion

As the positron research in InN and AlN is still rather scarce, it is difficult to make a full comparison of the three nitride materials. However, many properties can be compared separately between InN and GaN, and AlN and GaN. In the case of bulk GaN and AlN crystals, it is fascinating to observe that similar kind of interplay of vacancies and negative ion type defects (most likely impurities) is found, in spite of the difference between the conductivities: GaN is quite heavily *n*-type, while the Fermi level in AlN is more than 1 eV below the conduction band minimum making it "almost" *n*-type. However, even if in GaN the in-grown vacancies are bound to O impurities, in AlN this might not be the case due to the possible negative charge of the O impurities at sufficiently high Fermi level positions. It should be noted that the formation energy of a V_{Al}–O_N pair is predicted to be lower than that of a single Al vacancy [19], but the lack of suitable direct formation sites for the pair makes it probable that these pairs would form in AlN in a similar manner as in GaN, i.e., by formation of isolated vacancies and pairing during cooling down. However, if both the constituents of the complex are negatively charged when isolated, it is likely that there is quite a high barrier to be overcome during migration and pair formation.

Irradiation experiments have been performed in InN and GaN, but for the time being not in AlN. These experiments have shown that GaN behaves "normally" in the particle irradiation experiments, i.e., the introduction rates of the defects on both sublattices are on an expected level, both based on theory and other semiconductor materials [12]. Naturally there are unresolved issues, such as why some of the irradiation-induced Ga vacancies survive very high annealing temperatures (up to 800°C), and what are the exact effects of hydrogen as it is the only impurity abundant enough in high-quality undoped material to create complexes with a significant fraction of vacancies. In InN on the other hand, the production of In vacancies in particle irradiation is much slower than that could be expected, and in light of other peculiarities of the material (among others, the Fermi level in InN moves away from the midgap instead of approaching it in the irradiation) it has been proposed that the In vacancies undergo a similar acceptor-to-donor transition as the $V_{Ga} \leftrightarrow V_{As}+As_{Ga}$ transition in GaAs [29]. In fact, the recent annealing experiments suggest that very efficient vacancy clustering takes place already at very low In vacancy concentrations, supporting the view that very high N vacancy concentrations are present after the irradiation, and that some of them might actually originate from the above-mentioned complexes.

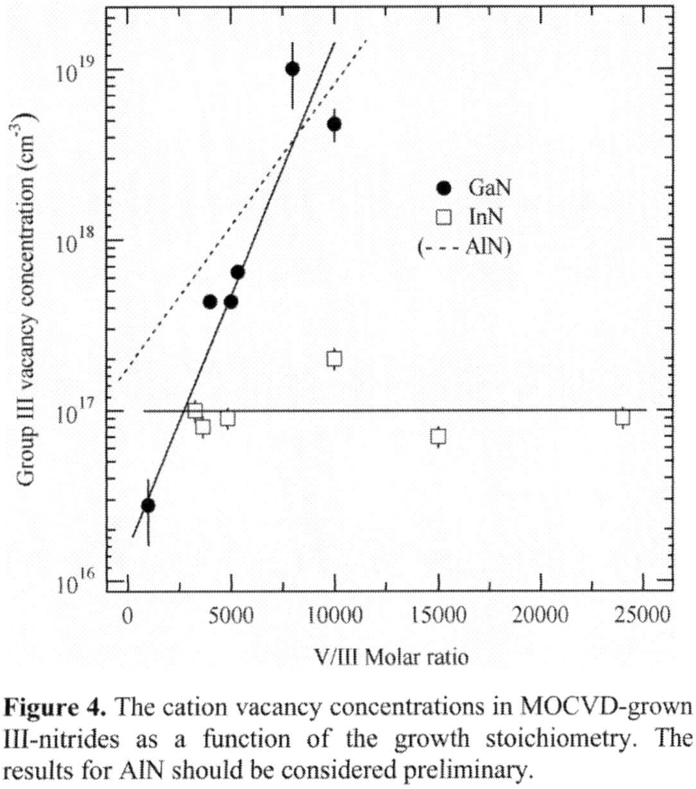

Figure 4. The cation vacancy concentrations in MOCVD-grown III-nitrides as a function of the growth stoichiometry. The results for AlN should be considered preliminary.

The three nitride materials can be compared in terms of vacancy formation in different MOCVD growth conditions. In all three materials, a higher growth temperature results in a higher vacancy concentration. On the other hand, other growth parameters can have different effects. Figure 4 shows the concentrations of Ga, In and Al vacancies in GaN, InN and AlN, respectively, as a function of the V/III ratio during growth. The difference in the behavior of the cation vacancies in GaN, InN and AlN is evident: the growth stoichiometry has no effect on the vacancy content in InN, while the effects are clear in GaN and AlN – the higher the V/III ratio the higher the III-vacancy concentration. A plausible explanation for the differences can be found in the vacancy formation energies and the temperatures of the MOCVD growth. In *n*-type (or "almost" *n*-type for AlN) material, the calculated formation energy of the Ga and Al vacancies is about 1 eV, while it is almost 3 eV for the In vacancy [19,37]. On the other hand, the growth temperature of AlN and GaN is around or above 1000 °C, but only around 500–600°C in the case of InN. As the concentrations of the Ga, Al and In vacancies are similar in the samples with low V/III ratios, the formation of the In vacancies must be dictated by other effects, such as strain or presence of dislocations, than the thermal formation (and subsequent stabilization by e.g. impurities) of an isolated In vacancy in an otherwise perfect lattice. On the other hand, the observed Al and Ga vacancy concentrations are of the same order of magnitude as could be expected from the growth temperature and the calculated formation energy, given that the vacancies (which are mobile already at relatively low temperatures) are stabilized by impurities (such as O in GaN) or other defects relatively close to the growth temperature. Hence it can be understood why the stoichiometric conditions affect the final Ga vacancy concentration in GaN and possible Al concentration in AlN more than the In vacancy concentration in InN.

The future directions in nitride research with positrons are varied. Concerning the fundamental material properties, the basic understanding of the vacancy defects and their interactions with impurities and other defects in all three materials needs to improved, especially in InN and AlN that have been studied much less than GaN. To this end systematic studies of effects of both *n*- and *p*-type doping (when possible) and of growth conditions, including polarity, need to be performed in addition to irradiation experiments also at low temperatures that provide valuable information on isolated intrinsic defects. A subject not touched upon in this paper is ion implantation and the defects induced by this kind of processing, although some studies already exist [38, 39]. As this is a standard processing technique in semiconductor technology, and is becoming an attractive way of making devices in nitrides as well, the produced defects and their impact should be better understood.

Up to now mostly binary alloys have been studied, with just a handful of papers published on ternary alloys such as $In_{1-x}Ga_xN$ or $Al_{1-x}Ga_xN$ [40, 41]. However, these are the materials that are of very high technological interest as they form the active layers in opto-electronic devices. The preliminary experiments that have been performed suggest that the vacancies in these ternary and eventually quaternary alloys do not behave in a similar manner as in the binary constituents. Hence these materials should and probably will be studied in the near future. Finally, closely related to the ternary alloys, quantum structures and possibly quantum dots will be the subject of future studies. The nitride quantum structures provide an interesting playground for both theoretical and experimental positron physics thanks to the polarization-induced in-grown saw-tooth-like electric fields in the wurtzite heterostructures [42–44]. Positrons are prone to be localized by the electric fields at the interfaces of such structures, and hence they could provide information about the local atomic configurations of the interfaces that may have a significant impact on device operation.

6. Summary

A short review and comparison of the results obtained with positron annihilation spectroscopy in studies of vacancy defects in AlN, GaN and InN are presented. The efficient use of positron lifetime and Doppler broadening techniques complemented by theoretical calculations has given the possibility to deduce both the identities and the concentrations of the vacancies in these materials. Measurements as a function of temperature have given information on the charge states of the detected defects. The III-sublattice vacancies are common defects in all the III-nitrides, and they compensate donors either by forming vacancy-impurity complexes or by providing deep states for electrons. In some cases also N vacancies have been observed. The formation of these defects under different growth conditions has been studied in the three materials. Also the effects of particle irradiation have been explored.

Acknowledgments
The author wishes to thank all those who have participated in the research presented in this paper, and in particular the past and present members of the positron group at Aalto University. The work has been partially funded by the Academy of Finland and the MIDE programme.

References
[1] Saarinen K, Laine T, Kuisma S, Nissilä J, Hautojärvi P, Dobrzynski L, Baranowski J M, Pakula K, Stepniewski R, Wojdak M, Wysmolek A, Suski T, Leszczynski M, Grzegory I and Porowski S 1997 *Phys. Rev. Lett.* **79** 3030
[2] Saarinen K, Hautojärvi P and Corbel C 1998 *Identification of Defects in Semiconductors* (*Semiconductors and Semimetals* vol 51A) ed Stavola M (New York: Academic Press) p 209
[3] Tuomisto F, Suski T, Teisseyre H, Krysko M, Leszczynski M, Lucznik B, Grzegory I, Porowski S, Wasik D, Witowski A, Gebicki W, Hageman P and Saarinen K 2003 *Phys. Status Solidi (b)* **240** 289
[4] Saarinen K, Seitsonen A P, Hautojärvi P and Corbel C 1995 *Phys. Rev. B* **52** 10932
[5] Corbel C, Pierre F, Saarinen K, Hautojärvi P and Moser P 1992 *Phys. Rev. B* **45** 3386
[6] Puska M J, Corbel C and Nieminen R M 1990 *Phys. Rev. B* **41** 9980
[7] Alatalo M, Kauppinen H, Saarinen K, Puska M J, Mäkinen J, Hautojärvi P and Nieminen R M: 1995 *Phys. Rev. B* **51** 4176
[8] Asoka-Kumar P, Alatalo M, Ghosh V J, Kruseman A C, Nielsen B and Lynn K G 1996 *Phys. Rev. Lett.* **77** 2097
[9] According to Web of Science® by Thomson Scientific in July 2008
[10] Tuomisto F, Saarinen K, Lucznik B, Grzegory I, Teisseyre H, Suski T, Porowski S, Hageman P R and Likonen J 2005 *Appl. Phys. Lett.* **86** 031915
[11] Hautakangas S, Ranki V, Makkonen I, Puska M J, Saarinen K, Xu X and Look D C 2006 *Phys. Rev. B* **73** 193301
[12] Tuomisto F, Ranki V, Look D C and Farlow G C 2007 *Phys. Rev. B* **72** 165207
[13] Oila J, Kivioja J, Ranki V, Saarinen K, Look D C, Molnar R J, Park S S, Lee S K and Han J Y 2003 *Appl. Phys. Lett.* **82** 3433
[14] Saarinen K, Hautakangas S and Tuomisto F 2006 *Phys. Scripta* **T126** 105
[15] Tuomisto F, Hautakangas S, Makkonen I, Ranki V, Puska M J, Saarinen K, Bockowski M, Suski T, Paskova T and Monemar B 2006 *Phys. Status Solidi (b)* **243** 1436
[16] Saarinen K, Suski T, Grzegory I and Look D C 2001 *Phys. Rev. B* **64** 233201
[17] Paskova T, Hommel D, Paskov P P, Darakchieva V, Monemar B, Bockowski M, Suski T, Grzegory I, Tuomisto F, Saarinen K, Ashkenov N and Schubert M 2006 *Appl. Phys. Lett.* **88** 141909
[18] Tuomisto F, Saarinen K, Bockowski M, Suski T, Paskova T and Monemar B 2006 *J. Appl. Phys.* **99** 066105
[19] Mattila T and Nieminen R M 1997 *Phys. Rev. B* **55** 9571
[20] Neugebauer J and Van de Walle C G 1996 *Appl. Phys. Lett.* **69** 503
[21] Tuomisto F, Paskova T, Figge S, Hommel D and Monemar B 2007 *J. Crystal Growth* **300** 251
[22] Tuomisto F, Paskova T, Kröger R, Figge S, Hommel D, Monemar B and Kersting R 2007 *Appl. Phys. Lett.* **90** 121915
[23] Oila J, Ranki V, Kivioja J, Saarinen K, Hautojärvi P, Likonen J, Baranowski J M, Pakula K, Suski T, Leszczynski M and Grzegory I 2001 *Phys. Rev. B* **63** 045205
[24] Hautakangas S, Oila J, Alatalo M, Saarinen K, Liszkay L, Seghier D and Gislason H P 2003 *Phys. Rev. Lett.* **90** 137402
[25] Hautakangas S, Saarinen K, Liszkay L, Freitas, Jr. J A and Henry R L 2005 *Phys. Rev. B* **72** 165303
[26] Rummukainen M, Oila J, Laakso A, Saarinen K, Ptak A J and Myers T H 2004 *Appl. Phys. Lett.* **84** 4887

[27] Saarinen K, Seppälä P, Oila J, Hautojärvi P, Corbel C, Briot O and Aulombard R L 1998 *Appl. Phys. Lett.* **73** 3253
[28] Oila J, Kemppinen A, Laakso A, Saarinen K, Egger W, Liszkay L, Sperr P, Lu H and Schaff W J 2004 *Appl. Phys. Lett.* **84** 1486
[29] Tuomisto F, Pelli A, Yu K M, Walukiewicz W and Schaff W 2007 *Phys. Rev. B* **75** 193201
[30] Pelli A, Saarinen K, Tuomisto F, Ruffenach S and Briot O 2006 *Appl. Phys. Lett.* **89** 011911
[31] Reurings F, Rauch C, Tuomisto F, Jones R E, Yu K M, Walukiewicz W and Schaff W J 2010 *Phys. Rev. B* **82** 153202; Reurings F, Tuomisto F, Gallinat C S, Koblmüller G and Speck J S, *Appl. Phys. Lett.*, in press
[32] Li S X, Yu K M, Wu J, Jones R E, Walukiewicz W, Ager III J W, Shan W, Haller E E, Lu H and Schaff W J 2005 *Phys. Rev. B* **71** 161201
[33] Tuomisto F, Mäki J-M, Chemekova T Yu, Makarov Yu N, Avdeev O V, Mokhov E N, Segal A S, Ramm M G, Davis S, Huminic G, Helava H, Bickermann M and Epelbaum B M 2008 *J. Crystal Growth* **310** 3998
[34] Bickermann M, Epelbaum B M and Winnacker A 2004 *J. Crystal Growth* **269** 432
[35] Koyama T, Sugawara M, Hoshi T, Uedono A, Kaeding J F, Sharma R, Nakamura S and Chichibu S F 2007 *Appl. Phys. Lett.* **90** 241914
[36] Mäki J-M, Tuomisto F, Bastek B, Bertram F, Christen J, Dadgar A and Krost A 2009 *Phys. Status Solidi (c)* **6** 2575
[37] Stampfl C, Van de Walle C G, Vogel D, Krüger P and Pollmann J 2000 *Phys. Rev. B* **61** R7846
[38] Nakano Y and Kachi T 2002 *J. Appl. Phys.* **91** 884
[39] Uedono A, Ito K), Nakamori H, Mori K, Nakano Y, Kachi T, Ishibashi S, Ohdaira T, Suzuki R 2007 *J. Appl. Phys* **102** 084505
[40] Onuma T, Chichibu S F, Uedono A, Sota T, Cantu P, Katona T M, Keading J F, Keller S, Mishra U K, Nakamura S and DenBaars S P 2004 *J. Appl. Phys.* **95** 2495
[41] Slotte J, Tuomisto F, Saarinen K, Moe C G, Keller S and DenBaars S P 2007 *Appl. Phys. Lett.* **90** 151908
[42] Bernardini F, Fiorentini V and Vanderbilt D 1997 *Phys. Rev. B* **56** 10024
[43] Fiorentini V, Bernardini F, Della Sala F, Di Carlo A and Lugli P 1999 *Phys. Rev. B* **60** 8849
[44] Lefebvre P, Morel A, Gallart M, Taliercio T, Allegre J, Gil B, Mathieu H, Damilano B, Grandjean N and Massies J 2001 *Appl. Phys. Lett.* **78** 1252

Journal of Physics: Conference Series **265** (2011) 012004 doi:10.1088/1742-6596/265/1/012004

Determination of defect content and defect profile in semiconductor heterostructures

A Zubiaga[1], J A Garcia[2], F Plazaola[2], J Zúñiga-Pérez[3] and V Muñoz-Sanjosé[3]

[1] Laboratory of Physics, HUT, P.O. Box 1100, 02015 TKK, Espoo, Finland

[2] Zientzia eta Teknologia Fakultatea, Euskal Herriko Unbertsitatea, P. K. 644, 48080, Bilbao, Spain

[3] Universitat de Valencia, Departamento de Fisica Aplicada i Electromagnetisme, Dr. Moliner 50, 46100 Burjassot, Valencia, Spain

E-mail: fernando.plazaola@ehu.es

Abstract. In this article we present an overview of the technique to obtain the defects depth profile and width of a deposited layer and multilayer based on positron annihilation spectroscopy. In particular we apply the method to ZnO and ZnO/ZnCdO layers deposited on sapphire substrates. After introducing some terminology we first calculate the trend that the W/S parameters of the Doppler broadening measurements must follow, both in a qualitative and quantitative way. From this point we extend the results to calculate the width and defect profiles in deposited layer samples.

1. Introduction

The actual technology requires high quality and very well characterized semiconductor layers and interfaces for the design of devices for different type of applications. The characterization of ZnO layers grown over sapphire; lead us to develop the way of obtaining further information from W and S parameters.

The physical and chemical properties of ZnO and the progress in crystal growth processes have renewed the interest on this material due to the potentials it presents in different technological fields, particularly in optoelectronic devices working at blue and ultraviolet wavelengths. On the other hand, due to the relative low cost of sapphire substrates, its availability in large area wafers and its transparency up to about 6 μm irradiation wavelength, sapphire is a potential substrate for mass production of ZnO films and it has been largely used for investigating that material. However, the heteroepitaxy of ZnO on sapphire presents several problems due to differences in their chemical nature, structure (wurtzite or corundum), and lattice parameters. Therefore, a necessary step, prior to the technological and practical use of ZnO in optoelectonics, is the knowledge of the ZnO/sapphire interface microstructure in order to control the defects of the deposited films and to improve the quality of the final devices. Even though, defects in ZnO are characterized by different experimental methods, photoluminescence being one of the most relevant, positron annihilation spectroscopy had already shown to be a promising technique for studying the defect structure in ZnO films [1–6].

The study of interfaces by positron annihilation spectroscopy can be performed using slow positron beams. Once one has access to a beam of monoenergetic positrons the most usual technique to get information of the sample under study is the Doppler broadening of the annihilation peak measured using a Ge detector. We used one monoenergetic slow positron beam, in the 0–40 keV range, of the Laboratory of Physics of the Helsinki Tecnological University. The Ge detector had an energy resolution of 1.24 keV at 511 keV. We used the following energy windows to define the S and W parameters of the annihilation line: $|E_g - 511\text{ keV}| < 0.8$ keV ($p_L/m_0c \leq 3 \times 10^{-3}$, where m_0 is the electron mass) for the central S parameter and 2.9 keV $< |E_g - 511\text{ keV}| < 7.4$ keV ($11 \times 10^{-3} \leq p_L/m_0c \leq 29 \times 10^{-3}$) for the wing W parameter. The absolute values of the parameters are meaningless, because they depend on the widths and positions of the windows. Only the changes of the parameters are important. To facilitate comparisons between various experiments, it is customary to report relative values like S/S_{ref} and W/W_{ref}, where a good reference is provided by free annihilation in the lattice. These relative values are rather independent of the energy windows and of the small variations in the energy resolutions of different Ge detectors.

The line shape parameters have characteristic values for free positrons in each material, depending on the materials electron momentum distribution. When positrons are trapped, the line shape is characteristic of the trapping defect. For a vacancy-type defect, the density of valence electrons is reduced. This leads to a narrowing of the momentum distribution, which is seen as an increase in S. On the other hand, the localized positron in a vacancy-type defect has less overlap with core electrons than a free positron, leading to a decrease in the core annihilation parameter W.

If the fraction η_D of positrons are trapped and annihilate at a defect D, then the parameters can be expressed as superposition

$$S = (1 - \eta_D)S_B + \eta_D S_D \tag{1}$$

$$W = (1 - \eta_D)W_B + \eta_D W_D \tag{2}$$

where the subscript B refers to the free state in the lattice. The changes $S - S_B$ and $W - W_B$ are proportional to the trapping fraction η_D. A new parameter, called R_D, can be defined if the ratio of the changes are taken into account [7–9]

$$R_D = |\Delta S/\Delta W| = |(S - S_B)/(W - W_B)| = |(S_D - S_B)/(W_D - W_B)| \tag{3}$$

which is independent of η_D and thus characteristic of the defect D.

A further illustration of the usefulness of the parameters is given in an (W, S) plot. Equations 1 and 2 define a segment of a straight line in the (W, S) plane. The slope of the line is R_D. If there is no trapping ($\eta_D = 0$), the point (W_B, S_B) represents the free positron state. With saturation trapping ($\eta_D = 1$) we get the point (W_D, S_D) corresponding to the trapped state. A partial trapping is just an intermediate state of the straight line that is between those two points. Thus any sample, which contains an unknown concentration of the defect D should have its (W, S) value on the straight line from (W_B, S_B) to (W_D, S_D), provided that no other positron traps are present.

2. Bilayer structures

The ZnO films that triggered our interest in the matter of the paper were grown at atmospheric pressure in a horizontal MOCVD reactor (MR Semicon 102) with a two-inlet configuration, avoiding pre-reactions in the gas phase between the zinc and oxygen precursors: dimethylzinc-triethylamine (DMZn-TEN) and tertiary-butanol (t-butanol). The total nitrogen flow rate through the reaction chamber was kept constant at 5.6 l/min, while the partial pressures of the precursors had a pressure ratio $R_{VI/II} = 5$.

Flat sapphire substrates of different orientations (1 1 -2 0), (0 0 0 1), (1 0 -1 0) and (1 -1 0 2) (referred as A-, C-, M-, and R-planes, respectively), were used as received. For sapphire planes A and M all experiences were carried out at 420 °C without any prior heat treatment of the substrates. The deposition time was 1000 or 7200 seconds. For C and R planes of sapphire another strategy was used too: a growth of a ZnO buffer layer, grown at 350 °C during 4 min (the estimated buffer layer thickness is about 60–70 nm) and a posterior growth at 420 °C. For more details on the MOCVD growth, see [10].

An un-doped single crystal from Eagle Picher, which did not present traps for positrons at room temperature and has a positron lifetime of 170 ps [11], was studied as reference.

Figure 1 shows W/S plots obtained after W and S parameter measurements performed with monoenergetic positrons ranging from 0–40 keV, in the ZnO bulk reference sample and in two ZnO films grown over two different faces of sapphire. In the reference sample (figure 1a) positrons penetrating the sample with the lowest energy have the smallest W parameter and the largest S one (bottom right corner of the figure). W parameter increases and S parameter decreases as positron implantation energy increases forming a line in the W/S plot. As in the bulk of the ZnO reference sample there is only one state for positrons, the straight line in the W/S plot indicates that positrons are annihilating at two states, one corresponding to the ZnO surface and the other to ZnO bulk [12]. The line is quite straight for W values above 0.75 and S values below about 1.05. At very low positron implantation energy, where many positrons annihilate in surface states, there is a small deviation related to surface inhomogeneities that may be present in the sample.

As the bulk of the ZnO reference sample has only one positron state, the straight line in the W/S plot indicates that positrons are annihilating from 2 states one corresponding to annihilations from ZnO surface and the other to ZnO bulk [12]. Increasing positron energy the amount of positrons annihilating from the bulk of ZnO is increasing in respect to the ones annihilating from the surface. Indeed, when positrons are implanted with large energy most of positrons are annihilating from bulk states, and the (W, S) value corresponding to the largest W and smallest S is the one corresponding to bulk states (left upper corner in figure 1a). Therefore, the distance between a certain (W, S) point and the one indicated as bulk is related to the positron trapping fraction in bulk state; on the contrary, the distance between that point and the surface point shown in figure 1a is related to the positron trapping fraction at surface state. This is due to the fact that S and W parameters corresponding to a particular implantation energy are a linear superposition of positrons annihilating from either of the two possible states [12].

The W/S plot corresponding to ZnO films grown on sapphire are completely different. They both show a clear cusp. Indeed, at very low implantation energies positrons are annihilating mainly from surface states (bottom right corner in figure 1b). At higher implantation energies the W/S plot follows a straight line towards a cusp whose maximum depends on the measured film. The cusp position corresponds to a relative maximum (minimum) of the W parameter (S parameter) versus implantation energy. At positron implantation energies just above the cusp (W, S) points follow back the previous line, but at further energies the measured values do not follow the previous straight line and they tend toward a new position. The position obtained at the highest implantation energies must correspond to positron annihilation at the bulk of the sapphire. The W/S plot corresponding to the cusp positions in all the studied ZnO layers fall within a straight line that contains the bulk position measured in the ZnO reference sample and the saturated trapping at Zn vacancies, which has been estimated from simultaneous lifetime and Doppler measurements in electron irradiated single crystal ZnO [11] (see figure 2). It indicates that the same type of defect, the Zn vacancy, is responsible for the trapping at the ZnO films presented here. Moreover, it shows also that all positrons implanted in the sample are annihilating from ZnO layer positron states. We will call "vacancy line" to the line formed by the (W, S) values of the cusps, the (W, S) point of the bulk and the one of the Zn vacancy, which in this particular case represent the linear superposition of positrons annihilating from

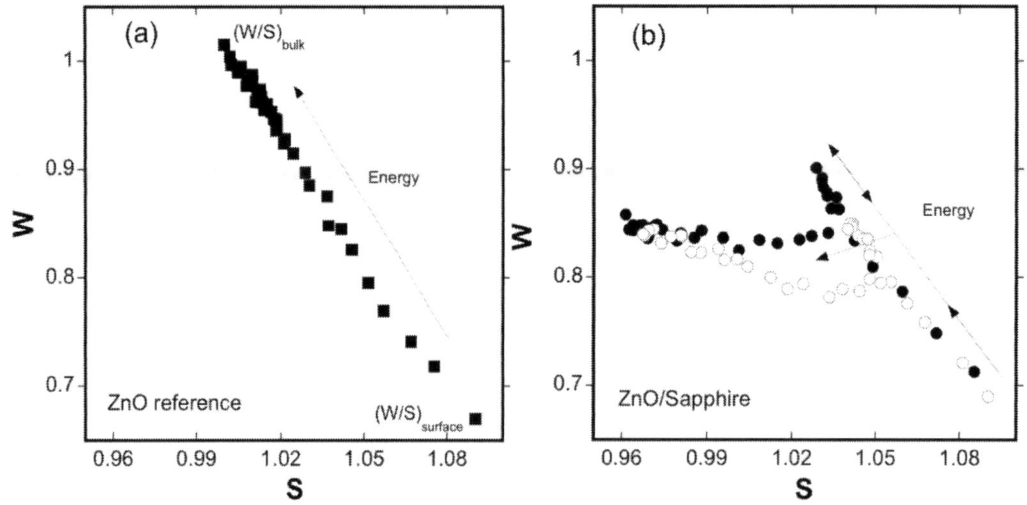

Figure 1. W/S plot of (a) the reference sample and (b) two ZnO layers deposited on sapphire substrate.

Zn vacancy and from ZnO bulk states. It is worth noting, the length between the cusp and the position where (W, S) points leave the straight line depends on the sample, and such points fall within the so called "vacancy line". It indicates there should exist a profile of Zn vacancy concentration in the studied films, and such a profile varies form sample to sample.

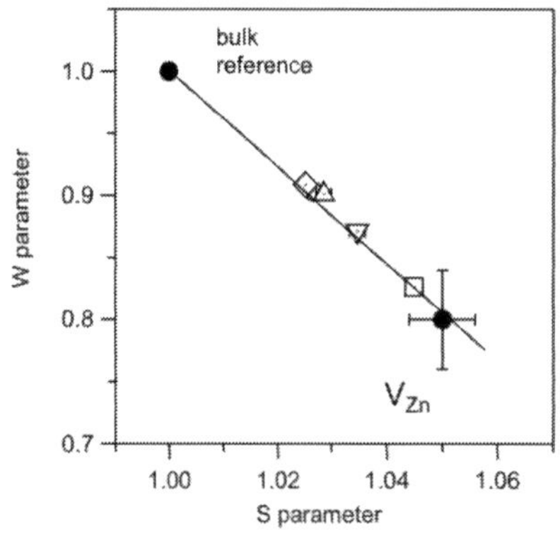

Figure 2. W/S plot of the cusp positions in all the studied bilayers.

2.1. Implantation profile

In order to obtain further information from W/S plots like the ones in figure 1b it is necessary to analyze the implantation process. Characteristics such as positron penetration and its implantation depth, positron diffusion length (in the layer and the substrate), and its probability distribution after the diffusion have to be considered. The positron implantation energy and the mean implantation depth of the positron into the sample, $\langle z \rangle$, are related by the equation

$$\langle z \rangle = \frac{BE^n}{\rho} \tag{4}$$

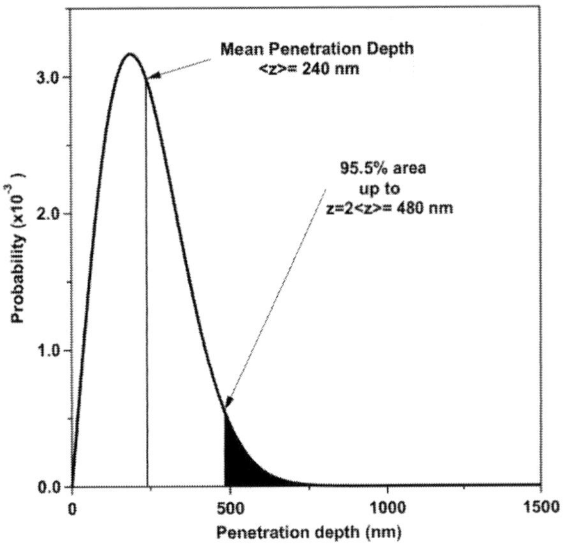

Figure 3. Shape of the positron implantation profile.

where ρ is the material density (5.6 g/cm^3 for ZnO) and n and B are 1.6 and 4×10^{-6} g/cm^{-2} keV$^{-1.6}$, respectively [12]. Once the positron penetrates into a sample its stopping profile is described by the following Makhovian function

$$I(z) = 2\frac{z}{{z_0}^2}e^{-\left(\frac{z}{z_0}\right)^2} \tag{5}$$

where z_0 is related to the mean penetration depth by $\langle z \rangle = 0.886z_0$ ([12] and references therein). As figure 3 shows the stopping profile function extends from the surface towards higher penetration values inside the sample. The stopping profile has a long tail, which extends towards larger depths. Indeed, the probability of implantation at depths larger than $2\langle z \rangle$ is small (4.5%). Neglecting the 4.5% area of the implantation profile tail above $2\langle z \rangle$, the resulting function is quasi-symmetrical and centered at the mean penetration depth value $\langle z \rangle$. The maximum penetration depth defined in this way is $2\langle z \rangle$, twice the mean penetration depth. The stopping profile in a ZnO layer deposited on a sapphire substrate is slightly different to that given by equation (5), because both materials have different densities. But, if both materials have similar density values, the shapes of the stopping profiles are qualitatively similar. At high penetration depths the implantation profile will have a slightly longer tail inside the sapphire, since its density is slightly lower than that of ZnO.

Taking into account that from the location where they have been implanted, positrons may diffuse inside the material until they annihilate, we must consider the mean diffusion length (L_{diff}) of the positron in the layer and in the substrate for a correct interpretation of the results. Considering the diffusion equation [12] and using the VEPFIT program [13], we can adjust the variation of the S parameter versus the implantation energy. Fitting the studied samples with VEPFIT gives values of 22 ± 5 nm and 80 ± 10 nm for the positron mean diffusion length in the layer and in the substrate, respectively. The obtained values have to be considered when evaluating the mean penetration depth. When L_{diff} in the ZnO layer is much smaller than the layer thickness, we can neglect it. As this is the case of the ZnO layers presented in this work, we can consider $2\langle z \rangle$ as the maximum penetration depth for the positrons in the layer.

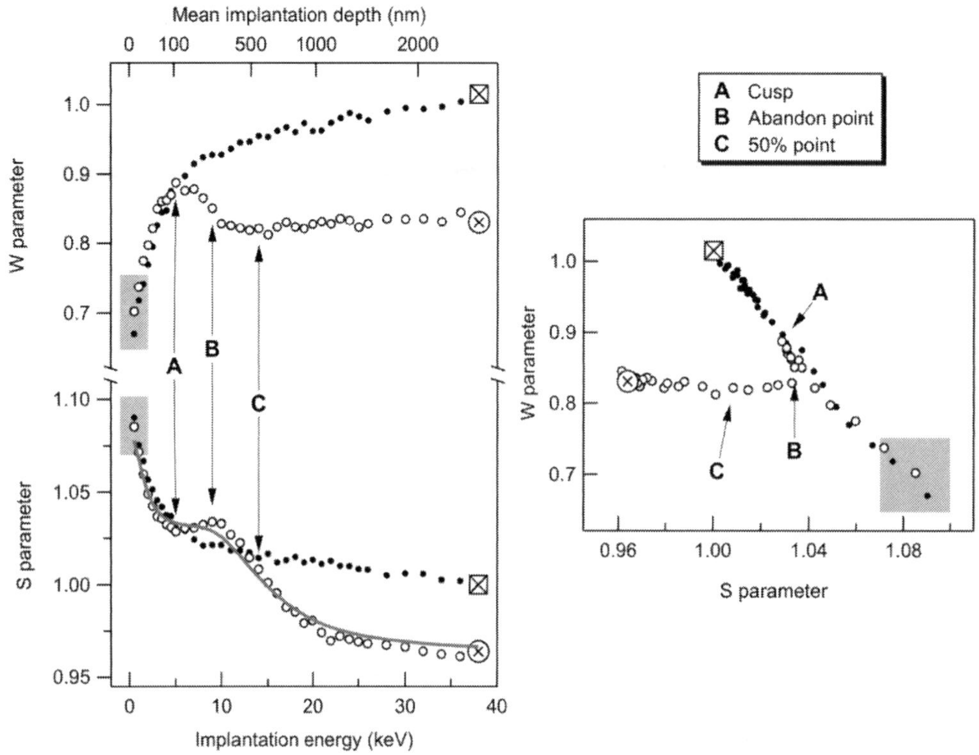

Figure 4. W and S parameters versus positron penetration depth for ZnO reference sample and a bilayer. A indicates the cusp position, B the abandon point and C corresponds to the 50% percentage. The continuous line corresponds to the VEPFIT S-parameter curve.

2.2. Determination of the layer thickness

The cusps mean implantation depths for the measured samples range between 70 and 240 nm. Therefore, the maximum penetration depth corresponding to the cusps is, as we will see, smaller that the layer thickness in all the samples. At implantation energy higher than the one corresponding to the cusp, the W/S measured values still lay over the "vacancy line". The position at which the (W, S) points start to leave the "vacancy line" will be called "abandon point". In the studied samples the "abandon point" is attained for mean penetration depths between 160 $\pm$ 30 nm and 440 $\pm$ 30 nm. Leaving the "vacancy line" indicates that positrons are starting to annihilate from another state, different to the ones that gave the "vacancy line" (delocalized state at ZnO and Zn vacancy). This indicates that for energies larger than the one corresponding to the "abandon point" positrons are annihilating outside the layer too. Taking into account this fact and the shape of the implantation profile (Figure 2), the "abandon point" must correspond to mean penetration depths in the order of half of the sample thickness. Therefore, under these conditions the stopping profile function extends from the surface of the layer up to the interface between the layer and the substrate, see figure 1. For larger energies, positrons cross the interface and start annihilating inside the substrate.

The previous discussion indicates that a simple and non-destructive way for determining the thickness of a layer grown on a substrate consists on determining the "abandon point" in the W/S plot. The thickness of the layer corresponds to twice its mean implantation depth ($2\langle z_{abandon}\rangle$). Table 1 shows the thickness of the studied samples obtained by two different

Table 1. Layers thickness measured by scanning electron microscopy (SEM) and by the method described in this work.

SEM (nm)	This work (±60 nm)
540±50	480
410±40	480
450±70	320
860±150	880
660±70	660
760±50	720
630±60	660

methods: the non-destructive method based on determining the abandon point, as introduced in the present work, and a destructive one based in scanning electron microscope images of the cleaved samples. Indeed, there is quite a good agreement between both methods. The proposed criterion yields good results in the studied samples because the positron diffusion is small inside the film. The measured spectra show that the abandon point is well defined in all the studied samples, indicating the very small influence diffusion causes in determining the thickness of the layer.

2.3. Determination of the defect profile

What does indicate the presence of a cusp in ZnO layers grown over sapphire? Between the cusp position and the "abandon point" the W/S curve is directed towards the zinc vacancy value following the "vacancy line". It indicates, clearly, that the concentration of zinc vacancies close to the interface is significant. At higher energies, the probability of positrons annihilating in the substrate increases and the measured W and S values tend to the annihilation values of sapphire. The distance in the W/S plot between the cusp position and the "abandon point" is proportional to the variation of the zinc vacancy concentration. From figures 1 and 4 it can be inferred that the vacancy concentration is minimum inside the layer at depths lower than one third of the layer thickness and increases up to the interface, that is, for mean penetrations around half of the thickness of the samples.

For mean penetration depths larger than half of the layer thickness the probability that positrons annihilate at the interface and inside the substrate increases, and from this point on the results cannot be easily separated into the different contributions. Moreover, a long positron diffusion length in the substrate might influence the amount of positrons annihilating in the substrate and trapping at interface states. However, taking into account the short positron diffusion length within the layers, only the number of positrons annihilating in the interface would increase. The results can be explained assuming only trapping in the layer and the substrate, so the effect of the interface states is minor in the measured samples.

In figure 4 point C is indicated by an arrow. Point C corresponds to a mean penetration depth of 480 nm, and although this point is outside the straight line (vacancy line), it keeps some information on the annihilation parameters of the layer but with some influence from the substrate. The corresponding mean penetration depth is close to the thickness of the layer, so, 50% of positrons are implanted in the substrate (50 % percentage) and the rest in the layer and it brings information about the annihilation characteristics near the interface. In the following we will analyze this general behavior in order to understand it and to obtain further information from these data.

The method we propose for the determination of the defect profile in semiconductor layers grown heteroepitaxially, can be generalized to other hetero(homo)structures if they fulfill the following conditions:

(i) The positron diffusion length is much smaller than the layer thickness.

(ii) The layer has only one defect trapping positrons at the measurement temperature.

(iii) Positron annihilation in the substrate comes from only one state.

The three conditions are not very restrictive and the method could be, thus, employed in many systems.

2.3.1. Definitions in the W, S plane We can divide the W/S curves of all the samples in the W/S plot into three different zones according to the positron implantation (and annihilation) depths into the sample: the layer zone, which corresponds to the "vacancy line" (all positrons annihilate within the layer), the intermediate zone, which starts at the abandon point and extends up to the third zone (positrons annihilate from both, the layer and the substrate), and the third zone, which is representative of the W/S substrate values (most of the positrons annihilate at the substrate). In figure 1b these three zones can be easily distinguished.

The W/S value corresponding to the sapphire substrate can be obtained directly from measurements of the substrate by the other side of the layer sample. On the other hand, all points in the W/S plot are enclosed by three lines, which establish the limits of the W/S plots. These lines are: the "vacancy line", already defined, the "bulk line", which runs from the ZnO bulk to the substrate W/S values, and the "saturation line", which joins the Zn vacancy and the substrate W/S values, see figure 5a.

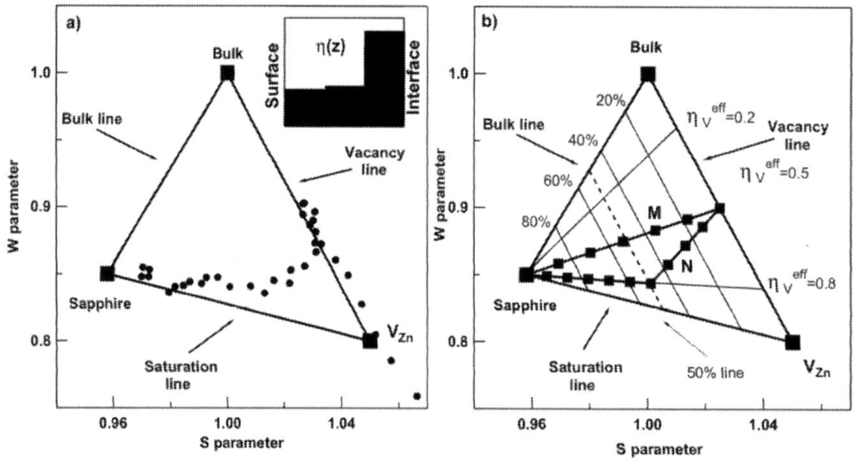

Figure 5. a) definition of the W/S plane. The inset shows the defect profile $\eta(z)$ of the layer, b) representation of percentage lines, effective constant trapping lines and 2 qualitative positron trends of W/S points.

2.3.2. W and S annihilation parameters as a function of penetration depth Now, lets analyze the trend of a point, which individual contributions to the W and S parameters will be written as (W/S), for a layer grown over a sapphire substrate (subst subscript) once it abandons the vacancy line. That is for mean penetration depths larger than approximately the layer half width. The individual contributions to the W and S parameters of a layer with a vacancy profile along the layer thickness and a substrate can be deduced from the next equation:

$$(W/S)_{total} = (W/S)_{layer}[0,d]P_{layer} + (W/S)_{subst}[d, D = \infty]P_{subst} \tag{6}$$

where,

$$(W/S)_{layer} = (W/S)_{bulk}\eta_{bulk}^{eff} + (W/S)_v\eta_v^{eff} \tag{7}$$

and

$$\eta_{bulk}^{eff} + \eta_v^{eff} = 1 \tag{8}$$

P_{layer} and P_{subst} correspond to the annihilation probability in the layer and the substrate, respectively. d and D are the layer and the substrate thicknesses, respectively. η_v^{eff} is the effective trapping at vacancies. It fulfils a relation similar to the trapping fraction, and allows for non constant trapping fraction:

$$\eta_v^{eff} = \frac{\int_0^d \eta_v(z)I(z)dz}{\int_0^d I(z)dz} \tag{9}$$

We can generalize equation (6), that is, the contribution to the experimental W and S values of the sample, $(W/S)_{total}$, as follows:

$$(W/S)_{total} = \int_0^d (W/S)_{layer}I(z)dz + \int_d^\infty (W/S)_{subst}I(z)dz \tag{10}$$

and,

$$(W/S)_{layer} = (W/S)_{bulk} + [(W/S)_v - (W/S)_{bulk}]\eta_v(z) \tag{11}$$

where $I(z)$ is the implantation profile or probability function along the whole sample, and $\eta_{bulk}(z)$ and $\eta_v(z)$ are the bulk and vacancy trapping fractions, respectively.

The ratio η_v/η_{bulk} is proportional to the defect concentration $C(v) = K\eta_v/\eta_{bulk}$, where $K = N_{at}\lambda_b/\mu$ is the proportionality constant. N_{at} is the atom density of the material, λ_b the annihilation rate at delocalized states and μ the defect specific trapping rate. η_v/η_{bulk} is dimensionless and it will be represented by the function $f(z)$. We can then write

$$\eta_v = f(z)\eta_{bulk} \tag{12}$$

and taking into account that

$$\eta_{bulk}(z) + \eta_v(z) = 1 \tag{13}$$

The trapping fraction of vacancies is

$$\eta_v(z) = \frac{f(z)}{(1+f(z))} \tag{14}$$

The $(W/S)_{total}$ can be written as:

$$(W/S)_{total} = (W/S)_{subst} \times e^{-\left(\frac{d}{z_0}\right)^2} + (W/S)_{bulk}\left(1 - e^{-\left(\frac{d}{z_0}\right)^2}\right) + \int_0^d [(W/S)_v - (W/S)_{bulk}]\left[\frac{f(z)}{1+f(z)}\right] I(z)dz \tag{15}$$

where $I(z)$ is [14]:

$$I(z) = \begin{cases} \frac{2z}{z_0^2} e^{-\left(\frac{z}{z_0}\right)^2} & ; z < d \\ \frac{2}{z_1} \frac{\left[d + \frac{\rho_1}{\rho_0}(z-d)\right]}{z_0} e^{-\left[\frac{d+\frac{\rho_1}{\rho_0}(z-d)}{z_0}\right]^2} & ; z > d \end{cases} \tag{16}$$

In the particular case of ZnO, $z_0 = 40/\rho_0/0.886E^{1.6}$ and $z_1 = 40/\rho_1/0.886E^{1.6}$. ρ_0 and ρ_1 are the densities of the layer and the substrate, respectively. It depends on the density of the material and the energy of positrons like a Makhovian function.

If the density of the substrate is much larger than the layer one, most positrons will preferentially annihilate in the layer, independently of their energy. On the other side, if the density of the substrate is much lower than the layer one, positrons with enough energy will enter easily into the substrate and the penetration depth will get largely increased. In the following we will consider that the densities of layer and substrate are similar and, therefore, the probability distribution in the substrate will be similar to the layer one. In the analysis above we have estimated the expression for $I(z)$ in a simple way, but a more exact expression could be obtained from experimental measurements or Monte-Carlo calculations.

2.3.3. W/S curve behaviour for energies larger than the one of the "abandon point" We can obtain the behavior of W/S points easily for positron energies larger than the one of the "abandon point", assuming $\eta_v = constant$. Such a constant corresponds to a layer with a constant defect concentration profile along its thickness. In this case $f(z) = f(0) = constant$, and the W and S parameters can be easily calculated.

In this case equation (15) becomes:

$$(W/S)_{total} = e^{-\left(\frac{d}{z_0}\right)^2}(W/S)_{subst} + \left(1 - e^{-\left(\frac{d}{z_0}\right)^2}\right)[(W/S)_{bulk}\eta_{bulk} + (W/S)_v\eta_v] \tag{17}$$

$$(W/S)_{total} = e^{-\left(\frac{d}{z_0}\right)^2}(W/S)_{subst} + \left[1 - e^{-\left(\frac{d}{z_0}\right)^2}\right](W/S)_{layer} \tag{18}$$

$$(W/S)_{total} = (1 - P_0)(W/S)_{subst} + P_0(W/S)_{layer} \tag{19}$$

where P_0 is the probability of the positron to annihilate inside the layer. P_0 is obtained integrating the probability $I(z)$ from 0 to the entire layer thickness. Its value depends on the layer thickness, the densities of the layer and the substrate, and the positron implantation energy.

$$P_0 = 1 - e^{-\left(\frac{d}{z_0}\right)^2} \tag{20}$$

In this case, equation (19) defines a straight line joining the vacancy line with $(W/S)_{subst}$ value. Other values of η_v give other straight lines joining $(W/S)_{subst}$ value with the corresponding η_v that is located on the vacancy line. Any of such lines, corresponds to constant defect profile density along the sample thickness. Lets take a value for $f(0)$ which leads to a value of 0.5 for η_v in equation (14). For this case, the W/S value will follow a line as the straight line M in figure 5b [15]. Therefore, we can define constant trapping fraction lines (η^{eff}), which join points with the same effective trapping fraction. They extend between the "vacancy line" and the substrate value.

On the other hand, the analyses above allows to define and draw parallel lines to the "vacancy line" in figure 5b, too. Each line corresponds to a mix of the layer and substrate W and S parameters in different proportions, that is, for different mean penetration depths or P_0. So we can draw for instance, the called "50% line" in figure 5b. Assuming that the W/S values, for the annihilation of positrons which penetrate twice the width of the layer sample, can be obtained as approximately a contribution of 50% of the W/S parameters of the film plus another 50% contribution of the W/S parameters of the substrate, equation (16).

Other different lines, for different values of the probability of the positron in the film, P_0, can be drawn parallel to the "vacancy line" as in figure 5b. We will call them the hundred percent or P_0 lines (%lines). The η^{eff} trapping fraction line is the one connecting the trapping fraction value at the vacancy line and the (W/S) substrate value. The effective trapping fraction in a particular point of each hundred percent line is given by the point where the above line cuts the corresponding hundred percent line.

2.3.4. Non constant defect profiles In heterostructures due to the interface the defect concentration profile is not constant. Indeed, it usually increases towards the interface. We can qualitatively predict the W/S behavior or trend for this case using the hundred percent lines.

In this situation once the (W/S) point abandons the "vacancy line", at mean penetration depths from the sample half width to approximately the width of the sample, the point will descend to higher trapping fraction values along each "% line" in figure 5b. At the same time it moves towards the W/S point representative of the sapphire substrate. This will happen for mean penetration depths up to the layer width 50% line in figure 5b), because at that mean penetration the maximum of the positron implantation profile is located at the interface. For higher mean penetration depths the substrate influence will increase and the W/S point will move along a straight line towards the W/S value of sapphire. In this case we can expect a trend with two lines of different slopes (N line in figure 5b) in the W/S plot. The first line of the above trend connects the abandon point and the point in the 50% line. The second line connects the 50% line point and the W/S value of the substrate.

We can deduce the defect concentration profile of a particular layer by a pair of values (P_0, η^{eff}) in the W/S plot. For any experimental point, the η^{eff} value and its associated P_0 value can be calculated and, using these values, the defect profile can be approximately obtained.

We can perform the analysis of the W/S plot for a varying defect concentration distribution, for what we only need to know the vacancy concentration $f(z)$ or the trapping fraction η_v as a function of the sample depth. Using equation (15) we can calculate the evolution of the W/S parameters versus implantation energy. We need only to solve the integral of equation (15) for the appropriate values of $f(z)$. The equation can be solved numerically for any distribution of zinc vacancies.

Considering the previous discussion we can also calculate the variation of the trapping fraction η_v along the sample fitting equation (15) to the experimental variation of the W and S parameters for different penetration depths [16].

2.4. Three layers structures

We can extend easily the equations obtained in the case of two layers for samples composed by two layers plus one substrate. In this case the two layers equation (6) becomes

$$(W/S)_{total} = (W/S)_{layer_1}[0, d_1]P_{layer_1} + (W/S)_{layer_2}[d_1, d_2]P_{layer_2} + (W/S)_{subst}[d_2, D = \infty]P_{subst} \quad (21)$$

where,

$$(W/S)_{layer_i} = (W/S)_{bulk_i}\eta^{eff}_{bulk_i} + (W/S)_{v_i}\eta^{eff}_{v_i} \quad (22)$$

and

$$\eta^{eff}_{bulk_i} + \eta^{eff}_{v_i} = 1 \quad (23)$$

In this case an equation for $\eta_{v_i}^{eff}$ can be obtained equivalent to equation (9),

$$\eta_{v_i}^{eff} = \frac{\int_{d_{i-1}}^{d_i} \eta_{v_i}(z) I(z) dz}{\int_{d_{i-1}}^{d_i} I(z) dz} \tag{24}$$

where $i = 1, 2$ and $d_0 = 0$. Following a similar discussion than in the case for two layers a similar function as the one described in equation (15) can be obtained. So,

$$(W/S)_{total} = (W/S)_{subst} \times e^{-\left(\frac{d_2}{z_0}\right)^2} + (W/S)_{bulk_1}(1 - e^{-\left(\frac{d_1}{z_0}\right)^2}) +$$
$$+(W/S)_{bulk_2}(e^{-\left(\frac{d_1}{z_0}\right)^2} - e^{-\left(\frac{d_2}{z_0}\right)^2}) + \int_0^{d_1} [(W/S)_{v_1} - (W/S)_{bulk_1}] \left[\frac{f_1(z)}{1+f_1(z)}\right] I(z) dz +$$
$$+ \int_{d_1}^{d_2} [(W/S)_{v_2} - (W/S)_{bulk_2}] \left[\frac{f_2(z)}{1+f_2(z)}\right] I(z) dz \tag{25}$$

As in the case of a bilayer, we can first obtain the behavior of (W/S) points assuming η_{v_1} and η_{v_2} constants, that is, considering that the two layers have constant defect concentration profile along their thicknesses (constants $f_1(z)$ and $f_2(z)$). In that case equation (19) becomes

$$(W/S)_{total} = [1 - (P_1 + P_2)] (W/S)_{subst} + P_1 (W/S)_{layer_1} + P_2 (W/S)_{layer_2} \tag{26}$$

where P_1 and P_2 are the probabilities of positron to annihilate inside layer 1 and layer 2 respectively.

Let's analyze qualitatively the above equation that corresponds to constant vacancy concentration profile in the two layers. If $P_1 \gg P_2$ and P_1 close to one, equation (26) yields the vacancy line of the first layer. If $P_1 \ll P_2$ and P_2 close to one, equation (26) yields the vacancy line for layer two. When P_1 very small and $P_{subst} \simeq 1 - P_2$ the equation yields a line that goes from the vacancy line of the second layer to the W/S value of the substrate. However, due to the relation between P_1 and P_2 the abandon points corresponding to the different lines are not as acute as in the bilayer case.

In the following we present in figure 6 the W and S parameters obtained after monoenergetic positrons implantation in a three layer formed by $ZnCdO/ZnO/Al_2O_3$ and we give a preliminary discussion of the results. The thin films were deposited on $R-$plane sapphire substrates. A relatively thick ZnO buffer layer (of about 800 nm, estimated from growing conditions) was deposited prior to ZnCdO alloy growth. The ZnO buffer was grown at 376 °C (see [17] for more details).

Figure 6a presents W and S parameters versus positron implantation energy and its corresponding W/S plot, figure 6b. The W/S plot shows four straight lines with different slopes for monoenergetic positrons in the range 0–40 keV. At low positron energies, most of positrons annihilate from surface states of ZnO:Cd(4.5%) layer (bottom right corner). Increasing energy (W/S) points follow a straight line towards a position where there is a relative large concentration of points with relatively similar W and S parameters. Those points correspond to the plateau of W and S parameters versus implantation energy observed in figure 6a, which ends at about 1.1 μm penetration depth. However, increasing positron energy, a new slope change occurs for positron penetrations up to 1.8 μm. From this point a short straight line runs up to positron penetration depths of 2.7 μm. After, this point and for larger positron penetration depths (W/S) points follow another straigth line that it is clearly directed toward (W/S) values of sapphire (not shown in the figure). Following the previous discussion, it is clear that the straight line running between 1.8 μm and 2.7 μm penetration depths must correspond to positron annihilation preferentially happening at the ZnO buffer, indicating that 2.7 μm corresponds to the "abandon

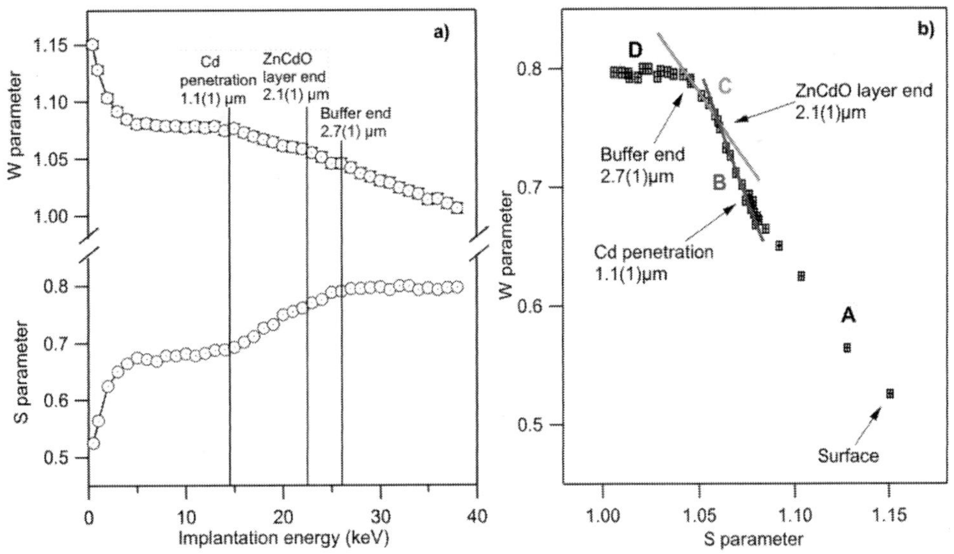

Figure 6. (a) W and S parameters versus positron implantation energy for a heterostructure formed by two ZnO layers deposited over a sapphire substrate and (b) its corresponding W/S plot.

point" between annihilations in the two layers and the substrate. Therefore, (W/S) points that run between surface and 1.8 μm positron penetration depths must correspond to W and S parameters of the overlayer of ZnCdO. However, the W/S plot indicates clearly the presence of two straight lines in such region and the piling up of many points for positron implantation energies in the range 5–15 keV. This behaviour suggest that Cd content along the layer is not uniforme. Indeed, from the surface to 1.1 μm depths Cd content uniformity is indirectly deduced for the piling up of (W/S) points corresponding to implantation energies in the range 5–15 keV. It indicates that the presence of Cd induces the presence of positron traps, which trap positrons preferentially for penetration depts upto 1.1 μm; however, for larger penetration depths, closer to the ZnCdO/ZnO the Cd induced defects decrease; which suggest the lower Cd presence in the neighbourhood of the ZnO/ZnCdO interface.

3. Conclusions

We have presented a method to analyze W/S plots of Doppler broadening data obtained with slow positron beams. A detailed analysis of of the method for a system formed by a semiconductor layer over a substrate has been presented. The method allows to determine easily and nondestructively the thickness of the layer. The method allows to perform quatitative determinations of the defect profile inside the layer, too. The method can be extended to heterostructures formed by several semiconductor layers over a substrate. In this work a preliminary extension of the method for heterostructures having two different layers over a substrate has been presented.

Acknowledgments

The authors want to acknowldge for finantial support the Basque Government under project IT44310 and the Spanish MEC under projects MAT2006-12743, MAT2007-06481 and MEC872/2006. One of us (A.Zubiaga) wants to acknowledge the Basque Government for finantial support under a PosDoc Grant.

References

[1] Pi X D., Coleman P G., Tseng T L, Burrows C P, Yavich B, and Wang W N 2002 *J. Phys.:Condens. Matter* **14** L243
[2] Chen Z Q , Yamamoto S, Maekawa M, Kawasuso A, Yuan X L and Sekiguchi T 2003 *J. Appl. Phys* **94** 4807
[3] Uedono A , Koida T, Tsukazaki A, Kawasaki M, Chen Z Q, Chichibu S F and Koinuma H 2003 *J. Appl. Phys.* **93** 2481
[4] Koida T, Chichibu S F, Uedono A, Tsukazaki A, Kawasaki M, Sota T, Segawa Y and Koinuma H 2003 *Appl. Phys. Lett.* **82** 532
[5] Zubiaga A, Tuomisto F, Plazaola F, Saarinen K, Garcia J A, Rommeluere J F, Zúñiga-Pérez J and Muñoz-Sanjosé V 2005 *Appl. Phys. Lett* **86** 042103
[6] Zubiaga A, Garcia J A, Plazaola F, Tuomisto F, Saarinen K , Zúñiga-Pérez J and Muñoz-Sanjosé V 2006 *J. Appl. Phys.* **99** 053516
[7] Mantl S and Triftshäuser W 1978 *Phys. Rev.* B **27** 1645
[8] Saarinen K, Hautojärvi P, Keinonen J, Rauhala E, Risanen J and Corbel C 1991 *Phys. Rev.* B **43** 4249
[9] Krause-Rehberg R and Leipner H S 2003 in *Positron Annihilation in Semiconductors* edited by M. Cardona et al (Springer Series in Solid-State Sciences, Berlin)
[10] Munuera C, Zúñiga-Pérez J, Rommeluere J F, Sallet V, Triboulet R, Soria F, Muñoz-Sanjosé V, Ocal C 2004 *J. Crystal Growth* **264** 70
[11] Tuomisto F, Ranki V, Saarinen K and Look D C 2003 *Phys Rev Lett.* **91** 205502
[12] Saarinen K, Hautojärvi P and Corbel C 1998 in *Identification of defects in Semiconductors* edited by M. Stavola (academic press, New York)
[13] Van Veen A, Schut H, Clement M, de Nijs J M M, Kruseman A and Ijpma M R 1995 *Appl. Surf. Sci.* **85** 216
[14] V. J. Ghosh, D. O. Welch and K. G. Lynn 1994 *AIP Conf. Proc.* **303** 37
[15] Zubiaga A, Garcia J A, Plazaola F, Tuomisto F, Zúñiga-Pérez J and Muñoz-Sanjosé V 2007 *phys. stat. sol.* (c) **4** 3973
[16] Zubiaga A, Garcia J A, Plazaola F, Tuomisto F, Zúñiga-Pérez J and Muñoz-Sanjosé V 2007 *Phys. Rev.* B **75** 205305
[17] Zúñiga-Pérez J and Muñoz-Sanjosé V, Lorenz M, Benndorf G, Heitsch S, Spemann D and Grundmann M 2006 *J. Appl. Phys.* **99** 023154

Journal of Physics: Conference Series **265** (2011) 012005 doi:10.1088/1742-6596/265/1/012005

Detection of vacancy-like defects during Cu diffusion in GaAs by positron annihilation

M Elsayed[1], V Bondarenko, K Petters and R Krause-Rehberg

Department of Physics, Martin Luther University, 06099 Halle, Germany

E-mail: mohamed.elsayed@physik.uni-halle.de

Abstract. The positron annihilation spectroscopy is a method for direct characterization of vacancy-type defects by measuring the positron lifetime. It provides information about open volume and concentration of defects. Such measurements were carried out to study the defect properties of semi-insulating GaAs after copper diffusion. A 30 nm layer of Cu was deposited by evaporation to undoped GaAs samples. The diffusion of Cu was performed during an annealing step at 1100 °C under different arsenic vapor pressures. The samples were quenched into room temperature water. The initial semi-insulating (SI) undoped GaAs sample shows no positron traps in that state. After gentle annealing, a vacancy-type defect complex in addition to shallow positron traps was observed to be an efficient positron trap. After Cu in-diffusion during the annealing process, the shallow positron trap is believed to be the Cu_{Ga} double acceptor. The exact nature of the vacancy-like defects could not be determined unambiguously. The concentration of these defects exhibits inverse relationship to the arsenic vapor pressure. Thus, the arsenic vacancy is believed to be part of this complex. The temperature-dependent Hall-effect measurements have revealed the presence of an acceptor level at $E_V + 0.5$ eV that is usually attributed to Cu_{Ga}.

1. Introduction

Despite extensive studies on GaAs during the recent decades, we do not yet have a full understanding of native point defects in this material. The thermodynamic properties of these defects in GaAs are of great technological interest. The thermodynamic analysis helps predicting the concentrations of the point defects incorporated into crystal under equilibrium conditions [1, 2]. The subject of the present work is the investigation of point defects quenched from different equilibrium states by means of positron annihilation lifetime spectroscopy (PALS). The positron has proved to be a valuable nondestructive probe for vacancy-type defects. During the past decades, PAL has been intensively applied to characterize defects in various semiconductors [3]. But in many cases it is difficult to conclude from the annihilation parameters alone which defect is responsible for the positron trapping. Basic thermodynamic considerations displayed in this paper helped us a lot in characterizing the origin of the observed vacancy-complex. The native point defect concentration can be expressed as a function of temperature and arsenic vapor pressure according to the law of mass action [1].

Copper is found as an unintentional impurity in most semiconductors. This is owing to the fact that Cu is a rapidly diffusing contaminant already at low temperatures. Cu exhibits an unusually large

[1] Corresponding author: Tel.: + 49-345-552-8502; Fax: + 49-345-552-7390.

diffusion coefficient in many semiconductor crystals. In GaAs, it was found to be as high as $D = 1.1\times10^{-5}$ cm^2 s^{-1} at 500 °C and $D = 1.8\times10^{-9}$ cm^2 s^{-1} at 100 °C [4]. Cu diffuses very fast by interstitial diffusion (kick-out process) [5]. Figure 1 presents the experimentally obtained solubility of Cu as a function of temperature as reported in Ref. 4. Moreover, the solubility was calculated to be 5×10^{9} cm^{-3} at 100 °C [4]. Depending on the cooling speed after a diffusion process, only a small fraction of the total Cu concentration is electrically active as acceptors. The portion of Cu that remains electrically inactive forms Cu-Ga precipitates [6].

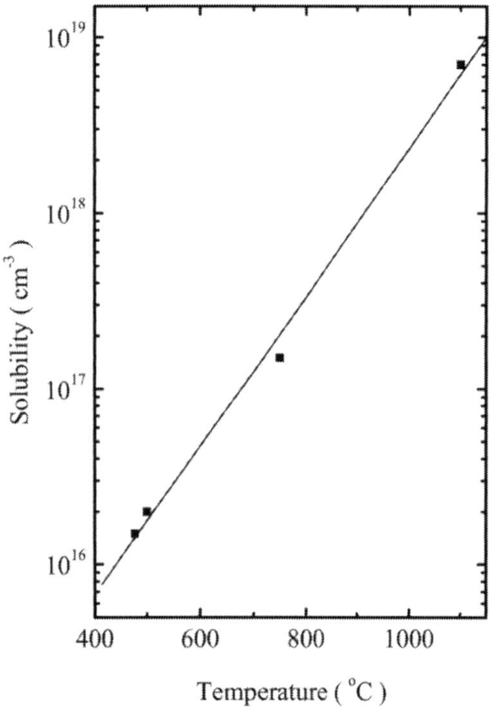

Figure 1. Solubility of Cu in undoped GaAs as a function of temperature (Ref. 4).

2. Experimental

The investigated samples were cut from the semi-insulating undoped LEC grown GaAs wafer ($5\times5\times0.5$ mm^3). The resistivity of initial material was about $10^6 - 10^7$ Ωcm. The samples were covered by 30 nm Cu by evaporating it under UHV conditions. This amount of Cu corresponds to a volume concentration of 6×10^{18} cm^{-3} which is approximately the upper solubility limit of Cu in GaAs at 1100 °C [4]. The deposited layer thickness was controlled by a thickness measurement device (frequency shift of a crystal oscillator) which was calibrated before by Atomic Force Microscopy. High purity copper-free quartz ampoules HSQ300 (Heraeus Quarzglas GMBH&Co) were used for the Cu diffusion annealing. Pure As (99.999%) was used as an arsenic source. The samples and the arsenic source were sealed in quartz ampoules under high vacuum. Annealing was performed for three hours in a two-zone temperature furnace at 1100 °C (sample temperature). The temperature of the arsenic source was varied in the region of 550 – 740 °C, which corresponds to an As-pressure of 0.2 – 9.68 bar [7]. After annealing, the ampoules were quenched into water at room temperature. According to the solubility, Cu is in oversaturated state and Cu atoms have the tendency to leave the lattice and start the out-diffusion, e.g. by forming precipitates. Hall-effect measurements were applied to measure the samples in the as-quenched state. Thereafter, the samples were isochronally annealed in the temperature range up to 850 K. The samples were cooled down relatively slowly after each annealing step. Between the annealing steps, PALS measurements in the temperature range of 20 – 500 K were carried out using a conventional fast-fast coincidence system with a time resolution of 225 ps. The

^{22}Na positron source was sandwiched between two identical 1.5 μm thick Al foils and placed between two identical samples. The spectra were analyzed with the two-component trapping model (one defect type) after source and background correction. The annealed SI GaAs samples were investigated by temperature-dependent Hall-effect measurements (T = 293...373 K). The samples annealed at 0.2 and 9.68 bar of As vapor pressure were chosen for chemical analysis by titration measurements.

3. Result and Discussion

The semi-insulating undoped GaAs sample without any Cu deposition (reference sample) did not show any positron trapping. After Cu in-diffusion, the average lifetime increases only slightly in the high-temperature region, which indicates detection of a small number of vacancy-type defects. Figure 2 represents the average positron lifetime versus measurement temperature after different annealing steps performed after Cu in-diffusion. A distinct decrease of the average lifetime at low temperatures is clearly shown for all curves. This is a typical dependence for shallow positron traps (negatively charged, non-open volume defects, such as ionized acceptors), which tend to trap positrons in the extended region of the Coloumbic potential, reflecting thereby the properties of the bulk as the annihilation characteristics of the positrons [3, 8]. Shallow traps are effective only at low temperatures due to the small binding energy of positron, while at higher temperatures positrons are detrapped because of the high detrapping rate. Here, after Cu in-diffusion, the shallow traps must be ionized Cu acceptor. Their concentration is up to 3×10^{17} cm^{-3} according to the Hall-effect measurements. During annealing up to 750 K, the average positron lifetime increases strongly up to the value of 260 ps, indicating the presence of vacancy-type defects. With a further increase of the annealing temperature, a rapid decrease of the average positron lifetime was observed. With annealing at temperatures higher than 800 K the vacancy clusters grow and the distance between them becomes larger than the positron diffusion length. Thus, they become invisible for positrons [9].

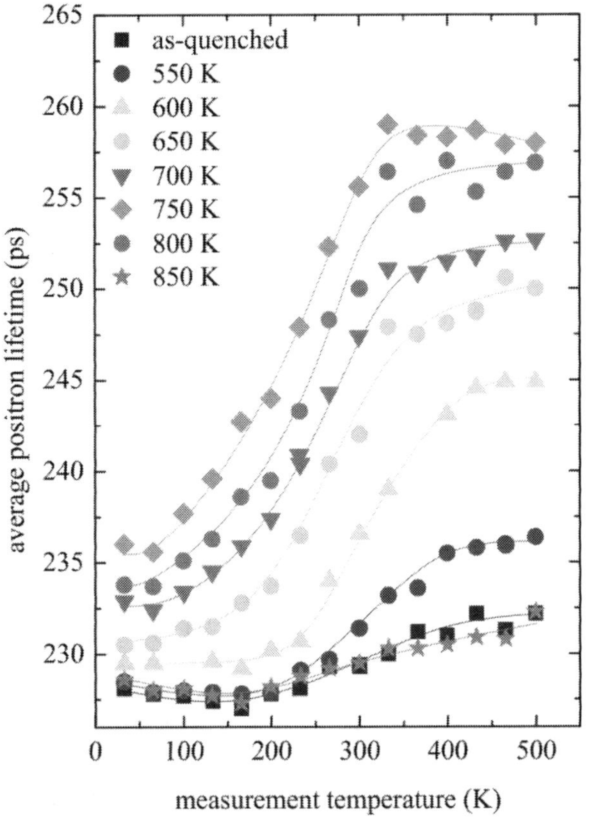

Figure 2. Average positron lifetime as a function of sample temperature in undoped SI GaAs. Prior to the experiment, about 6×10^{18} Cu atoms were introduced by evaporating a layer of 30 nm Cu onto the sample surface and by subsequent annealing at 1100 °C under 5.57 bar of As pressure (3h, then quenched into water). The temperature-dependent lifetime experiment was carried out after each annealing step as illustrated in the figure.

Figure 3 shows the annealing behavior of the average and defect-related lifetimes and its intensity. It can be shown that the open volume of the detected vacancy-type defect increases during annealing. The defect-related lifetime increases with increasing the annealing temperature but lies in the monovacancy region until the annealing temperature 750 K. The defect-related lifetime reaches the value of 332 ps at 800K but with a low intensity, which corresponds to divacancies (upper panel of figure 4).

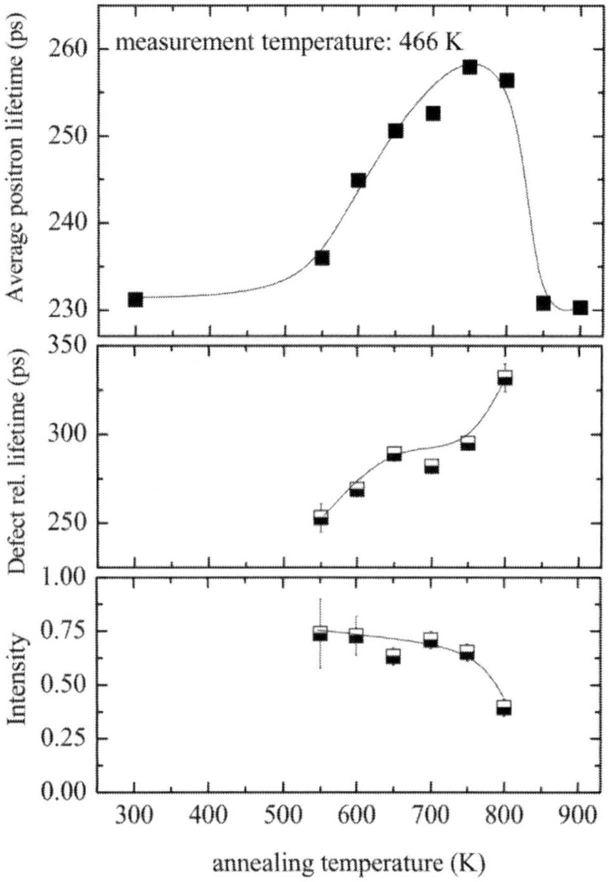

Figure 3. Positron lifetime results of the annealing experiment of undoped semi-insulating GaAs after in-diffusion of 6×10^{18} cm^{-3} Cu atoms at 1100 °C under 5.57 bar of As pressure. The average lifetime is shown in the upper panel. The defect-related lifetime and its intensity versus the annealing temperature are plotted in the lower two panels. The spectra were measured at a sample temperature of 466 K to diminish the influence of the shallow traps.

The lower panel of figure 4 represents the defect concentration versus the annealing temperature where the defect concentration increases from 6.97×10^{16} cm^{-3} at 550 K up to 1.49×10^{17} cm^{-3} at 750 K and decreases again at 800 K. The defect concentration is determined according to equation (1):

$$K_d = \mu C = \frac{1}{\tau_b}\left(\frac{\tau_{av} - \tau_b}{\tau_d - \tau_{av}}\right). \tag{1}$$

K_d is the trapping rate, μ is the trapping coefficient and taken as 10^{15} s^{-1} at 300 K [10,11], τ_b is the bulk lifetime 230 ps [12,13], τ_{av} is the average positron lifetime and τ_d is the defect-related lifetime. The upper panel in figure 4 represents the number of vacancies versus the annealing temperature where the number of vacancies is one vacancy in the temperature range up to 750 K and increase to be 2 vacancies at 800 K. This was concluded according to the calculation in Ref. 14 which is based on the superimposed-atom model by Puska and Nieminen [15].

To show the effect of Cu, we would like to compare the results above to the sample which was not treated with Cu and annealed under very similar conditions. As shown in figure 5, the as-quenched sample shows a higher value of the average lifetime. This can be attributed to the trapping of positrons in vacancies. The pronounced decrease of the average positron lifetime at temperatures below 200 K, may be explained by positron trapping at negative ions (shallow traps), C_{As}^{-}. Where semi-insulating high resistive crystals are produced by intentional doping of GaAs with carbon atoms, which are incorporated exclusively on the As sublattice forming shallow acceptor-like defects, C_{As} [1]. But this cannot be identified from the results of positron lifetime alone. With increasing the annealing temperature the average lifetime decreases. This is owing to the fact that the vacancies disappear. In contrast, in the case of GaAs samples after Cu in-diffusion (figure 2), almost no change of the average positron lifetime was observed in the as-quenched state. However, during the annealing steps until 750 K the average positron lifetime increases strongly, and at annealing temperatures higher than 800 K the vacancy cluster signal almost disappears.

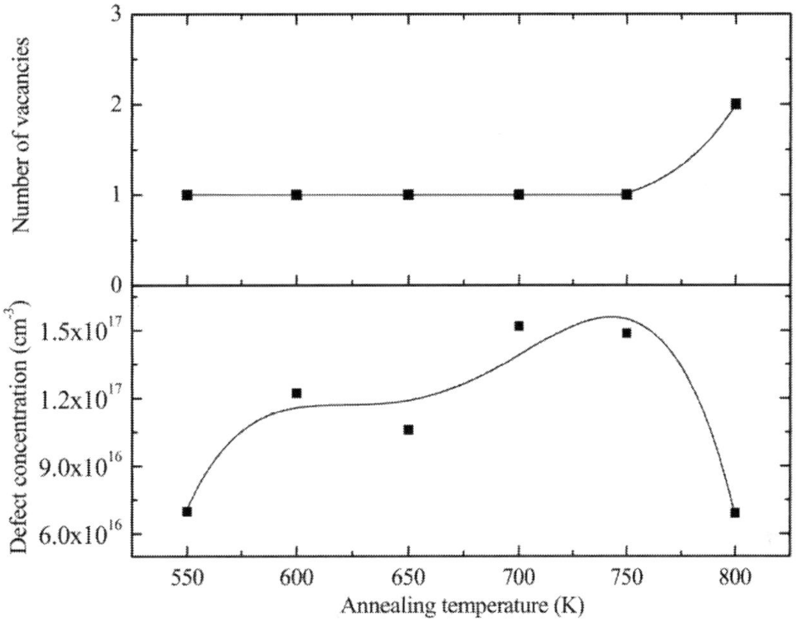

Figure 4. Defect concentration and the number of vacancies as a function of the annealing temperature in undoped SI GaAs after in-diffusion of 6×10^{18} cm^{3} Cu atoms at 1100 °C under 5.57 bar of As pressure. The number of vacancies in the observed clusters is shown in the upper panel. The defect concentration versus the annealing temperature is plotted in the lower panel. These Data were calculated using the positron lifetime results presented in figure 3.

Figure 6 shows the temperature dependence of the average and defect-related lifetime on the sample temperature for undoped semi-insulating GaAs annealed at 1100 °C for 3 hours under different arsenic vapor pressures compared with an unannealed reference sample. No Cu was incorporated into the samples. As shown in the lower part, no vacancy defects were observed in as-grown semi-insulating GaAs. The average positron lifetime of the reference sample is close to the defect-free bulk value of 230 ps at all measurement temperatures. After annealing, the average positron lifetime is strongly enhanced, which indicates the generation of vacancy-like defects during the annealing treatment.

The decrease of the average positron lifetime at low temperatures is due to positron trapping at shallow positron traps. With increasing the arsenic vapor pressure during annealing, the average positron lifetime decreases. The maximum average lifetime is observed after annealing at 0.2 bar. Since the average positron lifetime determines the magnitude of the positron trapping rate and thus the concentration of the vacancy-like defects, the pressure dependence of τ_{av} reflects an inverse relation between the concentration of the vacancy defects at 1100 °C and the arsenic vapor pressure. The upper part in figure 6 presents the defect-related lifetime obtained from the two-component decomposition of the spectra as a function of the measurement temperature. The defect-related lifetime for all the samples at 300 K is 293±10 ps, which is distinctly higher than the lifetimes in Te- and Si-doped GaAs (254 and 262 ps respectively) where Ga vacancies were found to be responsible for the positron trapping. But this value still lies in the region for a monovacancy, because the value calculated for the V_{Ga}-V_{As} divacancy defect in GaAs is 332 ps [16]. Since the defect-related lifetime for As vacancies in GaAs is about 290-300 ps [17], then based on the measured defect- related lifetime the observed vacancy-like defects attributed to As vacancies. In addition, this is found to be in a good agreement with the thermodynamic considerations.

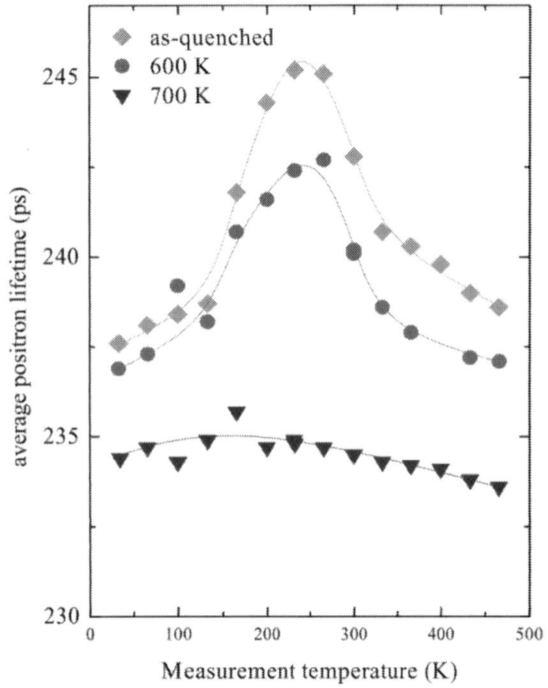

Figure 5. Average positron lifetime as a function of sample temperature in undoped SI GaAs. The samples were annealed at 1100 °C under 5.57 bar of As pressure. The samples were not treated with copper as a reference experiment to the results shown in figure 3. The temperature-dependent lifetime experiment was carried out after each annealing step as illustrated in the figure.

Let us consider the basic thermodynamic reactions for the vacancy formation in GaAs as:

$$1/4As_4^{gas} \leftrightarrow As_{As} + V_{Ga}. \qquad (2)$$

for the Ga vacancy and

$$As_{As} \leftrightarrow V_{As} + 1/4As_4^{gas}. \qquad (3)$$

for the As vacancy. The arsenic tetramer was chosen because it is the dominating As vapor component [18]. Thus, the concentrations of the Ga and As vacancies at a certain temperature should depend on the arsenic vapor pressure. According to the mass action law, the concentrations of these defects can be derived as:

$$[V_{Ga}] = K_{VGa} \times P_{As}^{1/4}. \qquad (4)$$

$$[V_{As}] = K_{VAs} \times P_{As}^{-1/4}. \qquad (5)$$

where K_{VGa} and K_{VAs} are mass action constants for gallium and arsenic vacancies at a certain temperature and P_{As} is the ambient arsenic vapor pressure. From equations (4) and (5), it is clear that the concentrations of V_{Ga} and V_{As} should have an opposite behavior with respect to As vapor pressure and the V_{As} concentration is inversely proportional to the arsenic vapor pressure, which gives us the possibility to discriminate between these two vacancies in both sublattices.

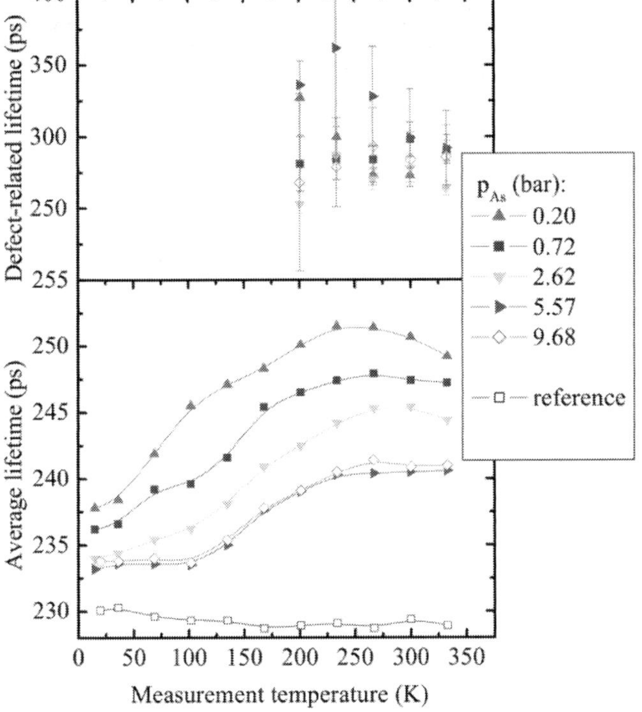

Figure 6. Average and defect-related positron lifetime versus measurement temperature for undoped semi-insulating GaAs annealed at 1100 °C for 3 hours under different arsenic pressures compared to a not annealed reference sample. Lines are to guide the eye only.

Figure 7 reveals an opposite behavior of the concentrations of the vacancy-like defects with increasing P_{As} compared to the data for *n*-type Si-doped GaAs ([Si] = 2×10^{19} cm^{-3}) [19]. The two-component trapping model and the specific trapping coefficient of 10^{15} s^{-1} at 300 K were used for calculating the vacancy concentrations. The fits to the experimental data (solid lines in figure 7) represent the power law and yield an exponent close to 0.25 for GaAs:Si and −0.25 for SI GaAs. As shown, for the GaAs:Si the vacancy concentration increases with increasing arsenic pressure, which refers to the Ga vacancy and is in accordance with the well-known formation of the V_{Ga}-Si_{Ga} defect complex in this material [11]. The same pressure dependence was also found in Te-doped GaAs [20]. The vacancy concentration in SI GaAs exhibits an opposite behavior and decreases with increasing arsenic pressures, which is characteristic for the V_{As}. Based on these results, the origin of observed vacancy-like defects in annealed semi-insulating GaAs is ascribed to V_{As}. On the other hand, it cannot be the isolated As vacancy. Arsenic vacancies should be positive (and thus not detectable with positrons) in semi-insulating or in *p*-type GaAs, where the position of Fermi level is in the middle or in the lower part of the band gap. The Hall-effect measurements showed that all investigated annealed samples became slightly *p*-type with a concentration of [*p*] = 10^{11}-10^{12} cm^{-3} that corresponds to the position of Fermi level at 0.4-0.5 eV above the valence band. Also all theoretical calculations agree that the arsenic vacancy is always positive in SI or at least in *p*-type GaAs and thus it should be invisible for positrons [3]. This leads us to suppose that we are dealing with a V_{As} defect complex which is not any more positively charged. Since Cu is the most common impurity in annealing studies [6, 21, 22], Cu is the first candidate that can be responsible for the formation of such a complex.

The concentration of the Cu impurities for the two samples annealed at 0.2 and 9.68 bar was determined by titration measurements (full magenta circles in figure 7). As can be seen, the Cu concentration was about 10^{16} cm^{-3} only; that is one order of magnitude lower than the measured number of the vacancy-complex. This means that copper is not a constituent of the observed defect complex. The observed V_{As}-like defect should be related to a native defect-complex. The exact nature of this complex cannot be determined from the results of positron annihilation alone. More likely, the electrically active part of Cu impurities acts as shallow positron traps. It causes the decrease of average positron lifetime in the low temperature region. Indeed, the temperature-dependent Hall-effect measurements have revealed the presence of an acceptor level at $E_V + 0.5$ eV that is usually attributed to Cu_{Ga} related defects [21, 22].

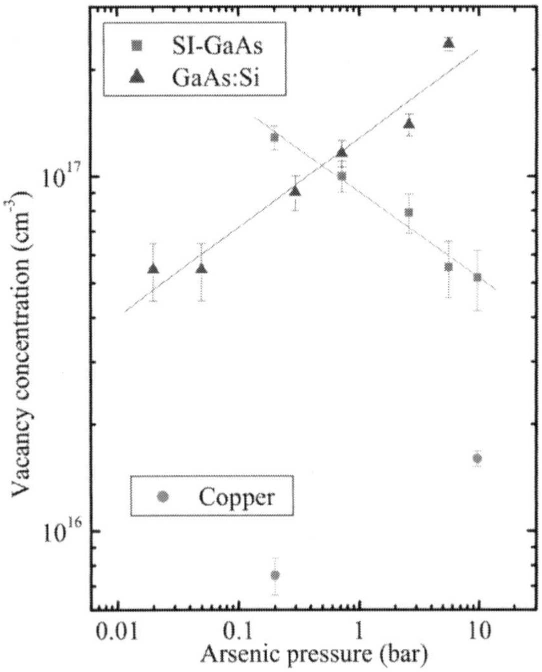

Figure 7. Vacancy defect concentrations in semi-insulating and Si-doped GaAs versus As vapor pressure during annealing at 1100 °C (Ref. 18). Solid lines are the power law fits to the data points. Closed circles present the concentration of Cu impurities obtained with the help of titration measurements for the samples annealed at 0.2 and 9.68 bar.

4. Summary

Positron annihilation lifetime was used to investigate the formation of point defects in undoped semi-insulating GaAs during the diffusion of copper. Our experiment findings are summarized below:

Almost no positron trapping is found after quenching from diffusion temperature (1100 °C), when Cu is distributed all over the crystal, i.e. after in-diffusion. During a subsequent annealing up to 750 K after the diffusion treatment (out-diffusion), the average positron lifetime increases strongly indicating the detection of vacancy-type defects. With a further increase of the annealing temperature to 850 K, a rapid decrease of the average positron lifetime was observed. It is probably due to the fact that vacancy clusters grow and the distance between them becomes larger than the positron diffusion length. Another possible reason for the disappearance of the vacancy signal is that vacancy clusters are dissolved. This cannot be distinguished by the obtained data alone.

It can be shown from the annealing behavior of the average lifetime that the open volume of the detected vacancy-like defect increases during annealing, in contrast to the semi-insulating undoped GaAs samples annealed under very similar conditions but not treated with Cu. The average lifetime decreases with increasing the annealing temperature. The defect concentration increases with increasing the annealing temperature up to 750 K. The number of vacancies incorporated in the observed clusters is one vacancy in the annealing temperature range up to 750 K and increases to reach 2 vacancies at 800 K. Vacancy-like defects and shallow positron traps were observed.

On the basis of positron annihilation parameters and thermodynamic considerations, it was concluded that the observed vacancy-like defect contains an arsenic monovacancy. The results obtained at undoped SI GaAs were compared to the results of similar experiments done on *n*-type Si-doped GaAs to confirm the reliability of such thermodynamic considerations, where the existence of a donor-gallium vacancy complex is well known.

The difference of the arsenic vapor pressure dependence of the vacancy concentration for undoped SI and *n*-type Si-doped GaAs was clearly shown. Taking into account that the isolated V_{As} in a *p*-type sample is positive and accordingly invisible to positrons, the presence of a vacancy complex containing an As vacancy was assumed. The charge of this complex must be neutral or negative in our *p*-type samples. Due to the high solubility and high diffusion coefficient of copper in GaAs, it was the first candidate that could be responsible for this complex. The contamination of Cu atoms was confirmed with means of titration measurements but the concentration was almost one order of magnitude lower than the vacancy concentration that is calculated from positron lifetime measurements.

We believe that the observed vacancy complex is not bound to Cu impurities and represents a native defect complex. But the structure of the complex cannot be exactly determined from positron annihilation parameters alone. The observed shallow traps in Cu-diffused GaAs can be explained by copper contamination. Cu atom placed on a Ga sublattice forms a double acceptor Cu_{Ga}^{2-} (energy level E_V + 0.5 eV) that acts as a positron shallow trap. The temperature-dependence Hall-effect measurements confirmed the existence of this acceptor level. But in SI GaAs samples which were not treated with Cu, shallow positron traps could be explained by negative ions, C_{As}^{-}.

Acknowledgments
We would like to thank B. Gruendig-Wendrock (TU Bergakademie Freiberg) for performing Hall measurements and S. Eichler (Freiberger Compound Materials GmbH) for providing the samples materials.

References

[1] Hurle D T J 1999 *J. Appl. Phys.* **85** 6957
[2] Cohen R M 1997 *Mater. Sci. Eng. R.* **20** 167
[3] Krause-Rehberg R and Leipner H S 1999 *Positron Annihilation in Semiconductors* (Berlin: Springer)
[4] Hall R N and Racette J H 1964 *J. Appl. Phys.* **35** 379
[5] Frank F C and Turnball D 1956 *Phys. Rev.* **104** 617
[6] Leon R, Werner P, Yu K M, Kaminska M and Weber E R 1995 *Appl. Phys.* A **61** 7
[7] Gokcen N A 1989 *Bulletin of Alloy Phase Diagrams* **10** 11
[8] Saarinen K, Hautojärvi P, Vehanen A, Krause R and Dlubek G 1989 *Phys. Rev.* B **39** 5287
[9] Bondarenko V, Petters K, Krause-Rehberg R, Gebauer J and Leipner H S 2001 *Physica* B **308-310** 792
[10] Krause-Rehberg R and Leipner H S 1997 *Appl. Phys. A: Mater. Sci. Process.* **64** 457
[11] Gebauer J, Krause-Rehberg R, Domke C, Ebert Ph and Urban K 1997 *Phys. Rev. Lett.* **78** 3334
[12] Le Berre C, Corbel C, Saarinen K, Kuisma S, Hautojarvi P and Fornari R 1995 *Phys. Rev.* B **52** 8112
[13] Gebauer J, Börner F, Krause-Rehberg R, Staab T E M, Bauer-Kugelmann W, Kögel G, Triftshäuser W, Specht P, Lutz R C, Weber E R and Luysberg M 2000 *J. Appl. Phys.* **87** 8368
[14] Staab T E M, Haugk M, Sieck A, Frauenheim Th and Leipner H S 1999 *Physica B* **273-274** 501
[15] Puska M and Nieminen R 1983 *J. Phys. F: Metal Phys.* **13** 333
[16] Gebauer J, Lausmann M, Staab T E M, Krause-Rehberg R, Hakala H and Puska M J 1999 *Phys. Rev.* B **60** 1464
[17] Saarinen K, Hautojärvi P, Lanki P and Corbel C 1991 *Phys. Rev.* B **44** 10585
[18] Tan T Y, You H M and Goesele U M 1993 *Appl. Phys.* A **56** 243
[19] Bondarenko V, Gebauer J, Redmann F and Krause-Rehberg R 2005 *Appl. Phys. Lett.* **87** 161906
[20] Gebauer J, Lausmann M, Redmann F and Krause-Rehberg R 1999 *Physica* B **273-274** 705
[21] Tin C C, Teh C K and Weichman F L 1987 *J. Appl. Phys.* **62** 2329
[22] Kuriyama K, Tomizava K, Uematsu S and Takahashi H 1994 *Appl. Phys. Lett.* **65** 746

Journal of Physics: Conference Series **265** (2011) 012006 doi:10.1088/1742-6596/265/1/012006

Calculation of positron characteristics for elements of the periodic table

J M Campillo Robles[1], E Ogando[2] and F Plazaola[2]

[1] Oinarrizko Zientziak Saila, Goi Eskola Politeknikoa, Mondragon Unibertsitatea, 20500 Arrasate, Basque Country

[2] Elektrizitatea eta Elektronika Saila, Zientzia eta Teknologia Fakultatea, UPV/EHU, 48080 Bilbao

E-mail: fernando.plazaola@ehu.es

Abstract. Positron characteristics have been calculated in bulk and monovacancies for most of the elements of the periodic table. Self-consistent and non-self-consistent schemes have been used for the calculation of the electronic structure in the solid, and different parametrizations for the positron enhancement factor and correlation energy. As it is known, positron lifetimes in bulk show a periodic behaviour with atomic number. These calculations also confirm that monovacancy lifetimes follow the same behaviour. The results obtained have been compared with selected experimental lifetime data, which confirms the calculated theoretical trends. Positron binding energies to a monovacancy have been calculated also for most of the elements of the periodic table. The binding energy shows a periodic behaviour with atomic number too.

1. Introduction

At the end of XIX century Mendeleev stated the periodic arrangement of the elements [1]. Since then, a lot of research has been made on elements and their properties. As a result of this, periodic behaviours have been tested in most of physical and chemical properties of the elements [2, 3]. Electronic properties of the elements show periodic trends too, and for this reason, positron annihilation characteristics are related to these periodic trends. MacKenzie et al discovered that experimental positron bulk lifetime shows a periodic behaviour with atomic number [4]. However, this periodicity is not only reflected in positron bulk lifetime; Doppler broadening experiments [5, 6] and positron affinity of elemental metals [7] also show periodic behaviour.

In this work, a systematic Density Functional Theory (DFT) calculation of positron characteristics has been performed for bulk and monovacancies of most of the elements of the periodic table. The effort made to calculate systematically the annihilation parameters is important to deepen into the knowledge of the calculation methods [8, 9], improving the theoretical background required for a good interpretation of the experimental data.

2. Calculation Method

Positron characteristics have been calculated for most of the elements of the periodic table. Our computational methods are described in more detail in reference 10, here we will only explain the main features. We have used two schemes for treating the electronic densities in calculations. First, we have used a non-self-consistent scheme, the atomic superposition approximation of Puska and Nieminen (AT-SUP) [11]. Moreover, we have used a self-consistent scheme, the linear muffin-tin orbital method within the atomic-spheres approximation (LMTO-ASA) [12, 13]. The exchange and correlation effects of the positron-electrons system have been simulated using two different approximations, one, within the local density approximation (BN), and the other, within the general gradient approximation (GGA) [14]. For the monovacancy supercells no relaxation has been performed in the atomic positions.

As a result of the computation of the positron state, the positron energy eigenvalue for that state can be obtained. In bulk state, delocalized positron, it is correct to use the zero-positron-density ($n_+ \to 0$) limit of the two-component density-functional theory (TCDFT). In monovacancy state, the positron is localized in the defect, and the positron density is not zero. However, the previous approximation, "conventional scheme", works well because the positron and its screening cloud of electrons can be considered as a neutral quasiparticle, which does not affect the average electron density [15]. For this reason, the positron binding energy to the monovacancy can be defined as the difference between positron energy eigenvalues in bulk (ε^+_{bulk}) and monovacancy (ε^+_{vac}):

$$E = \varepsilon^+_{bulk} - \varepsilon^+_{vac}.$$

The trapping of a positron in a monovacancy reduces the energy eigenvalue of the positron, releasing binding energy to the host.

3. Results and discussion

For most of the elements of the periodic table the bulk and monovacancy positron lifetime have been calculated using AT-SUP and LMTO-ASA within BN and GGA approximations. Calculated lifetimes can be consulted in reference 10. It is known from the work of MacKenzie et al [4], that experimental positron bulk lifetime shows a periodic trend with atomic number. The calculations presented in this work support MacKenzie et al results. However, we have proved that the periodic behaviour of bulk lifetime also appears in monovacancy lifetime. These periodic trends are independent of the calculation scheme, and follow the same behaviour as atomic volume (see figure 1). Monovacancy lifetimes for actinides do not appear in Fig. 1 because of convergence problems with the LMTO-ASA code. On the other hand, we have collected experimental positron lifetimes from reference 16. We have fixed some conditions to select the experimental values from this data-base with a minimum of quality and coherence [10]. As a result, the collected experimental values confirm our calculated theoretical trends.

We have calculated the binding energy of positron to a monovacancy for each element using AT-SUP and LMTO-ASA within BN and GGA approximations. Table 1 shows the LMTO-ASA results of positron binding energies within BN and GGA approximations. AT-SUP results are given in table 2.

There are not many values of binding energies in the literature. In order to determine experimentally the positron binding energy for a particular defect type it is necessary to observe detrapping from the defect [17]. In this way, Dannefaer and Kerr measured the binding energy of monovacancies in GaAs and InSb [18]. To our knowledge Hautojärvi et al [19] gave the first theoretical binding energy for Mg (0.3-0.4 eV). This value range is much lower than the ones presented in tables 1 and 2. Puska et al calculated binding energies in semiconductors [20]. The values corresponding to C, Si and Ge, performed by atomic superposition calculations, are larger than the ones presented in table 2 (BN) and the difference increases with atomic number. Moreover, latter Puska presented binding energies in Si and GaAs calculated by atomic superposition and LMTO-ASA-Green's function model in charged defects [21]. The non-self-consistent value of Si is identical to that of table 2, but the self-consistent value is larger than the one in table 1, which gives the value for an ideal vacancy. The most recent work in binding energies is the one of Makkonen and Puska [22]. They present binding energies for relaxed monovacancies in several metals and semiconductors, where they describe the electron-ion interaction on the projector augmented-wave method,

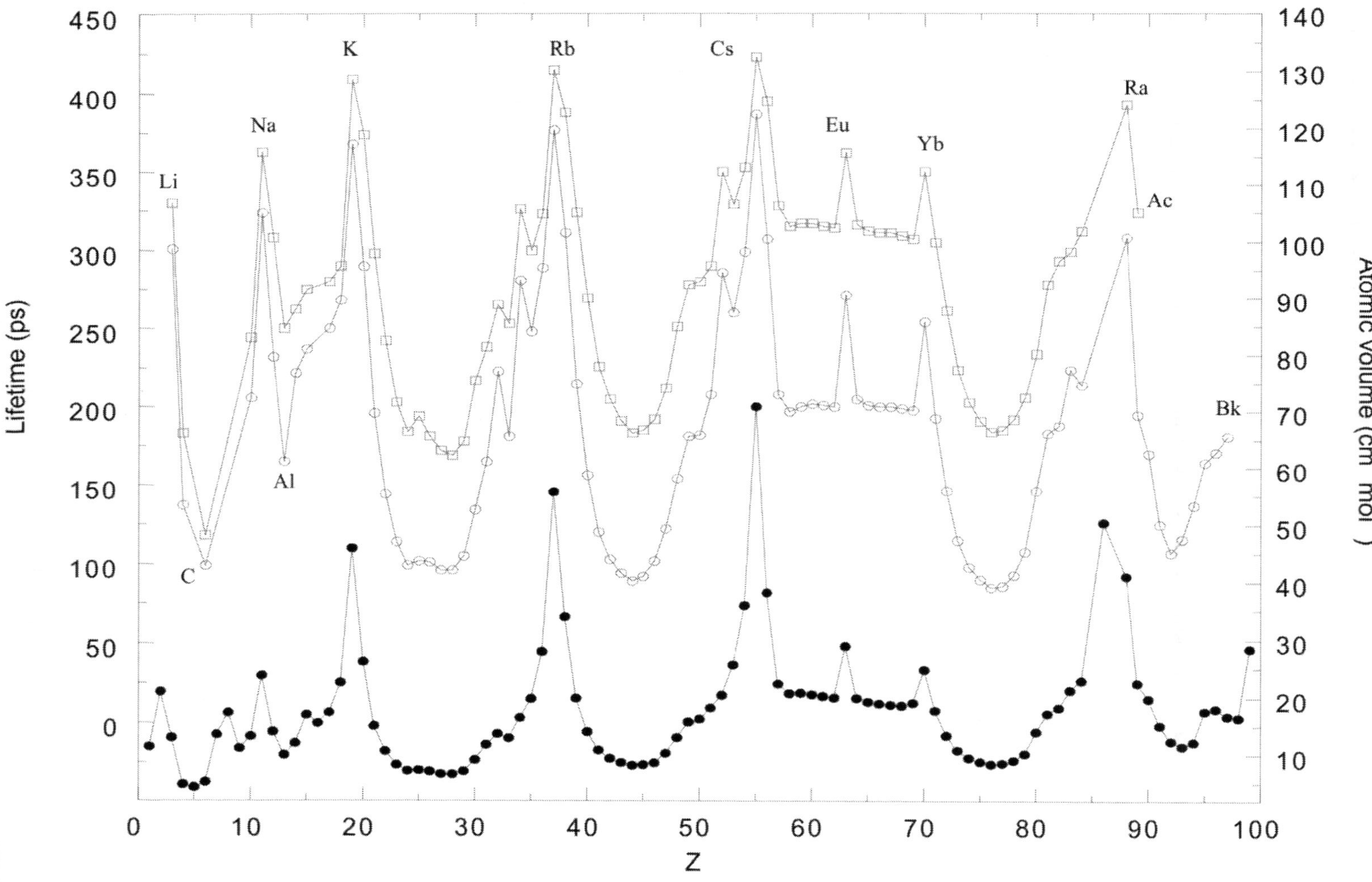

Figure 1. Atomic volume (filled circles) and positron lifetimes plotted against atomic number. Positron lifetimes are calculated in bulk (open circles) and monovacancy (squares) states within the LMTO-ASA method using the BN approximation.

Table 1. Positron binding energies calculated with the LMTO method within BN and GGA.

Key: Element — Nb
3.13 ← E_{BN}^{LMTO} (eV)
3.22 ← E_{GGA}^{LMTO} (eV)

H — —																	**He** — —
Li 0.17 1.07	**Be** 2.62 2.74											**B** — —	**C** 1.25 1.29	**N** — —	**O** — —	**F** — —	**Ne** 0.03 -0.02
Na 0.16 0.24	**Mg** 0.91 0.85											**Al** 1.88 1.71	**Si** 0.66 0.70	**P** 0.44 0.43	**S** — —	**Cl** -0.13 -0.26	**Ar** 0.07 -0.03
K 0.27 0.38	**Ca** 1.02 1.10	**Sc** 1.65 1.79	**Ti** 2.30 2.43	**V** 2.60 2.77	**Cr** 3.20 3.33	**Mn** 2.73 2.92	**α-Fe** 2.40 2.53	**α-Co** 2.41 2.56	**Ni** 2.07 2.24	**Cu** 1.67 1.83	**Zn** 1.56 1.58	**Ga** 0.96 0.85	**Ge** 0.67 0.72	**As** 1.11 1.06	**Se** 0.46 0.15	**Br** 0.43 0.37	**Kr** 0.13 -0.08
Rb 0.29 0.38	**Sr** 0.84 0.87	**Y** 1.70 1.79	**Zr** 2.57 2.64	**Nb** 3.13 3.22	**Mo** 3.99 4.00	**Tc** 3.84 3.87	**Ru** 3.94 3.99	**Rh** 3.29 3.35	**Pd** 2.34 2.43	**Ag** 1.61 1.71	**Cd** 1.50 1.50	**In** 1.16 1.10	**β-Sn** 1.30 1.14	**Sb** 1.16 1.03	**Te** 0.53 0.14	**I** 0.53 0.40	**Xe** 0.27 -0.01
Cs 0.39 0.44	**Ba** 1.14 1.29	**La** 1.84 1.98	**Hf** 2.79 2.85	**Ta** 3.39 3.44	**W** 4.57 4.51	**Re** 4.62 4.60	**Os** 4.94 4.86	**Ir** 4.44 4.34	**Pt** 3.46 3.38	**Au** 2.40 2.36	**β-Hg** 1.45 1.32	**Tl** 0.94 0.95	**Pb** 1.28 1.18	**Bi** — —	**Po** 1.37 0.93	**At** — —	**Rn** — —
Fr — —	**Ra** 0.96 0.98	**Ac** 2.57 2.60															
				γ-Ce 1.75 1.87	**Pr** 1.65 1.79	**Nd** 1.65 1.79	**Pm** 1.58 1.75	**Sm** 1.62 1.77	**Eu** 0.87 0.89	**Gd** 1.49 1.65	**Tb** 1.57 1.75	**Dy** 1.55 1.72	**Ho** 1.55 1.72	**Er** 1.56 1.73	**Tm** 1.58 1.75	**Yb** 1.03 1.07	**Lu** 1.84 1.92

Table 2. Positron binding energies calculated with the AT-SUP method within BN and GGA.

Element → **Nb**: 3.64 ← E_{BN}^{AT-SUP} (eV); 3.85 ← E_{GGA}^{AT-SUP} (eV)

H																	**He**
—																	—
—																	—
Li	**Be**											**B**	**C**	**N**	**O**	**F**	**Ne**
0.46	1.81											—	0.56	—	—	—	0.01
0.55	1.84											—	0.53	—	—	—	0.01
Na	**Mg**											**Al**	**Si**	**P**	**S**	**Cl**	**Ar**
0.22	0.72											2.13	0.36	0.13	—	0.01	0.02
0.32	0.71											2.10	0.32	0.09	—	0.00	0.00
K	**Ca**	**Sc**	**Ti**	**V**	**Cr**	**Mn**	**α-Fe**	**α-Co**	**Ni**	**Cu**	**Zn**	**Ga**	**Ge**	**As**	**Se**	**Br**	**Kr**
0.33	1.01	2.17	3.20	4.06	3.28	4.20	3.26	3.11	2.75	1.25	0.82	0.64	0.26	0.53	0.04	0.02	0.03
0.46	1.03	2.30	3.40	4.32	3.61	4.43	3.49	3.33	2.96	1.47	0.87	0.64	0.23	0.43	0.01	0.00	0.00
Rb	**Sr**	**Y**	**Zr**	**Nb**	**Mo**	**Tc**	**Ru**	**Rh**	**Pd**	**Ag**	**Cd**	**In**	**β-Sn**	**Sb**	**Te**	**I**	**Xe**
0.35	0.98	2.18	3.38	3.64	3.91	4.94	3.62	2.95	1.10	1.15	0.75	1.13	1.03	0.57	0.13	0.08	0.07
0.47	0.98	2.27	3.51	3.85	4.12	5.12	3.82	3.14	1.16	1.28	0.76	1.19	1.00	0.47	0.01	0.01	0.00
Cs	**Ba**	**La**	**Hf**	**Ta**	**W**	**Re**	**Os**	**Ir**	**Pt**	**Au**	**β-Hg**	**Tl**	**Pb**	**Bi**	**Po**	**At**	**Rn**
0.44	1.40	3.02	3.11	4.13	4.73	4.77	4.60	4.02	1.92	1.50	0.57	0.74	0.79	0.42	0.39	—	—
0.55	1.46	3.16	3.23	4.26	4.85	4.86	4.67	4.05	1.93	1.47	0.46	0.76	0.74	0.33	0.25	—	—
Fr	**Ra**	**Ac**															
—	1.23	3.34															
—	1.26	3.45															

γ-Ce	**Pr**	**Nd**	**Pm**	**Sm**	**Eu**	**Gd**	**Tb**	**Dy**	**Ho**	**Er**	**Tm**	**Yb**	**Lu**
3.20	2.95	2.80	2.71	2.60	1.23	2.35	2.31	2.23	2.14	2.08	2.03	1.04	2.04
3.43	3.18	3.02	2.93	2.82	1.30	2.48	2.53	2.44	2.35	2.28	2.22	1.10	2.15
Th	**Pa**	**U**	**Np**	**Pu**	**Am**	**Cm**	**Bk**	**Cf**	**Es**	**Fm**	**Md**	**No**	**Lr**
3.81	5.40	5.60	6.41	4.13	2.93	2.87	2.47	—	—	—	—	—	—
3.87	5.58	5.76	6.67	4.39	3.15	3.01	2.51	—	—	—	—	—	—

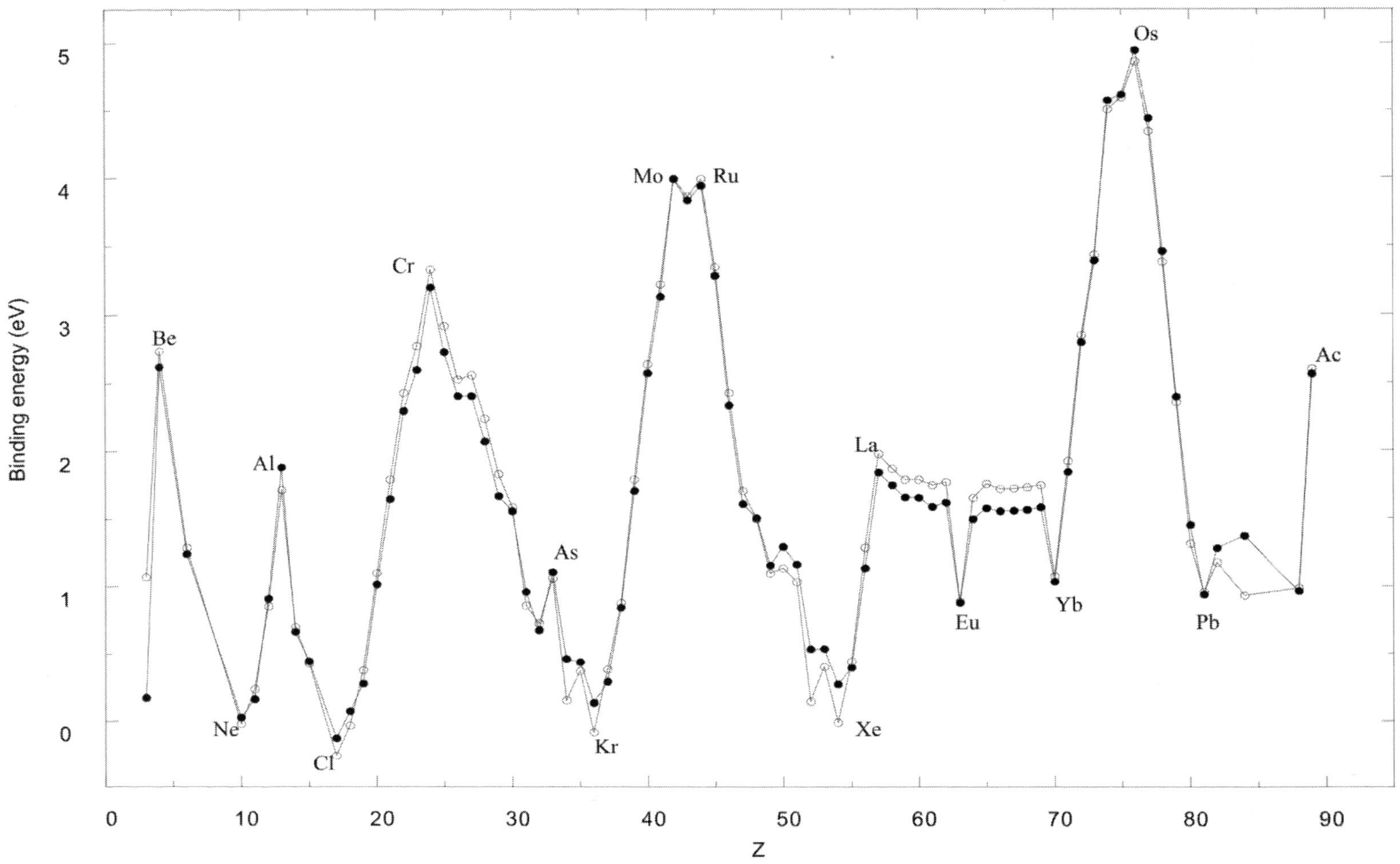

Figure 2. Positron binding energies as a function of the atomic number. Binding energies calculated with the LMTO method within BN (filled circles) and GGA (open circles).

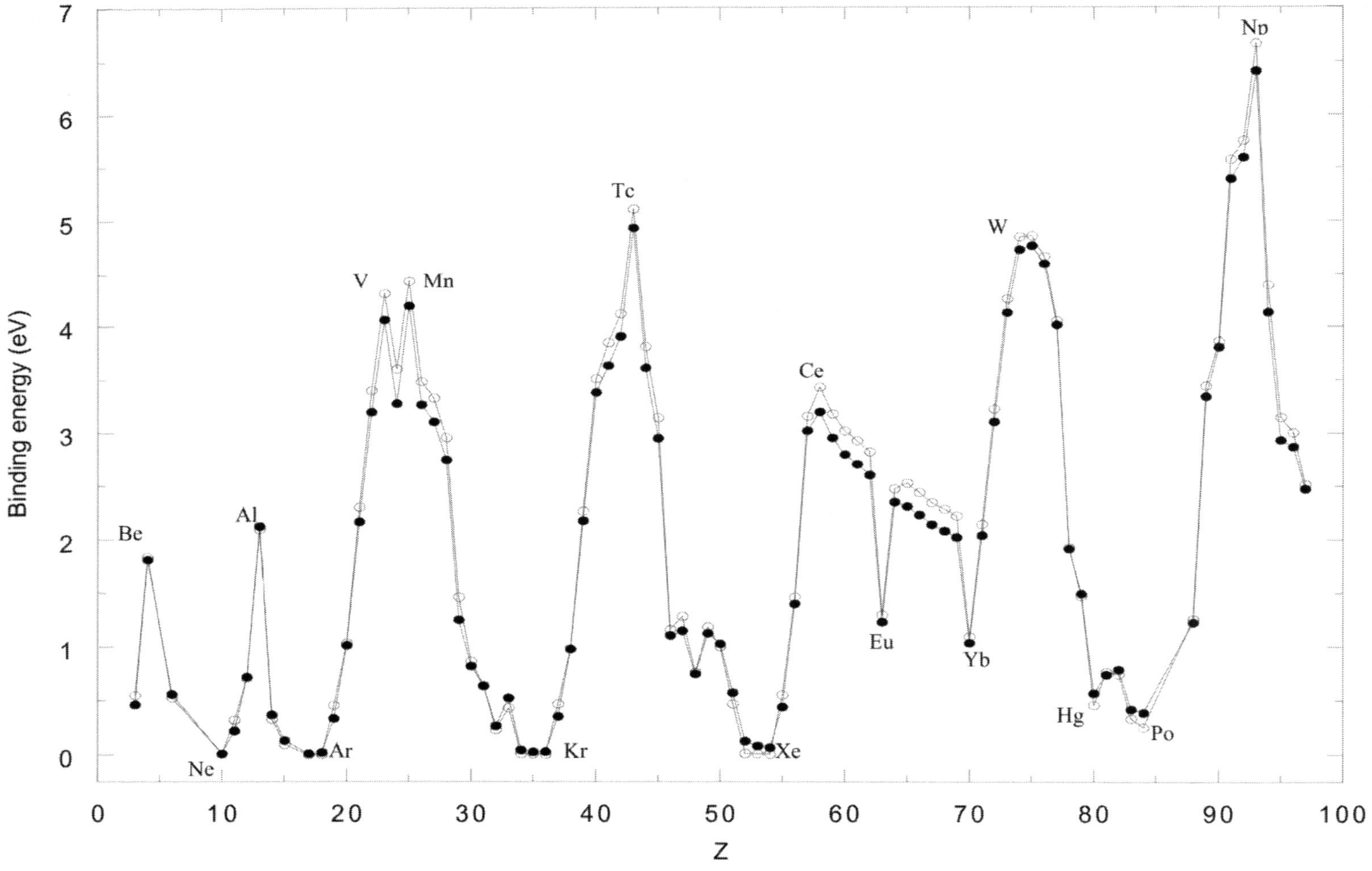

Figure 3. Positron binding energies as a function of the atomic number. Binding energies calculated with the AT-SUP method within BN (filled circles) and GGA (open circles).

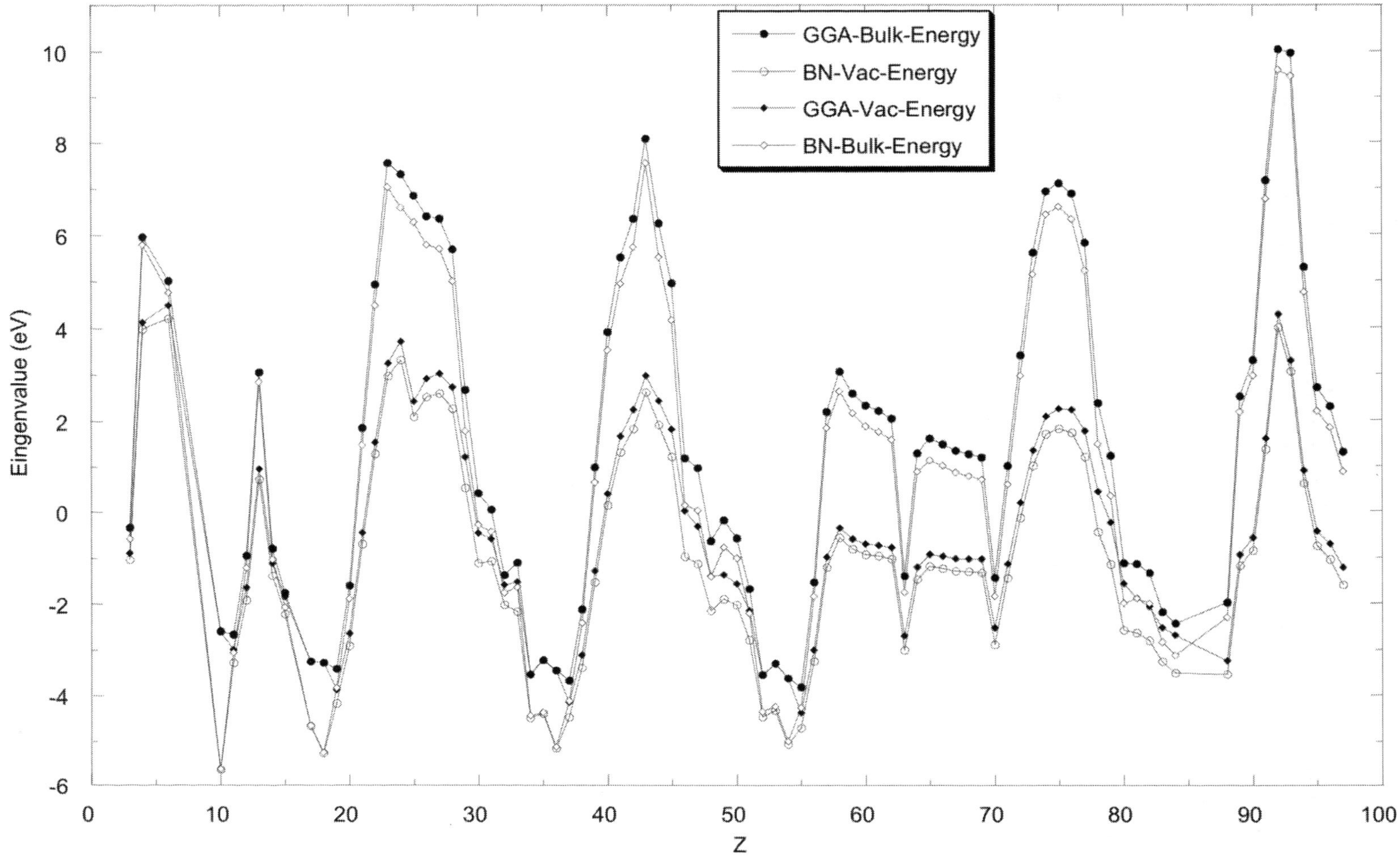

Figure 4. Positron energy eigenvalues of bulk (circles) and monovacancy (rhombuses) states plotted against atomic number. Positron energy eigenvalues are calculated with the AT-SUP method within BN (filled symbols) and GGA (open symbols).

implemented in the plane-wave code VASP. The values for Si and Fe are quite similar to the ones presented in table 1 (BN), but table 1 presents smaller values for Al and Cu and larger for Mg. This last work has been performed under the TCDFT, and this scheme has been showed to lead to rather localized positron states and too low core electron annihilation rates in comparison with experiments [15].

LMTO-ASA results of positron binding energies have been plotted in figure 2 as a function of atomic number. Figure 2 shows a very clear periodic trend. Comparing binding energy with atomic volume (lowest line in figure 1), it can be seen that binding energy shows an inverse behaviourcompared to that of atomic volume. Binding energy has minima near the noble gases, where the atomic volume gets its maxima, and maxima near the centers of "d" series, where atomic volume reaches its minima. In the case of lanthanides, binding energy is practically constant. Moreover, BN and GGA values are similar for most of the calculated elements. Anyway, GGA values are lower than BN ones in the previously cited minima of binding energy. The opposite happens in the case of 3d and 4d series and in lanthanides. However, for Eu there is no difference and in Yb the difference is very small.

Figure 3 shows binding energies calculated within AT-SUP plotted against the atomic number. All the periodic trends are similar to those of LMTO-ASA results. However, the difference in binding energies, E_{GGA}- E_{BN}, ranges between (–0.1 ± 0.3) eV in AT-SUP calculations, which is smaller than the one obtained with LMTO-ASA (–0.4 ± 0.2) eV. So, self-consistent calculations give larger values for BN binding energies. Moreover, in AT-SUP calculations there is a monotonous decrease of binding energies of lanthanides with atomic number increase, independently of the approximation (see figure 3). To understand this behaviour we have plotted in figure 4 energy eigenvalues of positron in bulk and monovacancy versus atomic number for AT-SUP. Figure 4 shows that monovacancy energy eigenvalue remains almost constant with atomic number, but in the case of bulk it decreases monotonically. It indicates that such variation is due to the inner f electrons, which increase with atomic number. Therefore, for lanthanides AT-SUP calculations give a relatively large annihilation with inner f electrons in bulk states. This behaviour is not observed in the case of binding energies calculations by LMTO calculations (see figure 2). Indeed, energy eigenvalues of lanthanides remain almost constant in bulk and monovacancy states (not shown). This last result indicates that non-self-consistent calculations give an enhanced contribution of f electrons in respect to self-consistent ones.

4. Conclusions

We have performed self- and non-self-consistent calculations of positron annihilation parameters in most of the elements of the periodic table, using BN and GGA approximations. The binding energies, as it has been shown for positron lifetimes, present a periodic behaviour too. However, the behaviour of binding energies is opposite to the one of positron lifetimes.

Acknowledgements

This work has been undertaken under CICYT and Basque government projects MAT2006-12743 and IT44310, respectively. The SGI/IZO-SGIker UPV/EHU is gratefully acknowledged for allocation of computational resources.

References

[1] Mendeleev D 1869 *Z. Chem.* **12** 405
[2] van Spronsen J W 1969 *The Periodic System of the Chemical Elements* (Amsterdam: Elsevier)
[3] Mazurs E G 1974 *Graphic Representation of the Periodic System during One Hundred Years* (Alabama: University of Alabama Press)
[4] MacKenzie I K, Jackman T E and Thrane N 1975 *Phys. Rev. Lett.* **34** 512
[5] Gosh V J, Alatalo M, Asoka-Kumar P, Nielsen B, Lynn K G, Kruseman A C and Mijnarends P E 2000 *Phys. Rev.* B **61** 10092
[6] Myler U and Simpson P J 1997 *Phys. Rev.* B **56** 14303

[7] Puska M J, Lanki P and Nieminen R M 1989 *J. Phys.: Condens. Matter* **1** 6081
[8] Takenaka H and Singh D J 2008 *Phys. Rev.* B **77** 155132
[9] Makkonen I, Hakala M and Puska M J 2006 *Phys. Rev.* B **73** 35103
[10] Campillo Robles J M, Ogando E and Plazaola F 2007 *J. Phys.: Cond. Matter* **19** 176222
[11] Puska M J and Nieminen R M 1983 *J. Phys. F: Met. Phys.* **13** 333
[12] Andersen O K, Jepsen O and Glötzel D 1985 *Highlight of Condensed-Matter Theory*, ed F Bassani, F Fumi and M P Tosi (New York: North-Holland)
[13] Andersen O K, Jepsen O and Šob M 1987 *Electronic Band Structure and Its Applications*, ed M Yossouff (Heildeberg: Springer-Verlag)
[14] Barbiellini B, Puska M, Torsti T and Nieminen R M 1995 *Phys. Rev.* B **51** 7341
[15] Puska M J, Seitsonen A P and Nieminen R M 1995 *Phys. Rev.* B **52** 10947
[16] Campillo Robles J M and Plazaola F 2003 *Deffect Diffus. Forum* **213-215** 141
[17] Corbel C, Pierre F, Saarinen K and Hautojärvi P 1992 *Phys. Rev.* B **45** 3386
[18] Dannefaer S and Kerr D1993 *Phys. Rev.* B **48** 9142
[19] Hautojärvi P, Johansson J, Vehanen A, Yli-Kauppila J, Hillariet J and Tzenétakis P 1982 *Appl. Phys.* A **27** 49
[20] Puska M J, Mäkinen S, Manninen M and Nieminen R M 1989 *Phys. Rev.* B **39** 7666
[21] Puska M J 1992 *Mat. Science Forum* **105-110** 419
[22] Makkonen I and Puska M J 2007 *Phys. Rev.* B **76** 54119

Journal of Physics: Conference Series **265** (2011) 012007 doi:10.1088/1742-6596/265/1/012007

Effect of electron- and neutron-irradiation on Fe-Cu model alloys studied by positron annihilation spectroscopy

Y Nagai[1,2], K Takadate[1], Z Tang[1,3], H Ohkubo[1], H Sunaga[4], H Takizawa[4] and M Hasegawa[1,5]

[1] The Oarai Center, Institute for Materials Research, Tohoku University, Oarai, Ibaraki 311-1313, Japan
[2] SCK•CEN, Nuclear Materials Science Institute, Boeretang 200, 2400 Mol, Belgium
[3] Department of Electronic Engineering, School of Information Science and Technology, East China Normal University, Dongchuan Road 500, Minhang, Shanghai 200241, P. R. China
[4] Takasaki Advanced Radiation Research Institute, Japan Atomic Energy Agency, 1233 Watanuki-machi, Takasaki, Gunma 370-1292, Japan
[5] Cyclotron and Radioisotope Center, Tohoku University, Sendai 980-8578, Japan

E-mail: nagai@imr.tohoku.ac.jp

Abstract. Electron- and neutron-irradiation effects on dilute Fe-Cu model alloys of nuclear reactor pressure vessel steels are studied by positron annihilation spectroscopy. We have found that, not only by high-dose neutron-irradiation but also by low-dose electron-irradiation, the aggregation of Cu atoms and vacancies takes place and the ultrafine Cu precipitates are formed after post-irradiation annealing at 400°C. In spite of large difference in the irradiation doses between the electron- and the neutron-irradiated samples, no significant difference is observed in the isochronal annealing behaviour above 400°C of positron annihilation and micro-hardness, indicating that small amount of extra vacancies enhance the aggregation of Cu atoms in Fe during the annealing-out process of the vacancies.

1. Introduction

Since light water reactors supply a large part of the electric power, their prolonged and safe operation is indispensable. In particular, the embrittlement of the reactor pressure vessel (RPV) steels due to long-term in-service exposure to neutron irradiation is one of the topical issues in the field of nuclear power energy. Therefore, the understanding of the detailed mechanism of the embrittlement is very important. The ultrafine precipitation of Cu impurities contained (~0.3%) in old RPV enhanced by the neutron irradiation is considered to be one of the main origin of the embrittlement of RPV steels (low alloy steels: mainly A533B) [1-6]. In spite of extensive studies, however, little is known about the irradiation enhanced Cu precipitation because it is very difficult to observe the ultrafine Cu precipitates even by current high-resolution transmission electron microscopes.

Recently, we have revealed the formation and recovery process of the ultrafine Cu precipitates in Fe-Cu model alloys formed by neutron irradiation using positron annihilation technique [7]. Vacancy-Cu complexes are formed by neutron irradiation and aggregate into nanovoids. The inner surface of the nanovoids is covered with Cu atoms. The dissociation of vacancies from the nanovoids by annealing at ~400°C leads to the formation of the ultrafine Cu precipitates. The Cu precipitates are coherent with Fe matrix and they are annealed out at ~650°C. In this paper, we study electron irradiated Fe-Cu alloys as well as neutron irradiated ones using positron annihilation spectroscopy (coincidence Doppler broadening (CDB) and positron lifetime) and Vickers micro-hardness, and compare the results for both irradiations.

Positron is the only probe that is able to detect vacancy-type defects in metals sensitively [8-11]. In addition, it is found that positron is also a sensitive tool to defect-free ultrafine Cu precipitates in Fe due to their positron affinity trapping [12]. Positron lifetime spectroscopy provides information on the size of the vacancy clusters and their number density. The CDB enables us to identify the chemical environment surrounding the site where the positron is trapped by measuring the electron momentum distribution. The high momentum region is namely the result of the positron annihilation with the inner shell electrons [13, 14]. Thus, the CDB method can reveal the formation of vacancy-Cu complexes, their aggregations and (defect free) ultrafine Cu precipitates.

In this paper, we report that, in spite of the much lower electron dose (10^{-4} dpa: displacement probability per atom) than the neutron dose (10^{-2} dpa), Cu atoms aggregate around vacancies and the ultrafine Cu precipitates are formed after post-irradiation annealing at 400°C even in the electron-irradiated dilute Fe-Cu alloys. This indicates that small amount of extra vacancies enhances much the aggregation of Cu atoms in Fe during the annealing-out process of the vacancies.

2. Experimental procedure

The prepared samples were Fe-0.3wt%Cu and -0.05wt%Cu. They were made from high purity Fe (4N) and Cu (5N) by arc melting and cold rolled to 0.3 mm in thickness. The high and low Cu samples are the model alloys of the practical old RPV steels (~0.3wt%) and recent RPV steels (~0.05wt%), respectively. The samples were heated to 825°C and kept for 4h, followed by quenching into iced water. Some of the samples were bombarded with 3 MeV electrons below 50°C using a dynamitron accelerator of the JAEA-Takasaki Institute. In this condition, only Frenkel pairs are introduced and no collisional cascade is formed. The electron fluence was 2.0×10^{18} e/cm^2, which corresponds to the irradiation dose of 9.9×10^{-5} dpa (displacement probability per atom). Other samples were irradiated with fast neutrons in the Irradiation Facility of Hydraulic Rabbit II of the Japan Material Testing Reactor (JMTR) at ~100°C. The neutron fluence was 8.3×10^{18} n/cm^2, which corresponds to the irradiation dose of 1.2×10^{-2} dpa. In this condition, vacancy-rich regions are created in the small area of the collisional cascade of the primary knock-on atom [15].

Isochronal annealing experiment from 100°C to 750°C with 50°C step (30 min) of positron annihilation and Vickers micro-hardness were performed. Positron lifetime measurements were carried out using a conventional fast-fast spectrometer with a time resolution of 190 ps (FWHM). About 4×10^6 coincidence events were accumulated for each measurement for 12h. After subtracting the source component and background, the spectra were decomposed into two components (τ_1 and τ_2). In order to identify the size of the vacancy clusters (V_n, n: the number of vacancies), the experimental results were compared with the calculated positron lifetimes in the vacancy clusters by using a simple superimposed atom method [16]. The detail of the method is described in ref. [17].

The CDB spectra were measured using two Ge detectors. The details of this method are described by Asoka-Kumar *et al.* [14]. The overall energy resolution is ~1.1 keV in full width at half maximum (FWHM), which corresponds to the momentum resolution of $\sim4.3\times10^{-3}$ m_0c (FWHM), where c is the speed of light and m_0 is the electron rest mass. Total counts of more than 2×10^7 for each measurement were accumulated for 12 h. The CDB ratio spectrum was obtained by normalizing the momentum distribution of each spectrum to that of unirradiated (defect free) pure Fe. The shape of the spectrum in high momentum region (typically $>10\times10^{-3}$ m_0c) exhibits characteristic signals of the elements through the positron annihilation with their inner shell electrons. In the case that positron annihilates with Cu *3d* electrons, a broad peak around 25×10^{-3} m_0c appears [12]. The *S*- and the *W*-parameters are defined as the ratios of the counts in low momentum ($|p_L| < 4\times10^{-3}$ m_0c) and high momentum (18×10^{-3} $m_0c < |p_L| < 30\times10^{-3}$ m_0c) regions in the Doppler-broadening spectrum to the total counts, respectively. When the positron is trapped at vacancy-type defect, the *S*-parameter increases and the *W*-parameter decreases. In addition, the *W*-parameter increases when positron annihilates with Cu electrons corresponding to the broad peak around 25×10^{-3} $m_0\,c$ in the CDB ratio spectrum.

We also measured Vickers micro-hardness with conventional apparatus with a load of 200 g for 15 sec in each step of the isochronal experiments.

3. Results and discussion

Figure 1 shows the CDB ratio spectra of as-irradiated (a) Fe-0.3wt%Cu and (b) Fe-0.05wt%Cu, together with well annealed pure Cu as a reference. Broad peak around 25×10^{-3} m_0c characteristic of Cu *3d* electrons is clearly observed for both the electron- and neutron-irradiated Fe-0.3wt%Cu and Fe-0.05wt%Cu. Enhancement in the low momentum region and dehancement (except the broad peak around 25×10^{-3} m_0c) in the high momentum region due to positron trapping at vacancy-type defects are also observed. These results indicate that vacancy-type defects associated with Cu atoms are formed.

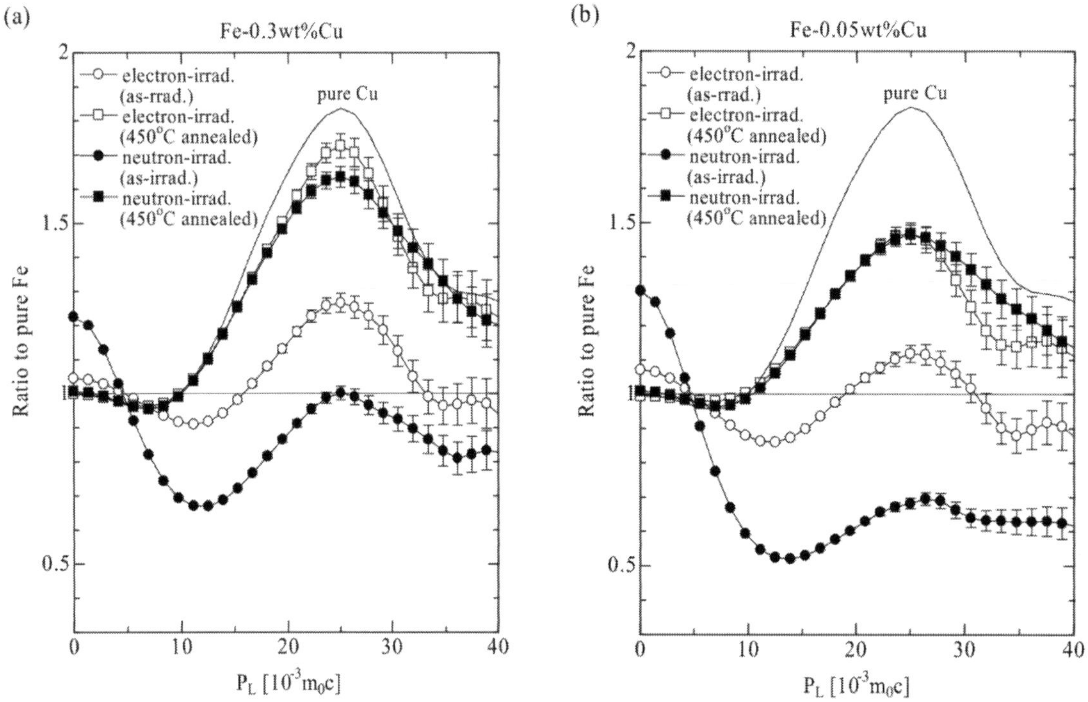

Figure 1. CDB ratio spectra of electron- and neutron-irradiated Fe-Cu alloys: (a) Fe-0.3wt%Cu, (b) Fe-0.05wt%Cu.

The results of positron lifetime measurements in the irradiated samples are shown in figure 2 ((a) electron- and (b) neutron-irradiation). The calculated positron lifetimes for some vacancy clusters in Fe are also shown by horizontal dashed lines. For the neutron-irradiated samples, the longer lifetime τ_2 of 400ps arising from three dimensional vacancy clusters (nanovoids) consisting of ~30 vacancies and the shorter lifetime τ_1 of ~170ps arising mainly from monovacancies (V_1) are consistent with our previous report [7], indicating that the irradiation-induced vacancies, binding with Cu solute atoms, aggregate into the nanovoids. The surfaces of the nanovoids are mostly covered with Cu atoms. The other parts of vacancies, as V_1, are surrounded by Cu atoms. On the other hand, for the electron-irradiated samples, the longer lifetime τ_2 are ~170ps for Fe-0.3wt%Cu with the intensity I_2 of 80% and ~220ps for Fe-0.05wt%Cu with I_2 of 60%. These values are close to the calculated positron lifetime in V_1 and V_3, respectively. Although large vacancy-clusters are not observed because of low energy (3MeV) electron irradiation, the irradiation induced vacancies certainly survive at room temperature as V_1 and V_3 surrounded by Cu atoms. It should be noted that the vacancy-Cu aggregation takes place by low dose (~10^{-4} dpa) electron irradiation even in the very dilute Fe-0.05wt%Cu where the Cu precipitation is hardly expected so far.

We have also performed three-dimensional atom probe (3D-AP) observations for the electron irradiated samples. 3D-AP is a unique method to map out the alloying elements in three dimensional space with nearly atomic scale. However, no Cu precipitate is observed. The vacancy-Cu aggregation should be too small to be detected even by 3D-AP.

The isochronal annealing behaviour of positron lifetime is shown in figure 2: (a) electron- and (b) neutron-irradiated Fe-Cu alloys. The longer lifetime components disappear by the annealing above 350°C for the electron irradiation and 400°C for the neutron irradiation, and positron average lifetime decreases to the value (~110ps) for bulk pure Fe or Cu, indicating that the vacancy-type defects are annealed out. However, the recovery is not complete recovery. The CDB ratio spectra of the irradiated Fe-Cu alloys annealed at 450°C are shown in figure 1. The shape of the spectra having a broad peak around 25×10^{-3} m_0c is very similar to the ratio spectrum of pure Cu. This means that defect-free ultrafine Cu precipitates are formed.

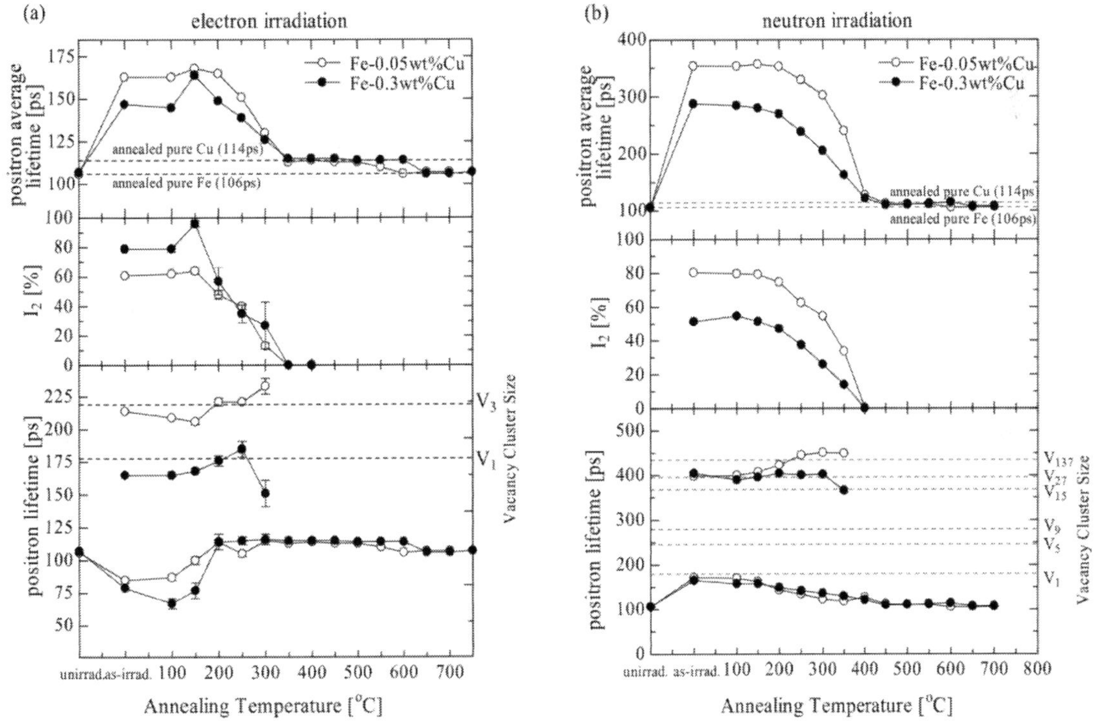

Figure 2. Isochronal annealing behaviour of positron lifetime of the irradiated Fe-Cu alloys: (a) electron-irradiation, (b) neutron-irradiation.

The isochronal annealing behaviour of *S*- and *W*-parameters, and Vickers micro-hardness are shown in figure 3. The *S*-parameters, sensitive to vacancy-type defects, decrease to the value for well annealed pure Fe or pure Cu up to 400°C annealing, consistent with the results of positron lifetime. Correspondingly, the *W*-parameters increase by the annealing up to the temperature where the vacancy-type defects are annealed out. They hold high values until high temperature values, up to 500°C for Fe-0.05wt%Cu and 550°C for Fe-0.3wt%Cu. They completely recover to the values of well annealed pure Fe at 600°C and 650°C for Fe-0.05wt%Cu and Fe-0.3wt%Cu, respectively. These results indicate the dissolution of the defect-free ultrafine Cu precipitates. The dissolution temperatures are consistent with the solubility limit of Cu in Fe [19].

The micro-hardness in the neutron-irradiated samples shows two-step recovery; the first is due to the annealing out of the vacancy-type defects and the second due to the dissociation of the Cu precipitates. On the other hand, the micro-hardness in the electron-irradiated samples increases with the recovery of the defects and spontaneous Cu precipitate formation, and then recovers with the dissociation of the Cu precipitates.

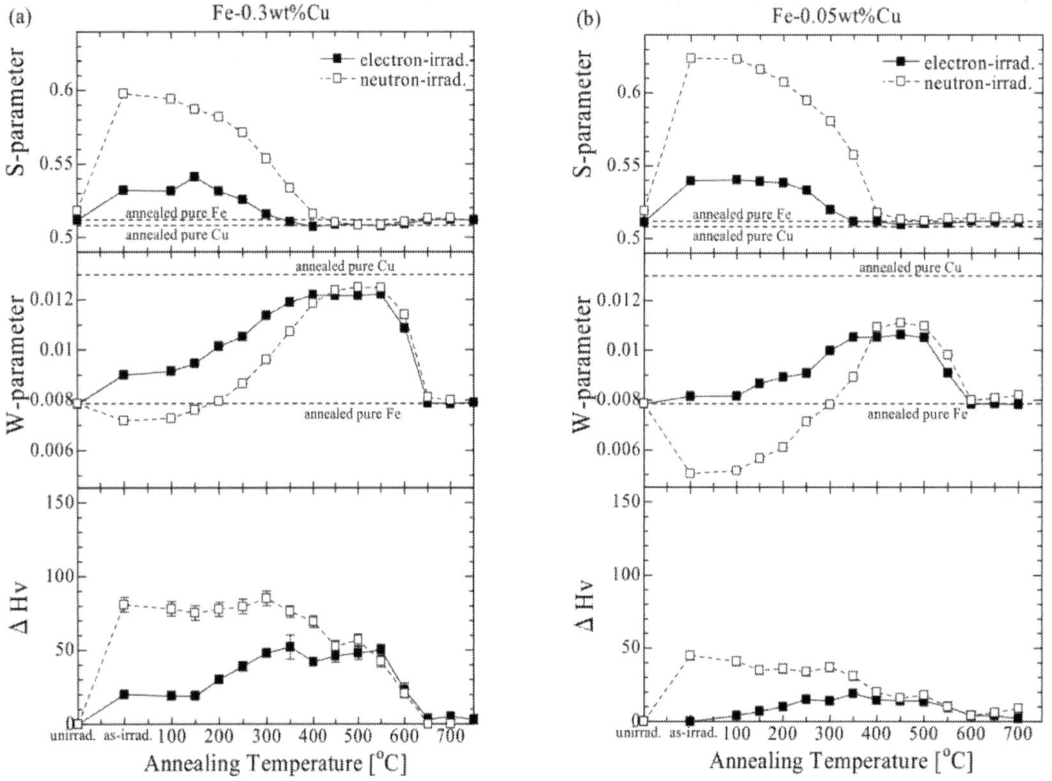

Figure 3. Isochronal annealing behavior of *S*- and *W*-parameter and Vickers micro-hardness of the irradiated Fe-Cu alloys: (a) Fe-0.3wt%Cu, (b) Fe-0.05wt%Cu.

It should be noted that the values of *W*-parameter and the micro-hardness for the electron irradiated and neutron irradiated samples are very similar to each other above the annealing temperature (400~450°C) where the vacancy-type defects are annealed out, in spite of the large difference in the irradiation dose. This indicates that small amount of extra vacancies enhance the aggregation of Cu impurity in Fe during the migration process of the vacancies. The difference in the *W*-parameter and micro-hardness between the electron- and neutron-irradiated samples below the annealing temperature dominantly comes from the Cu precipitates including nanovoids in addition to so-called matrix defects, considered to be dislocation loops formed by high dose neutron irradiation.

4. Conclusions

Electron- and neutron-irradiated Fe-Cu model alloys of nuclear reactor pressure vessel steels are studied by using positron annihilation spectroscopy. Not only by high dose neutron-irradiation but also by low dose electron-irradiation, vacancy-Cu aggregations are formed in dilute Fe-(0.05, 0.3wt.%)Cu model alloys and ultrafine Cu precipitates are formed after post-irradiation annealing at 400°C. In spite of large difference in the irradiation doses between the electron- and the neutron-irradiated samples, no significant difference is observed in the isochronal annealing behaviour above 400°C of both positron annihilation and micro-hardness results. These indicate that small amount of extra vacancies enhance the aggregation of Cu impurity in Fe during the annealing-out process of the vacancies.

Acknowledgments
This work is partly supported by Radioactive Waste Management Funding and Research Center, Grant-in-Aid for Scientific Research of the Ministry of Education, Science and Culture (No. 17002009 and 15106015). The electron irradiation was carried out using the JAEA (Takasaki) facility supported by the Inter-University Program for the Joint Use of JAEA Facilities. We thank Mr. T. Takahashi for preparing the samples by arc-melting.

References
[1] Phythian W J and English C A 1993 *J. Nucl. Mater.* **205** 162
[2] Othen P J, Jenkins M L and Smith G D W 1994 *Phil. Mag. A* **70** 1
[3] Buswell J T, Phythian W J, McElory R J, Dumbill S, Ray P H N, Mace J and Sinclair R N 1995 *J. Nucl. Mater.* **225** 196
[4] Stoller R E 1996 *Effects of Radiation on Materials: 17th International Symposium*, ASTM STP 1270, ed D S Gelles, R K Nanstad, A S Kumar and E A Little, American Society for Testing and Materials 25
[5] Odette G R and Lucas G E 1998 *Rad. Eff. Def. Sol.* **144** 189
[6] Carter R G, Soneda N, Dohi K, Hyde J M, English C A and Server W L 2001 *J. Nucl. Mater.* **298** 211
[7] Nagai Y, Tang Z, Hasegawa M, Kanai T and Saneyasu M 2001 *Phys. Rev. B* **63** 134110
[8] 1979 *Positron in Solids* ed P Hautojärvi (Berlin: Springer-Verlag)
[9] 1983 *Positron Solid-State Physics* ed W Brandt and A Dupasquier (Amsterdam: North-Holland)
[10] Puska M J and Nieminen R M 1994 *Rev. Mod. Phys.* **66** 841
[11] 1995 *Positron Spectroscopy of Solids* ed A Dupasquier and A P Mills Jr (Amsterdam: IOS Press)
[12] Nagai Y, Hasegawa M, Tang Z, Hempel A, Yubuta K, Shimamura T, Kawazoe Y, Kawai A, Kano F 2000 *Phys. Rev. B* **61** 6574
[13] Lynn K G, MacDonald J R, Boie R A, Feldman L C, Gabbe J D, Robbins M F, Bonderup E and Golovchenko J 1977 *Phys. Rev. Lett.* **38** 241
[14] Asoka-Kumar P, Alatalo M, Ghosh V J, Kruseman A C, Nielsen B and Lynn K G 1996 *Phys. Rev. Lett.* **77** 2097
[15] for example, 1976 *Radiation Damage in Metals* ed N L Peterson and S D Harkness (American Society for Metals)
[16] Puska M J and Nieminen R M 1983 *J. Phys. F.: Met. Phys.* **13** 333
[17] Ohkubo H, Tang Z, Nagai Y, Hasegawa M, Tawara T and Kiritani M 2003 *Met. Sci. Eng A* **350** 95
[18] Salje G and Feller-Kniepmeier 1977 *J. Appl. Phys.* **48** 1833
[19] Toyama T, Nagai Y, Tang Z, Hasegawa M, Almazouzi A, van Walle E and Gerard R 2007 *Acta Materialia* **55** 6852

Journal of Physics: Conference Series **265** (2011) 012008 doi:10.1088/1742-6596/265/1/012008

Positron annihilation techniques applied to reactor steels

Vladimír Slugeň, Róbert Hinca and Matúš Stacho

Slovak University of Technology, FEI, Ilkovičova 3, SK-81219 Bratislava, Slovakia

E-mail: Vladimir.Slugen@stuba.sk

Abstract. Different positron annihilation spectroscopy (PAS) techniques were applied in investigations of nuclear reactor steels. Summary of previous applications and evaluation of PAS potential contributions in the area of microstructural study of materials foreseen for fusion and fission technologies are discussed in detail.

1. General view to PAS applications in steels

If we consider about positron annihilation from the general point of view, we could say that the positron annihilation spectroscopy (PAS) is nowadays well recognized as:
- a powerful tool of microstructure investigations of condensed matter,
- a spectroscopic technique for the study of vacancy type defects,
- a technique in material study for registration also very low concentration of defects,
- one of suitable techniques for defects study in the near surface region [1-4].

The utility of positron annihilation studies of reactor pressure vessel steels (RPV) relies on the fact that the characteristics of the annihilation process depend almost entirely on the initial state of the positron-many-electron system. Since energetic positrons are rapidly thermalized after entering condensed matter, if they are not bound to electrons, the characteristics of the annihilation process in most cases depend on the initial state of the many-electrons system where positrons annihilate. **The energies, moment, and time of the gamma rays emitted during the annihilation may be measured** with high precision using the modern detector systems. Therefore, studying the positron annihilation process characteristics can identify the state of electrons in metallic substances.

The annihilation lifetime can be determined by lifetime measurement. Momentum distribution of the annihilation gamma rays can be measured using the method of angular correlation of the two collinear annihilation gamma rays, and energy distribution using the method of Doppler broadening.

Hence, while the lifetime method yields information regarding the electron density in the region of electron-positron overlap through the determination of positron decay rates from various states in the metal. The method of angular correlation and Doppler broadening (momentum technique) gives information regarding the electron moment in this overlap region through the shape of the angular (Θ) and energy (E) distributions of the annihilating γ-rays. These shapes are usually characterised by one or more "line shape parameters". The most commonly used ones are the normalised peak-counts determined in an interval around 0 (in angular correlation) or $E = 0$ (in Doppler broadening). These measure in various, but qualitatively equal, ways the change of annihilations with electrons of low momentum relative to those with high momentum. These changes in the annihilation characteristics yield information whether the positron annihilates in a defect-free region of the crystal or from a defect site.

2. PAS applications on RPV steels

Since mid 1980-s PAS has been extensively used in the study of RPV steels [5-19]. The positron annihilation technique can give essential information about the deterioration of the mechanical properties of RPV steels during their irradiation, which is known as neutron embrittlement. The lifetime of positrons trapped at radiation-induced vacancies, vacancy-impurity pairs, dislocations, microvoids, etc. is longer than that of free positrons in perfect region of the same material. As a result of the presence of open-volume defects, the average positron lifetime observed in structural materials is found to increase with radiation damage.

The embrittlement of steel is a very complicated process depending on many factors (thermal and radiation treatment, chemical compositions, preparing conditions, ageing, etc.). Properties of the RPV steels and influence of thermal and neutron treatment on these properties are routinely investigated by macroscopic methods such as Charpy V-notch and tensile tests. A number of semi-empirical laws, based on macroscopic data, have been established, but, unfortunately, these laws are not completely consistent with all data and do not provide the desired accuracy. Therefore, many additional test methods [20-22] have been developed to unravel the complex microscopic mechanisms responsible for RPV steel embrittlement.

RPV embrittlement poses one of limiting factors in the lifetime of vessels of today's nuclear power plant (NPP) This problem is very serious in Eastern (Russian) types of nuclear reactors (VVER). It is due to the narrower gap between the outside surface of the core barrel and the inside surface of the RPV than in Western RPV's. The neutron flux and consequently neutron fluency on the RPV wall is generally higher for VVER-440 type reactors than for other equivalent types. This influence of neutron flux (even neutrons of energy over 0.5 MeV) on RPV embrittlement is much more pronounced than other contributions, e.g., from a coolant temperature or from the operational pressure in the primary circuit.

According to previous reports [23-25] it seems to be generally accepted that even in the Western types of RPV steels containing more than 0.1 wt. % of Cu, the Cu- and P-rich precipitates play a dominant role in thermal and neutron embrittlement. In the case of Eastern-type RPV steels, comprehensive PAS studies [26-28] have suggested that carbide formation is an observable additional microstructural mechanism.

Positron annihilation can be used to recognize and study the following features and properties of RPV steels.

2.1. Differences in reactor pressure steel according to chemical composition

Comparison of 5 western and 3 Russian RPV steels using PAS LT techniques was published in [18] (see figure 1) and table 1. The increased amount of alloying elements causes larger grain boundaries and nonhomogeneity which result to increased positron lifetime.

Table 1. Chemical composition of studied RPV-steels.

Code	Type of steel	Contens of alloying elements in RPV specimens (wt.%)											
		C	Si	Mn	Mo	Ni	Cr	Cu	P	S	V	Co	Total
EGF	22NiMoCr37	0.22	0.23	0.88	0.51	0.84	0.39	0.080	0.006	0.004			3.160
D4	A508B Cl.3	0.20	0.28	1.43	0.53	0.75	-	0.055	0.008	0.008			3.261
CH	18MND5	0.18	0.26	1.55	0.50	0.65	0.18	0.140	0.007	0.002			3.469
JRQ	A533B Cl.	0.18	0.24	1.42	0.51	0.84	0.14	0.140	0.017	0.004			3.491
73W (weld)	Linde 0124	0.10	0.45	1.56	0.58	0.60	0.25	0.310	0.005	0.005			3.860
YA	15Ch2MFA	0.14	0.31	0.37	0.58	0.2	2.64	0.090	0.014	0.017	0.27	0.019	4.650
YTA (HAZ)	15Ch2MFA	0.14	0.24	0.40	0.72	0.14	2.93	0.110	0.013	0.017	0.31	0.011	5.031
XTA (HAZ)	15Ch2NMFA	0.18	0.24	0.52	0.62	1.26	2.22	0.080	0.010	0.013	0.08	0.008	5.231

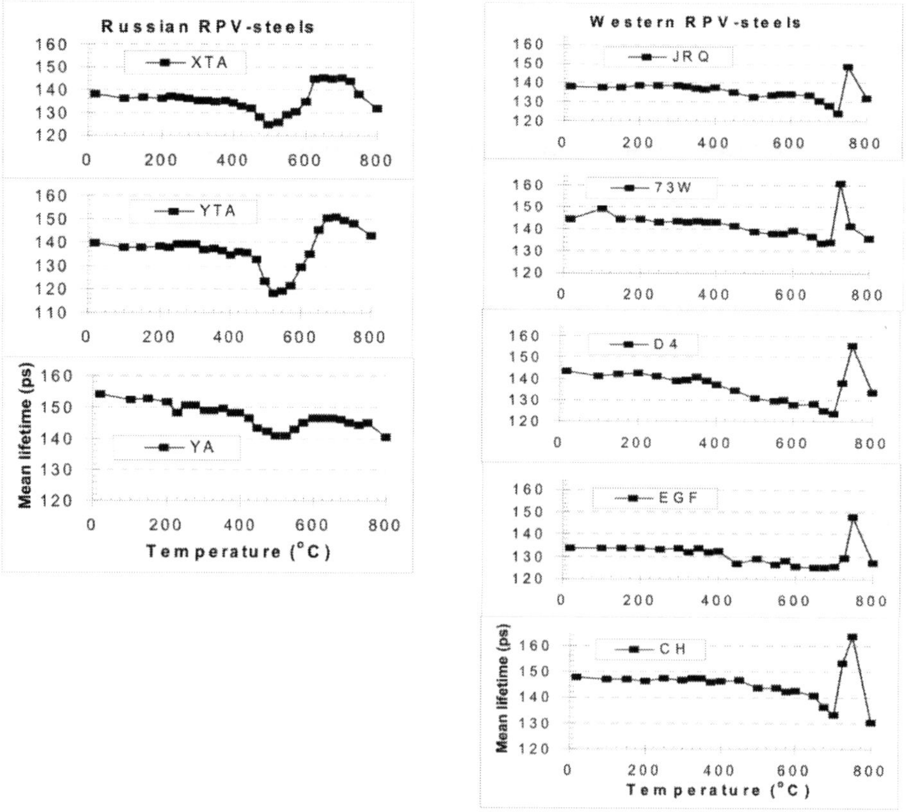

Figure 1. Comparison of positron mean lifetimes according to different chemical composition of RPV steels.

2.2. Differences in the same type of steels according to different thermal treatment

PAS techniques are very sensitive on annealing. Many works were focused on registration of optimal annealing procedures. One of the most appreciate picture in the "positron community" from industrial point of view was PAS confirmation of optimal annealing temperature of 15Kh2MFA RPV steel (see figure 2). Next picture shows the temperature region of annealing and informs that annealing at 500 °C causes creation of next large amount of small defects connected to hug probably carbide precipitation.

2.3. Irradiation effect

The positron annihilation can register an increase of vacancy type defects due to irradiation [5,12,15], however, the loads during RPV operating lifetime 10^{20} n/m^2 are too small to register large changes in positron annihilation parameters. Irradiation temperature at level of about 300 °C causes partial annealing of small (vacancy) defects, so that positron lifetimes stay practically unchanged or decrease after initial period (probably after Cu, P and/or carbides precipitation and connected re-structuralisation). For more information see [29] and figure 3.

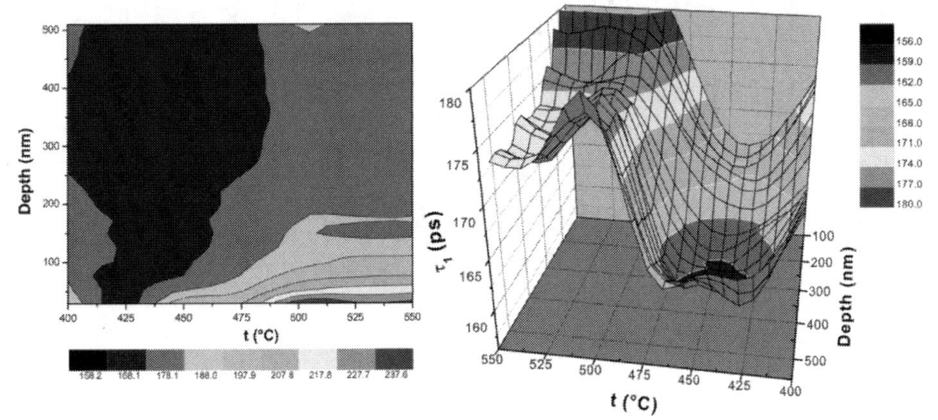

Figure 2. The 3D presentation of positron lifetime results of irradiated (neutron fluence 1.25×10^{24} m^{-2}) and annealed Sv-10KhMFT steel (WWER-440 weld metal).

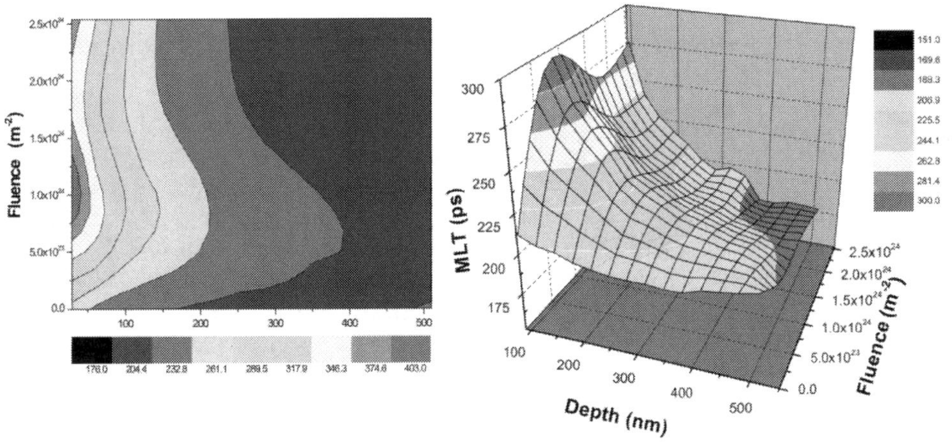

Figure 3. The 3D presentation of PLEPS results (MLT) of irradiated Sv-10KhMFT steel (weld metal). The first about 140 nm were neglected due to possible surface defects.

2.4. Depth profiling

Interesting depth profile was observed in the near surface region in base as well as weld metal in case of Russian WWER-440 RPV steels. The increased defects concentration after initial period of irradiation is much more observable in the near surface region. It decreases with the depth into the bulk. A plausible explanation is connected to the Cu precipitation and segregation on the grain boundaries.

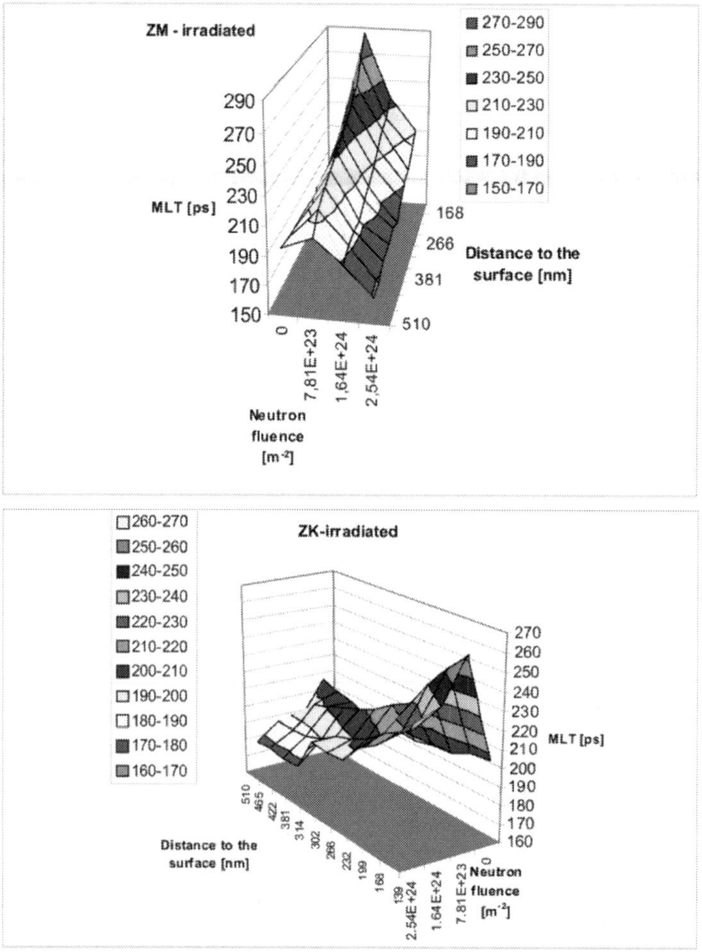

Figure 4. The 3D presentation of PLEPS results in form of the MLT dependence on neutron fluence in WWER-440 weld (ZK) and base metal (ZM). The first about 140 nm were neglected due to possible surface defects.

2.5. PAS parameters in comparison to hardness

Similar process (Cu, P and C precipitation) as in D can be confirmed also using measurements of WWER-440 RPV steels published in [18]. Figure 5 shows a comparison of hardness and PAS parameters versus Hollomon-Jaffe´s parameter which calculate temperature and time of annealing in one.

2.6. Recognition of nano-size defects

Positron annihilation can recognize also nanosize-defects. In figure 6 the positron lifetime is assigned not only to number of vacancies in defect but also to its diameter. The graph shows two curves obtained using two different enhancement factors: the one from Boronski-Nieminen [30] and the one from Stachowiak-Lach [31]. In the case of ref. [32] we can see a somewhat too slow saturation of the lifetime with the increasing size of the cluster. The reason is the incorrect behaviour of the corresponding enhancement for low electron densities (although the limit for zero electron density is correct). This is avoided in the approach of ref. [31], but the resulting lifetimes are a bit too short for small vacancy clusters. The positron surface state was not considered in these calculations [33]. These calculations were used also for positron study of ITER first wall materials [34].

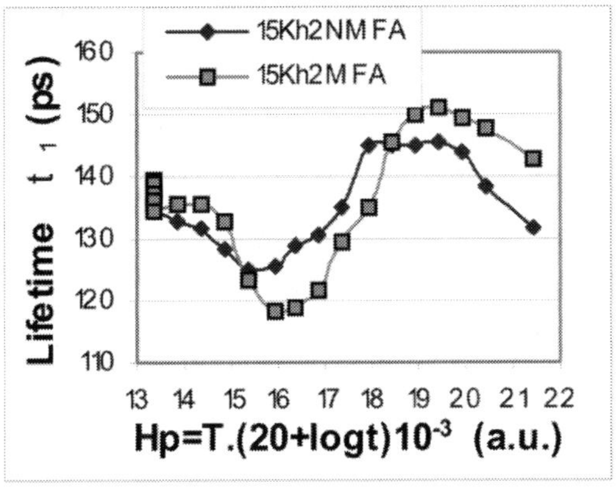

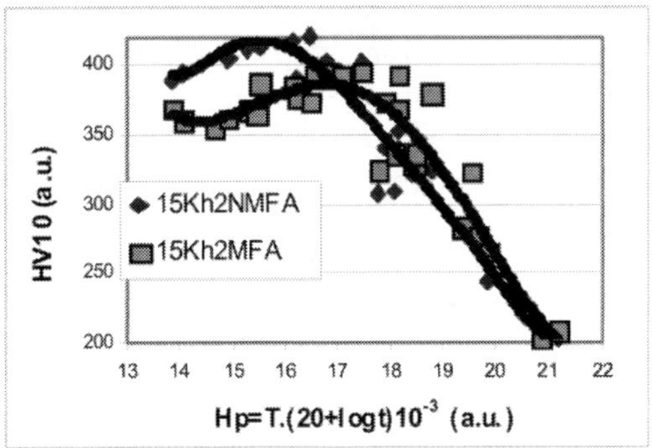

Figure 5. Comparison of positron lifetime results and hardness versus Hollomon-Jaffe´s parameter.

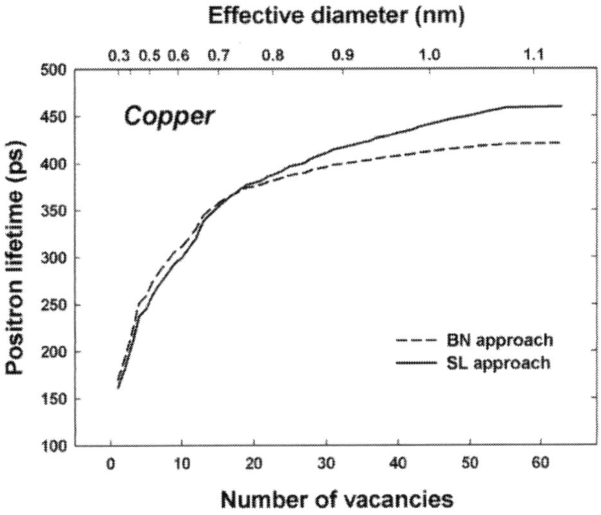

Figure 6. The calculated dependence of the positron lifetime on the size of the vacancy cluster in Cu. The full line was calculated according to Stachowiak-Lach approach. The dashed line was calculated according to Boronski-Nieminen approach.

2.7. Positron affinity calculations

From the point of view of PAS, coherent and incoherent particles exhibit different behaviour. In the latter case, there are free volumes at the matrix-particle interface and positrons may get trapped there. But this does not apply to a coherent particle, and positrons can be localised inside such a particle. Coherent particles may represent either a potential barrier or a potential well for positrons. An important quantity to evaluate the behaviour of the potential in the vicinity precipitates is the positron affinity. The change of the positron affinity at the matrix-particle interface determines the step of the positron potential.

In order to have some idea whether the positron trapping at precipitates may occur, the positron affinity calculations for Cu and possible precipitates which could exist in CuCrZr and CuAl25 systems were performed [34]. Table 2 presents the obtained results.

The necessary condition for positron trapping at an embedded particle is that the positron affinity of such a particle lies below the positron affinity of the matrix. One can see that this is not fulfilled for any considered particle except for Al_2O_3. In the case of Zr and Cu_5Zr there is no difference in affinities calculated using the GC computational scheme (within the precision of calculations). Thus, the positron trapping cannot be excluded, but it is not very likely. Another aspect is that the particle must have a certain radius (size) in order to hold at least one positron level [34]. When the difference in affinities is small, the necessary radius may become very large. Table 2 also contains calculated positron lifetimes together with their available experimental counterparts.

Table 2. Summary of affinity (A_+) and lifetime (τ) calculations for Cu and possible precipitates. A_+ and τ are in units of eV and ps, respectively. Available experimental results are also available.

System	Cu	Cr	Zr	Cu_5Zr	α-Cr_2Zr	α-Al_2O_3
BN A_+	-4.8	-2.8	-4.5	-4.6	-4.2	-7.3
GC A_+	-4.1	-2.3	-4.1	-4.1	-3.7	-6.8
Exp.	-4.2	-2.4				
BN τ	104	98	154	114	120	138
GC τ	118	104	154	126	124	152
Exp.	110	120				

2.8. Comparison of calculated and measured positron lifetimes

It is not easy to assign the positrons lifetime only to one type of defects. The calculated positron lifetimes for different types of defects in pure iron and different carbides in low alloy Cr-Mo-V steel are shown in table 3.

Table 3. Calculated or measured positron lifetimes in Fe or in carbides.

Material	Positron lifetime (ps)	Reference
Fe-bulk	110	[35]
Fe-dislocations	165	[36]
Fe-monovacancy	175	[35]
Fe-divacancy	197	[36]
Fe-3 vacancy cluster	232	[36]
Fe-4 vacancy cluster	262	[36]
Fe-6 vacancy cluster	304	[36]
VC	99	[37]
$V_{0.86}Cr_{0.09}Mo_{0.04}Fe_{0.01}C$	105	[37]
Mo_2C	112	[37]
$Mo_{1.4}Cr_{0.6}C$	116	[37]
Cr_7C_3	107	[37]
$Cr_{23}C_6$	112	[37]
$Mn_{26}C_6$	99	[37]
Fe_3C	101	[37]

However, due to FWHM on the level of 200 ps it is difficult to distinguish between all types of defects. The common approach in this area is grouping defects (mostly into 2 groups). One group comprise small defects like dislocations and mono-, di- vacancies together with bulk value and the second lifetime is mostly assigned to vacancies clusters. There were several trials to distinguish between edge and screw dislocations, which are characterised by lifetimes of 145 (screw) and 165 ps (edge) [38]. According to this work positrons are trapped mostly at dislocation lines (65-85 % of positrons in Russian RPV steel 15Kh2MFA had lifetime on the level 150 ps).

2.9. Calculation of defects concentration from PAS results

Of particular interest is the concentration of defects as a function of irradiation dose and thermal treatment. Generally, this effect can be obtained from positron studies only if the bulk lifetime can be resolved from the shortest lifetime, attributed to annihilation from defects. In the present case of saturation trapping this was impossible. Therefore, from the individual lifetime spectra we can only conclude a total trapping rate κ larger than about $10^{10}s^{-1}$.

However, from the variation of the mean lifetime τ_{av} as a function of positron implantation energy, we can estimate κ even in the case of saturation trapping. The problem was fully analyzed in [39,40]. Based on this theory the total defect concentration can be estimated as:

$$C_d = \frac{\kappa}{\kappa_{spec}}$$

$$\kappa = \mu C_d = \frac{I_2}{I_1}\left(\frac{1}{\tau_b} - \frac{1}{\tau_d}\right)$$

I_1, I_2 are intensities of positron lifetimes,
τ_b – positron lifetime in bulk
τ_d – positron lifetime in defect

In figure 7 and 8 results for the evaluation of κ (trapping rate) as obtained by the procedure described, are presented. For the specific trapping rate κ_{spec} the plausible value 10^{15} (s^{-1}) [41] has been assumed. Because of slight surface oxidation, this evaluation of κ results in a systematic underestimate. A lower limit of $10ns^{-1}$ for κ may be derived within the framework of the STM since we have observed saturation trapping at defects.

According to the results from our measurements performed on different irradiated RPV-steels, the total trapping rate κ in ns^{-1} as well as the total defect concentration c_d (the same values but in ppm) increases slightly for both base and weld materials as a function of the irradiation dose (see figure 8). The weld material (Sv10KhMFT) seems to be less sensitive to the changes caused by neutron-irradiation or by post-irradiation heat treatment than the base material (15Kh2MFA) (see figure 7). Nevertheless, the differences in the positron trapping rate κ are not too large. It seems reasonable to relate the observed trapping rates with the ones which have been derived for trapping into precipitated carbides from electron microscopic images [15]. Accordingly, in the 15Kh2MFA the trapping rate into chromium carbides (Cr_7C_3, $Cr_{23}C_6$) is predicted as $\kappa_{Cr}=1.8\times10^8$ s^{-1} and into vanadium carbides as $\kappa_{VC}=2.2\times10^{10}$ s^{-1} [15]. Thus precipitated vanadium carbide could indeed account for the observed trapping rates. But on the other hand, as shown by calculations [35], positrons experience a repulsive potential from carbides embedded in an iron matrix. Thus only the defects at the iron-carbide interface could provide an acceptable trapping site for positrons.

The total trapping rate κ and the defect concentration c_d are stable or increase slightly for 15Kh2MFA steel as a function of Hollomon-Jaffe´s parameter which calculates the influence of annealing temperature T [K] and time t [s] together [5,42]. Between 15.5 to 16.5 (corresponding to the temperature region 400 to 450°C) the defect concentration increases. On the other hand (figure 8), in the same range, there is a marked decrease in the lifetime of defects. A simple explanation could be the dissolution of precipitates and defect clusters which would reduce the average size of the defects, and, by the same process, would increase the concentration of the defects.

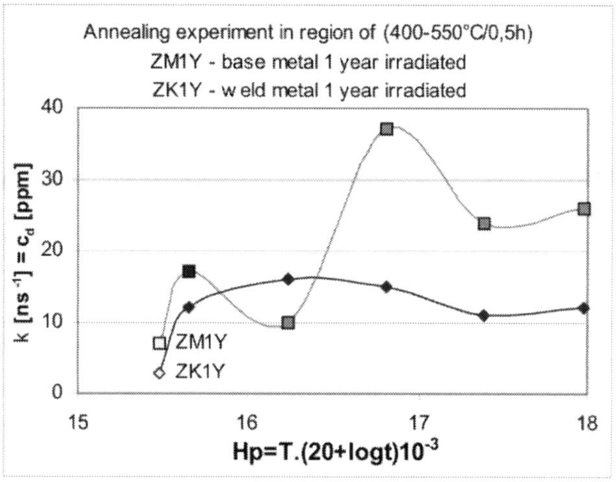

Figure 7. The total trapping rate κ versus Hollomon-Jaffe´s parameter at isochronal annealed (step 25°C) specimens of base (ZM) and weld (ZK) alloys after 1 year of irradiation in the reactor (fluency of about 7.8×10^{23} m^{-2}). The lower limits for κ, as derived from saturation trapping according to the STM, is 10 ns^{-1}.

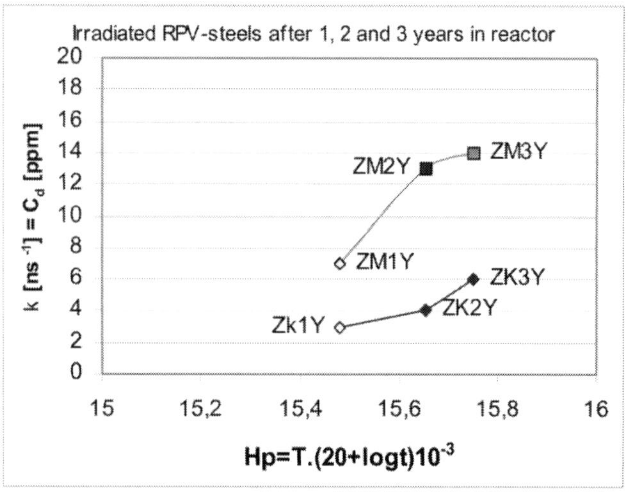

Figure 8. The total trapping rate κ versus Hollomon-Jaffe´s parameter at RPV specimens from base (ZM) and weld (ZK) alloys after 1, 2 and 3 years residence in reactor irradiation chambers (neutron fluency in the range from 7.8×10^{23} m^{-2} to 2.5×10^{24} m^{-2}). The lower limits for κ, as derived from saturation trapping according to the STM, is 10 ns^{-1}.

2.10. Applications of positron microscope

In the positron annihilation spectroscopy, some new lifetime techniques like positron microscope [42] seems to be very promising in defect study. Figure 9 shows a comparison of defect imagine by cathodoluminiscence (CL) and SEM versus pictures obtained from positron microscope in Munich (University of Bundeswehr). The most important problem is actually the relative low positron beam intensity. It can be significantly improved after moving this unique microscope to FRM II reactor in Garching.

2.11. Possible studies of near-surface region

Positron microscope enables very effective material study in the near surface region. For the enhancement of this a special approach was developed in PAS group in Munich based on the cutting of specimen in very small angle. Detailed explanation can be found in [43,44].

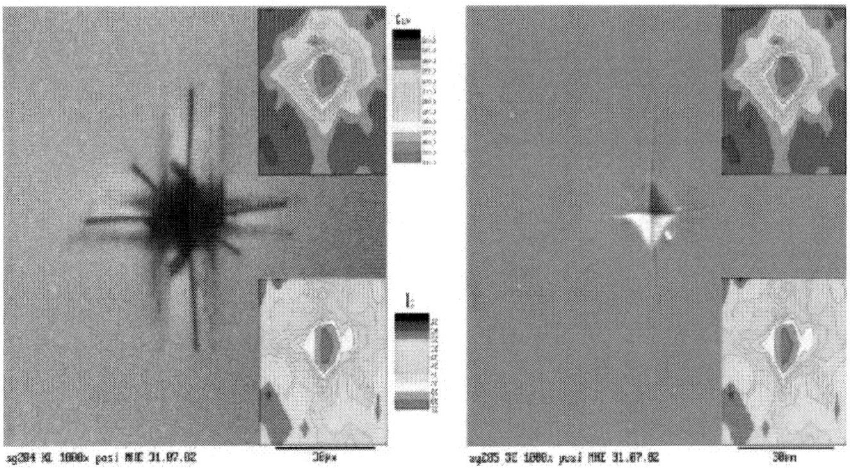

Figure 9. Comparison of defect imagine by cathodoluminiscence (left) and SEM (right) versus pictures obtained from positron microscope at Uniwersität der Bundeswehr, Neubiberg-Munich.

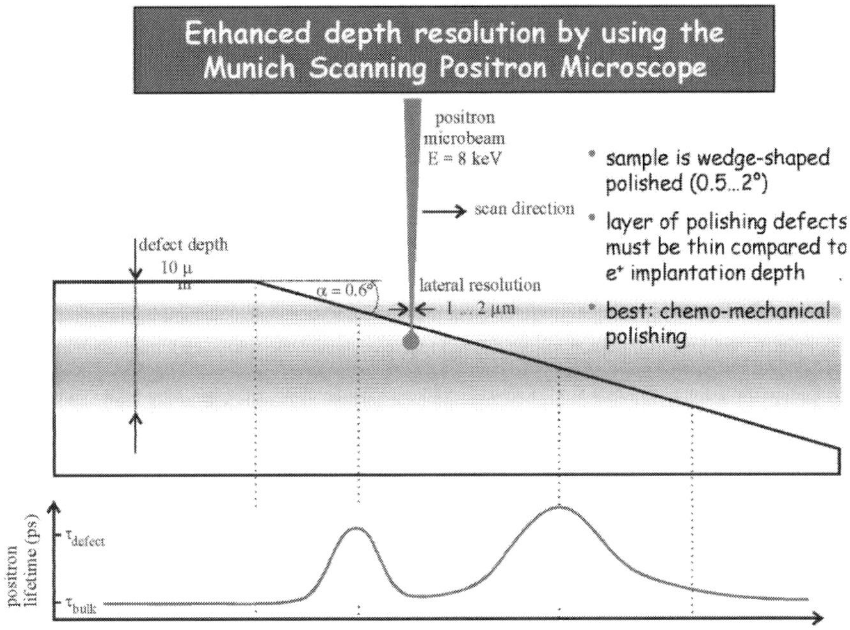

Figure 10. Abilities of positron annihilation lifetime technique based on positron microscopy for vacancy type defects study in material depth profile [44].

2.12. Evaluation of high implanted (irradiated) regions in nuclear materials

Defects in the form of vacancies (loops, voids, etc.) in selected Cu-alloys (99% of Cu) foreseen for applications in thermonuclear fusion technologies were studied using the improved Pulsed Low Energy Positron System (PLEPS), developed at the Institut für Nukleare Festkörperphysik, Universität der Bundeswehr, München [45]. This system enables the study the micro-structural changes in the region from 20 to about 500 nm (depth profiling).

For the simulation of the radiation damage due to neutrons, the ion implantation of protons has been applied. The protons were chosen having approximately the same mass as a neutron. We supposed that the ballistic influence of protons at the primary-knock-atoms (PKA) production could simulate the ballistic influence of neutrons at Cu-alloys in the fusion reactor ITER. The selected specimens were irradiated in the Ion Beam Laboratory of FEI STU Bratislava [46]. The energy of the implantation was EH=2×95 keV for the molecular H_2^+ ion beam (implantation dose up to 1.3×10^{19} ions/cm^2 (1.1 C/cm^2)). For the protons with the energy of 95 keV, the range of 480 nm was found by TRIM calculations. It optimal corresponds to the scope of the PLEPS equipment. Results enable not only to compare the implanted damage profile to positron lifetimes, but also to study the annealing behaviour of studied materials. By higher implantation doses the blistering effect can be studied, too [47].

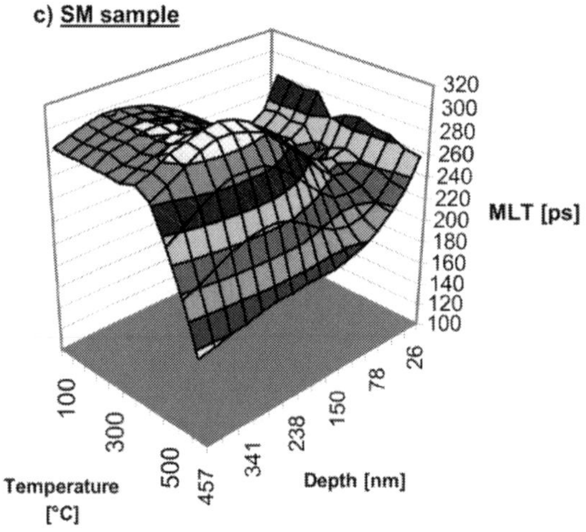

Figure 11. Annealing behavior of H_2^+ ions implanted 99CuZrCr alloys [48].

2.13. Evaluation of role of Cr in Fe-Cr alloys

Depth profiling of defects down to one micrometer below the surface has been performed on an untreated Fe11.62%Cr sample and three Fe11.62%Cr samples with different damage levels using positron implantation energies between 1 keV and 18 keV. The depth profiles were measured with the Pulsed Low Energy Positron System (PLEPS) [49] at the high intensity positron source NEPOMUC [50] at the research reactor FRMII at TUM.

To obtain cascade collisions in the microstructure of the studied materials without neutron activation, accelerated helium ions have been used. The depth profile of collision events can be seen in figure 12. The measured positron lifetimes spectra were analyzed using the LT 9.0 program [51] and a modified version of PosFIT [52]. The differences of these two analyses were negligible. All the lifetime spectra could be decomposed into three lifetime components with variances close to one.

Figure 13a shows the positron mean lifetime as a function of helium implantation dose and mean positron implantation depth. The positron mean lifetime (MLT) is increasing with the implantation dose, thus indicating the creation of defects due to implantation. The increase of the MLT close to the surface (< 200nm below the surface) is probably due to positrons annihilating in surface oxide layer. At higher depths the course of the MLT depth profile corresponds to the expected zone of maximum damage. In the zone of maximum damage in the implanted specimens two different defect lifetimes have been observed. The shorter lifetime between 240 ps and 300 ps could be assigned to small vacancy clusters < 6 vacancies or larger clusters filled with helium. The longer lifetime between 400 ps and 500 ps corresponds to annihilation in large voids (> 1 nm). The intensity of this longer component (I_3) increases dramatically with the helium implantation dose as can be seen from figure 13b. The course of the I_3 depth profile again corresponds to the expected zone of maximum damage.

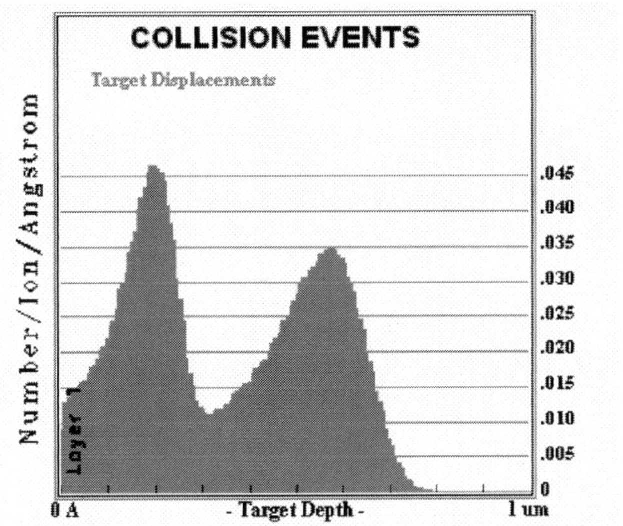

Figure 12. Depth profile of the helium implantation, E=250keV+100keV (SRIM simulation of 10^6 ions).

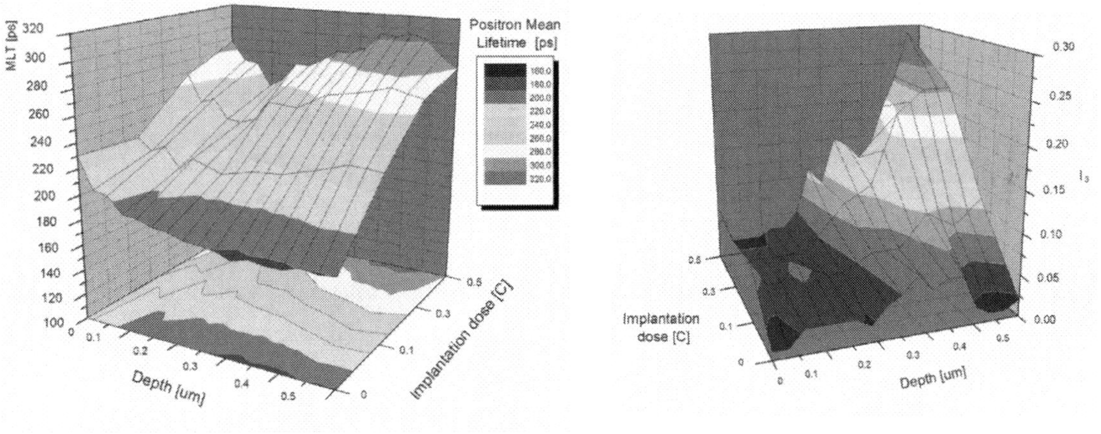

Figure 13. a) Positron mean lifetime in different treated Fe11.62%Cr alloy and b) intensity of annihilation in large defects (voids) measured in Fe11.62%Cr alloy.

3. Conclusions

In the last years several original irradiated RPV steel samples were investigated by PAS technique with the aim to observe some micro structural changes due to thermal and neutron treatment resulting from operating conditions in NPP. This investigation is aimed mainly at increasing the safety margin against the RPV brittle fracture. The only possibility to reduce this irradiation embrittlement is thermal annealing. Non-destructive test methods like PAS in combination with proved destructive test methods can contribute to a progress in optimisation of temperature-time regime for post-irradiation thermal treatment. This approach is especially useful in the case of highly radioactive materials and also in those cases where only small amounts of irradiated materials are available. Of course, in the long –

term point of view, PAS techniques can be effectively used in the testing and selection process of optimal reactor steels foreseen for Generation IV reactors and thermonuclear fusion facilities.

Acknowledgement
This study was supported by VEGA 1/3188/06 and *7RP*-Euratom/CU.

References

[1] Hautojärvi P 1979 *Positrons in Solids* (Berlin: Springer-Verlag)
[2] Krause-Rehberg R 1997 *Positrons in semiconductors* (Berlin: Springer-Verlag)
[3] Dauwe C, Dorikens M, Dorikens-Vanpraet L and Segers D 1974 *Appl. Phys.* **5** 117
[4] Sperr P and Kögel G 1997 *Mater. Sci. Forum* **255-257** 109
[5] Brauer G, Liszkay F, Molnar B and Krause R 1991 *Nuclear Engineering and Design* **127** 47
[6] Pareja R, De Diego N, De La Cruz R M and Del Rio J 1993 *Nucl. Technol.* **104** 52
[7] Lopes Gil C, De Lima A P, Ayres De Campos N J, Fernandez J V, Kögel G, Sperr P, Triftshäuser W and Pachur D 1989 *J. Nucl. Mater* **161** 1
[8] Valo M, Krause R, Saarinen K, Hautojärvi P and Hawthorne R 1992 *Effects of Radiation on Materials: 15th International Symposium, ASTM STP 1125* ed R E Stoller, A S Kumar and S G Gelles (Philadelphia: ASTM) p 172
[9] Slugeň V, Haščík J, Gröne R, Bartík P, Zeman A, Kögel G, Sperr P and Triftshäuser W 2001 *Mat. Sci. Forum* **363-365** 47
[10] Nagai Y, Tang Z and Hasegawa M 2000 *Radiation Physics and Chemistry* **58** 737
[11] Brauer G, Puska M, Sob M and Korhonen M 1995 *Nucl. Eng. Desg.* **158** 149
[12] Slugen V 2005 *Nucl. Eng. Desg.* **235** 1961
[13] Hartley J H, Howell R H, Asoka-Kumar P, Sterne P A, Akers D and Denison A 1999 *Appl. Surf. Sci.* **149** 204
[14] Becvar F, Cizek J, Lestak L, Novotny I, Prochazka I and Sebesta F 2000 *Nucl. Instr. Meth. A* **443** 557
[15] Cizek J, Prochazka I, Kocik J and Keilova E 2000 *Phys. Stat. Sol. (a)* **178** 651
[16] Van Hoorebeke L, Fabry A, van Walle E, Van de Velde J, Segers D and Dorikens-Vanpraet L 1996 *Nucl. Instr. Meth. A* **371** 566
[17] Ghazi-Wakili K, Zimmermann U, Brunner J, Tipping P, Waeber W B and Heinrich F 1987 *Phys. Stat. Sol. (a)* **102** 153
[18] Slugen V, Segers D, De Bakker P M A, DeGrave E, Magula V, Van Hoecke T and Van Waeyenberge B 1999 *J. Nucl. Mater.* **274** 273
[19] Goland A N 1976 *Proceedings of the International School of Physics "Enrico Fermi"* ed G Carlioti (Amsterdam: North-Holland) p 110
[20] Phythian W and English C A 1993 *J. Nucl. Mater.* **205** 162
[21] Amaev A D, Dragunov Yu G, Kryukov A M, Lebedev L M and Sokolov M A 1986 *IAEA Specialists Meeting Proceedings*, Plzen
[22] De Bakker P, Slugen V, De Grave E, Van Walle E and Fabry A 1997 *Hyperfine Interaction* **110** 11
[23] Koutsky J and Kocik J 1994 *Radiation Damage of Structural Materials* (Elsevier: Amsterdam)
[24] Törrönen K 1979 *Technical Research Centre of Finland, Materials and Processing Technology, publication* **22**, Espoo
[25] Ghoneim M M and Hammad F H 1997 *Int. J. Pres. Ves. Piping.* **74** 189
[26] Davies M 1999 *Int. J. Pres. Ves. Piping.* **76** 163
[27] Becvar F, Jirasková Y, Keilová E, Kocik J, Lestak L, Prochazka I, Sedlak B and Sob M 1992 *Mat. Sci. Forum* **105-110** 901
[28] Brauer G 1995 *Mat. Sci. Forum* **175-178** 303
[29] Slugen V, Zeman A, Lipka J and Debarberis L 2004 *NTD &E International* **37** 651

[30] Boronski E and Nieminen R M 1986 *Phys. Rev. B* **34** 3820
[31] Stachowiak H and Lach J 1993 *Phys. Rev. B* **48** 9828
[32] Dai G H, Moser P and Van Duysen J C 1992 *Mater. Sci. Forum* **105-110** 941
[33] Puska M J and Nieminen R M 1994 *Rev. Mod. Phys.* **66** 841
[34] Slugeň V, Kuriplach J, Ballo P, Domonkoš P, Kögel G, Sperr P, Egger W, Triftshäuser W, Domanková V M, Kováč P, Vávra I, Stanček S, Petriska M and Zeman A 2004 *Fusion Engineering and Design* **70** 141
[35] Puska M J, Lanki P, and Nieminen R M 1989 *J. Phys.: Condens. Matter* **1** 6081
[36] Hautojärvi P, Pöllönen L, Vehanen A and Yli-Kauppila J 1983 *J. Nucl. Mater.* **114** 250
[37] Vehanen A, Hautojärvi P, Johansson J, Yli-Kauppila J and Moser P 1982 *Phys. Rev. B* **25** 762
[38] Kocik J, Keilova E , Cizek J and Prochazka I 2003 *J. Nucl. Mater.* **303** 52
[39] Kögel G 1996 *Appl. Phys. A* **63** 227
[40] Slugen V, Kögel G, Sperr P and Triftshäuser W 2002 *Appl. Surf. Sci.* **194** (1-4) 150
[41] Brandt W 1967 *Positron Annihilation*, ed A T Stewart and L O Roelling (New York: Academic Press) p 155
[42] Magula V and Janovec J 1994 *Ironmaking and Steelmaking* **21** 64
[43] David A, Kögel G, Sperr P and Triftshäuser W 2001 *Phys. Rev. Lett.* **87** 067402
[44] Krause-Rehberg R 2003 *Material Research in Positron Annihilation* (oral presentation at Delft University)
[45] Bauer-Kugelmann W, Sperr P, Kögel G and Triftshäuser W 2001 *Mater. Sci. Forum* **363-365** 529
[46] Kovac P, Bortniansky A, Klopenkov M and Pavlovets M 2001 *Proc. Int. Conf. on Applied Charged Particle Accelerators in Medicine and Industry ACCELERATORS 2001* (Moscow: CNIIatominform) p 430
[47] Constantinescu B and Sarbu C 2000 *Fusion Engineering and Design* **49-50** 171
[48] Slugeň V, Kuriplach J, Ballo P and Domonkoš P 2004 *Nuclear Fusion* **44** 93-97
[49] Egger W, Sperr P, Kögel G and Dollinger G 2007 *Phys. Stat. Sol. (c)* **4** 3969
[50] Hugenschmidt C, Kögel G, Repper R, Schreckenbach K, Sperr P, Straßer B and Triftshäuser W 2004 *Nucl. Instr. Meth. B* **221** 160
[51] Kansy J 1996 *Nucl. Instr. Meth. A* **374** 235
[52] Kirkegaard P and Eldrup M 1972 *Comput. Phys. Commun.* **3** 240

Journal of Physics: Conference Series **265** (2011) 012009 doi:10.1088/1742-6596/265/1/012009

Positron annihilation in neutron irradiated iron-based materials

M Lambrecht[1,2] and A Almazouzi[1]

[1] SCK•CEN, Nuclear Materials Science Institute, Boeretang 200, B-2400 Mol, Belgium

[2] University of Ghent, Department of Materials Science and Engineering, Technologiepark 903, 9052 Gent, Belgium

E-mail: Marlies.Lambrecht@SCKCEN.BE

Abstract. The hardening and embrittlement of reactor pressure vessel steels is of great concern in the actual nuclear power plant life assessment. This embrittlement is caused by irradiation-induced damage, like vacancies, interstitials, solutes and their clusters. But the reason for the embrittlement of the material is not yet totally known. The real nature of the irradiation damage should thus be examined as well as its evolution in time. Positron annihilation spectroscopy has been shown to be a powerful method for analyzing some of these defects. In fact, both vacancy type clusters and precipitates can be visualized by positrons. Recently, at SCK•CEN, a new setup has been constructed, calibrated and optimized to measure the coincidence Doppler broadening and lifetime of neutron irradiated materials. To be able to compare the results obtained by the positron studies, with those of other techniques (such as transmission electron microscopy, atom probe tomography and small angle neutron scattering), quantitative estimations of the size and density of the annihilation sites are needed. Using the approach proposed by Vehanen *et al.*, an attempt is made to calculate the needed quantities in Fe and Fe-Cu binary alloys that were neutron irradiated to different doses. The results obtained are discussed highlighting the difficulties in defining the annihilation centres even in these simple model alloys, in spite of using both lifetime and Doppler broadening measurements in the same samples.

1. Introduction

For the safe operation of the nuclear power plants (NPPs), it is mandatory to ensure the mechanical integrity of the reactor pressure vessels (RPV). This irreplaceable component is made of steels that are prone to embrittlement due to neutron irradiation. In fact, these neutrons, when colliding with the material, produce microstructural defects, such as vacancies, interstitials and their clusters. The abundance of these defects is the origin of irradiation-induced precipitation and segregation. Thus, the study of defect formation under neutron irradiation and their thermal stability is of paramount importance for the understanding of radiation damage of structural materials. Moreover, to extend the lifetime of the NPPs, there is an urgent need for predictive models based simultaneously on experiments and numerical computer simulation.

Irradiation-induced embrittlement can be summarized by three main causes [1], i.e. precipitation, matrix-damage and grain boundary segregation. The nature of these defects has to be known for the better understanding of the interplay of hardening and embrittlement. However and up to now, it is still uncertain to assess the main mechanism that gives rise to the matrix-damage. Among other, Malerba [2] has shown that directly in the cascades some candidate cluster-solute complexes inducing small vacancy and interstitial clusters are produced. These defects aggregate to larger defects such as nano-voids and interstitial loops. At the same time, various solutes may diffuse to these clusters giving rise to complex defect-solute configurations.

PAS is found to give valuable information on these irradiation-induced defects [3]. In this technique, the positron is applied as probe. As anti-particle of the electron, the positron is trapped by defects with a different electron density than the bulk material, like vacancies, vacancy clusters, interfaces, second phase particles, dislocations, etc. [4]. The annihilation characteristics are different in the neighbourhood of defects compared to defect-free materials. Moreover, due to the difference in positron affinity of the different atoms, positrons annihilate with a different probability in the precipitates as compared to the bulk material [5]. The power of PAS is its possibility to find very small defects (> 0.1 nm) with very low concentrations (> 1 ppm), its "self-seeking" nature (the positron diffuses trough the matrix before it gets trapped [3]) and its non-destructiveness.

Significant insight into potential vacancy clusters has been made by PAS during the past decades [6]. Focus should now being put on the quantitative estimations of the average size and density of the annihilation sites, as this would lead to a better comparison with other techniques, such as transmission electron microscopy (TEM), atom probe tomography (APT) and small angle neutron scattering (SANS). On top of this, the quantitative determination of the defects is necessary for the production and validation of the predictive models.

Over the last 40 years a lot of investigations have been performed to determine the specific lifetime of positrons trapped at different types of defects in pure iron and Fe-based alloys. Experimentally, this has been done by recovery experiments of deformed iron [7], [8], as well as by the annealing of vacancies in electron irradiated iron [9], [10]. But, also models are produced to correlate a specific trapping site with the observed annihilation characteristics. While these calculations have been fairly successful for monovacancies in a wide variety of metals [11], they are not easily generalized to more complicated defect geometries.

In metallic systems, vacancies and small vacancy clusters show a well-defined relationship between the positron lifetime and the size of the open-volume region [12]. The local electron density determines the positron annihilation rate. Since the electron density is lower at an open-volume region compared to bulk material, the lifetime is increased for positrons trapped at vacancy clusters. The lifetime depends on the size of the open-volume region, so positron lifetime measurements can be used as a measure of the vacancy cluster size.

To interpret the experimental results, a theoretical model, which gives positron lifetime as function of micro-void size, is vital. Based upon a simple superimposed-atom model, Puska and Nieminen [13] have performed such calculations for positron lifetimes in micro-voids consisting of 1 (V_1) tot 15 (V_{15}) vacancies, using a fitting parameter to reproduce the experimental bulk lifetime. To examine larger micro-voids, however, ab-initio based superimposed-atom model calculations have been performed by Hempel *et al.* [14].

The above mentioned models are used to figure out how the size and density of the vacancy-type of defects that were produced in each of the neutron irradiated materials. Eldrup *et al.* [15] have already done such an attempt to extract the number density of microvoids present in pure Fe after neutron irradiation at 70 °C.

The present work consists on assessing the ageing mechanism of the reactor pressure vessel steels of old (generation II) nuclear reactors, which are in operation for more than 30 years. It is well known that these RPV-steels contain a considerable amount of copper impurities. Therefore, binary Fe-Cu alloys are often used to mimic their behaviour. Indeed, Fe-Cu model alloys show a similar increase in hardening during irradiation coming from the copper-rich precipitates [16]. Moreover, binary Fe-Cu alloys are chosen, to compare them with pure iron and iron containing a certain amount of carbon (as C plays an important role in steels). Finally, an attempt is made to calculate the needed quantities (i.e.

size and density of the irradiation-induced vacancy type of defects), using the approach proposed by Vehanen *et al.* [9], in pure iron and Fe-Cu binary alloys that were neutron irradiated to different doses.

2. Materials and irradiation conditions

The model alloys listed in table 1 were prepared using argon-arc melting and zone refinement methods, starting from electrolytic iron. The resulting ingots were cold worked after austenisation and tempering. A final heat treatment at 1075 K for 1 h was performed to release the stresses and to get well re-crystallized materials.

Table 1. The nominal composition of the investigated alloys in wt%.

Material	Nominal composition in wt%
Pure Fe	< 30 ppm C
Fe-C	> 30 ppm C
Fe-0.1% Cu	0.1 Cu (< 30 ppm C)
Fe-0.3% Cu	0.3 Cu (< 30 ppm C)

Samples from each alloy were irradiated in the test reactor BR2 at SCK•CEN. Table 2 lists the neutron fluxes and the reached fluences. During irradiation the temperature and the pressure were maintained constant respectively in between 290 and 295 °C and at 150 bar. The four of the irradiation conditions have the same flux of about 9.5×10^{13} n cm^{-2} s^{-1}, but different doses going from about 0.025 dpa to about 0.2 dpa. It is worth mentioning that at the highest flux of pressurised water reactors, 0.1 dpa corresponds to about 40 years of operation.

Table 2. The determined flux and fluence of the materials after neutron irradiation.

Dose (dpa)	Flux (10^{13} n cm^{-2} s^{-1})	Fluence (10^{19} n cm^{-2})
0.026	9.36 ± 0.52	1.5 ± 0.7
0.051	9.32 ± 0.55	3.0 ± 0.5
0.1	9.54 ± 0.54	6.9 ± 0.5
0.19	9.58 ± 0.52	12.5 ± 1.1

3. Measurements and analysis

As the positron affinity for both vacancies and copper atoms is higher than the one of pure iron [5], PAS will be sensitive to these types of defects. Therefore, positron annihilation lifetime spectroscopy (PALS) and coincidence Doppler broadening spectroscopy (CDB), both available in the hot cell laboratory of SCK•CEN, have been used in a complementary way. PALS measurements give information on the electronic density of the material at the place of the annihilation. While, CDB spectroscopy allows measuring the momentum distribution of the electron-positron annihilated pair, giving information of the chemical environment surrounding the annihilation sites.

3.1. Positron lifetime measurements

Non- and irradiated specimens of (10 × 10 × 1) mm^3, have been first surface polished to a mirror finish, and then chemically etched using a ($HF + H_2O_2 + H_2O$)-solution to remove the deformed surface layer. The specimens, taken two-by-two in a sandwich-like fashion, with the positron source in-between, were mounted in the specimen holder under biologically shielded environment. The sandwich was then automatically loaded (with a pneumatic loading system) in the measurement cell between three detectors working in anti-coincidence mode.

Measuring irradiated materials is proven to be more difficult, as care has to be taken to separate correctly the γ-peaks coming from the specimens and those related to the positrons. The setup used at SCK•CEN has been fully optimized for this purpose, as reported by Jardin *et al.* [17], [18]. Indeed, this triple coincidence setup was demonstrated to be very efficient in preventing the collection of the events due to ^{60}Co, which is the main isotope produced by neutron activation, during the irradiation of the alloys under investigation. On top of this, the detectors are mounted such that the distance separating the detectors can be adjusted to minimize the background, depending on the radioactivity of the samples.

After the collection of a statistically sufficient amount of triple coincidence events, the newly developed software analyses the raw data of each event, and allows a clear visualization of the results. Four characteristics of every signal such as peak height, peak intensity, base level and rising time are determined. The appropriate value ranges are then selected to filter the signals within the defined energy-range, avoiding any pileup. Thus, only the signals satisfying the selected ranges are used for the generation of a positron lifetime spectrum. In order to achieve a statistically relevant number of counts (i.e. at least one million counts), the acquisition of each spectrum has required in-between a few days and about one month.

The analysis of the obtained spectra has been performed using the LT-software [19]. The resolution of the setup was constantly checked to be about 175 ps, while the source contribution was set to 15 % with a single lifetime of 393 ps. All measurements were performed at room temperature, using a ^{22}Na source, with a strength of 2 MBq, wrapped in Kapton foils.

3.2. Coincidence Doppler broadening measurements

The CDB setup has been described with ample details by Verheyen *et al.* [20]. Here, both annihilation photons are detected in coincidence in order to minimize the background, [21], while the pileup is prevented, by making the detectors movable to larger distances from the sample (up to 1.5 m). The stability of the system was continuously checked, as a ^{57}Co source was installed inside the setup. This source emits photons with energies of respectively 122 keV and 136 keV. These peaks are used in combination with the 511 keV peak to monitor the calibration and resolution simultaneously.

The specimen preparation and the loading of the specimen in the setup were done in a similar way, as it has been explained for the PALS measurements. The overall energy resolution was ~1.07 keV full width at half maximum (FWHM), which corresponds to the momentum resolution of $\sim 4.2\times10^{-3}$ *mc* (FWHM). For each measurement, more than 1×10^{8} counts were accumulated what takes 2 to 3 days of acquisition.

The analysis of the results has been made based on the selection of a channel across the coincidence region of the 2D-spectra obtained from the measurements. Thus, the resulted one dimensional spectrum from each sample was normalized to the one obtained for the unirradiated defect-free pure Fe. The as such obtained CDB ratio curves allow a clear description of the momentum distribution. Two parameters, *S* and *W*, were extracted form each spectrum. These parameters are defined respectively as the ratio of low-momentum ($|cp_L| < 2.5 \times 10^{-3}\ mc$) and high-momentum ($15 \times 10^{-3}\ mc < |cp_L| < 25 \times 10^{-3}\ mc$) regions in the CDB spectrum to the total region (*c* is the speed of light and p_L is the longitudinal component of the positron-electron momentum along the direction of the γ-ray emission).

4. Determination of the average size of the defects

As it was shown in section 1, the determination of the average size of the defects has been performed by many authors. In the following, all the results from literature are given and used with equal value.

4.1. Pure Fe

In figure 1, the results found in literature are summarized. The available data can be separated in four groups. Each of them used a different approach to estimate the positron lifetime as function of the vacancy size. Group 1 contains papers from Vehanen, Hautojärvi, Puska and co-workers [7], [9], [10],

[13], [22], [23]. Group 2 gives the results published by Nagai, Tang and co-workers [3], [8], [14], [24], [25]. Group 3 shows the data given by Kuriplach *et al.* [26]. And the results of group 4 were published by Asoka-Kumar and co-workers [27], [28]. However, from the figure, it could be seen that the same trend is observed from all published data. Especially, the results from group 2 and 3 are quite identical as they use a similar approach for their calculations. In this work, a general trend line has been estimated based on all literature data points. As it is shown in figure 1, this trend is pretty close to the most recent results of group 2 and 3, that were obtained using ab-initio type of calculations.

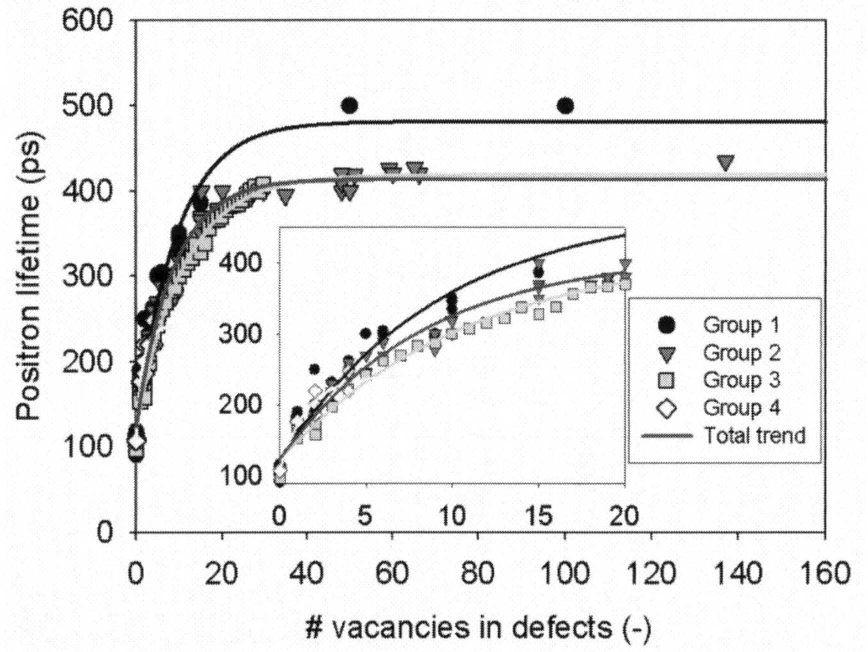

Figure 1. The results found in literature for the positron lifetime component of vacancy clusters with increasing amount of vacancies inside the clusters.

All calculations and experiments have of course been performed for pure iron. Adding alloying elements will lead to a shift of the above curve. As this shift depends on the amount of alloying elements at the annihilation sites, it is very difficult to determine the correct amount of vacancies inside the defect that gives rise to a particular lifetime component.

4.2. Fe-C

As for Fe doped with some C (the basic component of a steel), it is expected that the lifetime of vacancy clusters, combined with some carbon will change compared to the same pure vacancy clusters. In literature, however, two contradictory opinions have been found for the positron lifetime component of a carbon-vacancy pair. Hautojärvi *et al.* [23] found experimentally that this component should be equal to 160 ps (i.e. 15 ps less than the value of a monovacancy), while Asoka-Kumar *et al.* using his model [27] found that the addition of 1 carbon atom to 1 and 2 vacancies in Fe would lead to an increase of about 5 ps (i.e. 181 ps for the carbon-vacancy pair lifetime component).

As it is still questionable which of the two opinions is the most accurate, in this paper it is chosen to use the same values for the alloy containing some carbon as for the pure iron. Moreover, even for the pure iron differences of more than 10 ps have been found between the different groups for a cluster containing a certain amount of vacancies (see figure 1). On top of his, it should be mentioned that it is very difficult to determine experimentally if carbon is located near the annihilation site or not (see for instance [16]).

4.3. Fe-Cu

For the binary Fe-Cu alloys, it is expected that the lifetimes found for the vacancy clusters containing some copper will be different from the lifetimes found for the copper-free clusters. Indeed, Kuriplach *et al.* have shown that the lifetime component slightly decreases with increasing copper coverage of the vacancy clusters (see figure 2 of [26]).

As the presence of copper is well observed in the CDB curves, it is possible to estimate the amount of copper present at the annihilation sites, using the approach developed by Nagai *et al.* [29] and [25]. They found that the distribution in the binary Fe-Cu alloy can be approximately expressed in function of the distributions of pure iron and pure copper, as it is shown in (1), where $N_X(p_L)$ represents the high-momentum distribution of the CDB spectrum of the alloy. In this equation, x is the coverage of the defects by copper atoms (i.e. the fraction of Cu atoms that substitute Fe atoms in the first shell surrounding the annihilation site).

$$N_{Fe-Cu}(p_L) \propto (1-x)\cdot N_{Fe}(p_L) + x\cdot N_{Cu}(p_L) \quad (1)$$

As the ratio curve is given by $R_{X/Y}(p_L) = N_X(p_L)/N_Y(p_L)$, the ratio curve for the binary Fe-Cu alloy can be expressed with (2).

$$R_{(Fe-Cu)/Fe}(p_L) \propto (1-x) + x\cdot R_{Cu/Fe}(p_L) \quad (2)$$

Therefore, figure 2 plots the normalized CDB curves obtained for each irradiation condition of the two binary alloys investigated here. For comparison and illustration, the ratio curve for pure Cu normalized to pure Fe is also reported.

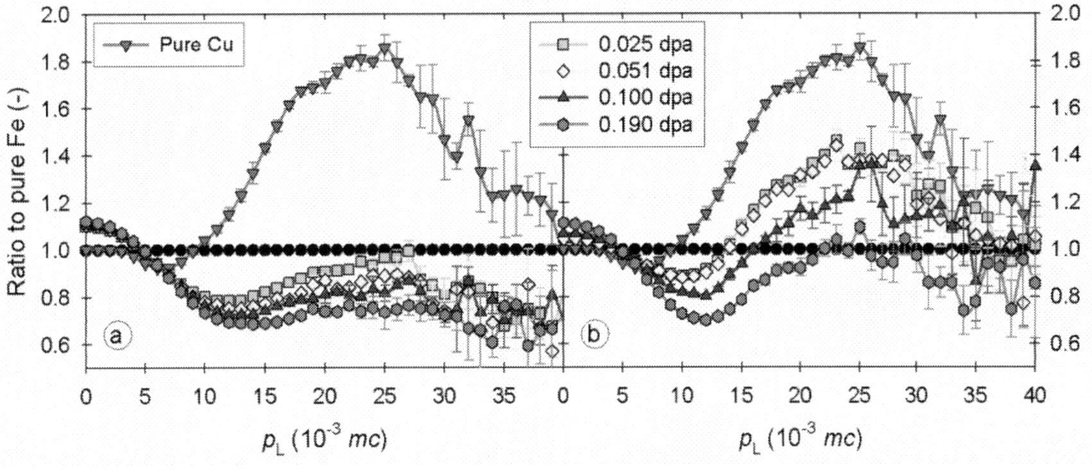

Figure 2. The CDB ratio curves for the binary Fe-Cu alloys at the different irradiation conditions. (a) shows the results for the Fe-0.1% Cu alloy and (b) those for the Fe-0.3% Cu alloy.

Figure 3 illustrates the estimated fractions of the positrons annihilating with copper electrons for the two Fe-Cu alloys after irradiation to the different doses. The estimated fractions are determined by fitting the Cu curve to the Fe-Cu curves, using (2) and taking into account the effect of vacancies who tend to increase the S parameter and decrease the high momentum region ($p_L > 7\times10^{-3}$ *mc*). Thus, the x-parameter is varying between zero when no Cu atoms are surrounding the annihilation site and 1 when this first shell is entirely made of Cu atoms.

It is found that for the alloy containing 0.1 % of copper, the degree of coverage is very low, i.e. lower than 50 %, while for the alloy containing 0.3 % of copper, it increases to about 100 %. This is easily understood, as more copper is available in the latter alloy.

The copper in the alloys combines with the relatively small amount of vacancies, produced at low doses. With increasing irradiation dose, the degree of coverage decreases steadily, as more and more vacancies are available that can reach the already formed clusters.

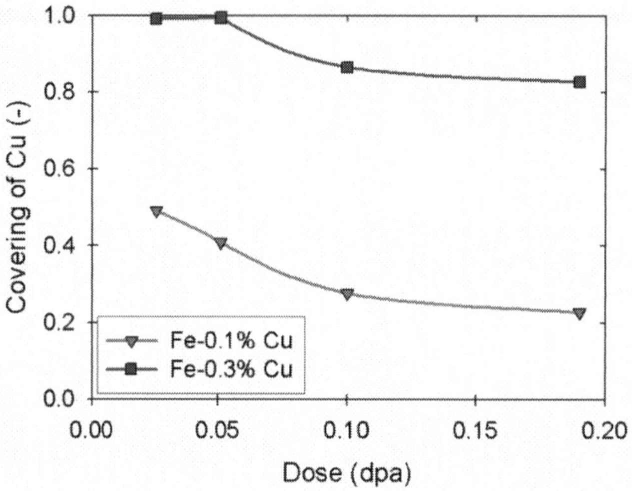

Figure 3. The estimated fraction of the coverage with copper on the inner surfaces of the defects.

Based on the obtained degree of coverage and using both the average lifetime of vacancies in pure iron from figure 1 and the results of Kuriplach *et al.* (see figure 2 of [26]), it becomes possible to identify the V-Cu clusters associated with the second component of the lifetime measured in the Fe-Cu alloys. In fact, these second lifetime component results of the PALS analysis can lead to an estimation of the average size of the clusters, including both the number of vacancies and those of copper atoms, co-clustering together. It should be noticed that an increasing amount of vacancies will lead to an important increase of the lifetime of the defect, while an increasing amount of copper atoms will slightly decrease this lifetime.

5. Determination of the concentration of annihilation sites

In order to extract the real properties of the defects, their size and density, from the experimentally obtained lifetime components and their associated intensities, the following well-known set of differential equations needs to be solved.

$$\frac{d}{dt}n_B(t) = -\left(\lambda_B + \sum_{i=1}^{k}\kappa_i\right)\cdot n_B(t) + \sum_{i=1}^{k}\delta_i \cdot n_{D_i}(t) \tag{3}$$

$$\frac{d}{dt}n_{D_i}(t) = \kappa_i \cdot n_B(t) - \left(\lambda_{D_i} + \delta_i\right)\cdot n_{D_i}(t) \tag{4}$$

Here, the quantities $n_B(t)$ and $n_{D_i}(t)$ are the densities of the positrons in respectively the bulk and the defect state at time t, while λ is the annihilation rate, κ is the trapping rate and δ is the detrapping rate. The indices B and D_i denote the free (bulk) and the trapped (defect) state, respectively. It should be mentioned at this point that the detrapping rate is considered to be negligible in most cases compared to annihilation and trapping rates [22].

In most cases, only one trapping state is used (i.e. k is equal to 1). But this situation cannot be applied for irradiated samples, as more than one trapping site is expected. Nevertheless, for the pure iron and the Fe-C alloy irradiated at low flux, reasonable values for τ_1 have been found with a single trapping model. In all other cases a second trapping state had to be included, as it is illustrated in figure 4. For this model, a 3-component analysis of the lifetime spectrum is required, however, in this work, a 2-component analysis gave a fairly sufficient fitting of the obtained spectra, while the analysis with 3 components failed to converge to singular results. Therefore, and in order to comply with the model, one of the experimentally obtained components should be considered as a combination of two components. In this work, it is assumed that the experimental first component fulfils this role. In fact, these first lifetime components are found to be in-between 90 and 120 ps, and therefore they are much higher than what is expected for a polycrystalline Fe alloy. Indeed, $\tau_{1,\exp}$ is expected to be equal to $\tau_{1,calc}$, given by (5). This equation shows that $\tau_{1,\exp}$ should be lower than the bulk lifetime, τ_B, which is equal to 107 ps.

$$\tau_{1,calc} = \left(\lambda_{1,calc}\right)^{-1} = \left(\lambda_B + \kappa_1 + \kappa_2\right)^{-1} \tag{5}$$

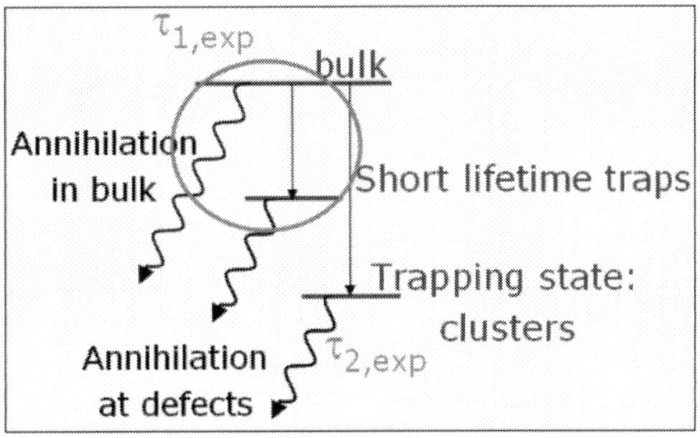

Figure 4. Two types of defect traps are assumed, i.e. short lifetime traps and clusters. The lifetime of the short lifetime traps is supposed to be experimentally measured in the first component.

It is thus supposed that the lifetime of the short lifetime traps is experimentally measured as part of the first component, as illustrated in figure 4. This has been firstly introduced by Vehanen *et al.* [9] and in the present work the formulae proposed by Vehanen *et al.* (6) and (7) will be applied. Also here, $\tau_{1,calc}$ is given by (5). For clarity, from now on the short lifetime traps will be denoted as sh, while the traps with longer lifetimes are named clusters (cl).

$$\begin{cases} I_{1,\exp} = I_{1,calc} + I_{sh} \\ I_{2,\exp} = I_{cl} \end{cases} \tag{6}$$

$$\begin{cases} \tau_{1,\exp} = \dfrac{I_{1,calc} \cdot \tau_{1,calc} + I_{sh} \cdot \tau_{sh}}{I_{1,calc} + I_{sh}} \\ \tau_{2,\exp} = \tau_{cl} \end{cases} \tag{7}$$

With the equations given by (6) and (7) and the fact that $I_{1,\exp} + I_{2,\exp} = 1$, only five equations are available to calculate six unknown parameters. Therefore, one of the parameters should be determined by another way. In this respect, a parametric study has been performed to determine the best value for τ_{sh}. Figure 5 shows the estimated defect densities for both the short lifetime traps (figure 5 (a)) and the clusters (figure 5 (b)), estimated using the equations given before, inserting the experimental results for irradiated pure iron, as well as a fixed value for τ_{sh}. These calculations have been performed for different values for τ_{sh}, changing from 120 ps to 175 ps, with an interval of 5 ps, using as input values those provided by Vehanen *et al.* [9]. The results shown here were carried out for the pure iron experiments, but they were also performed for the other alloys, with comparable results.

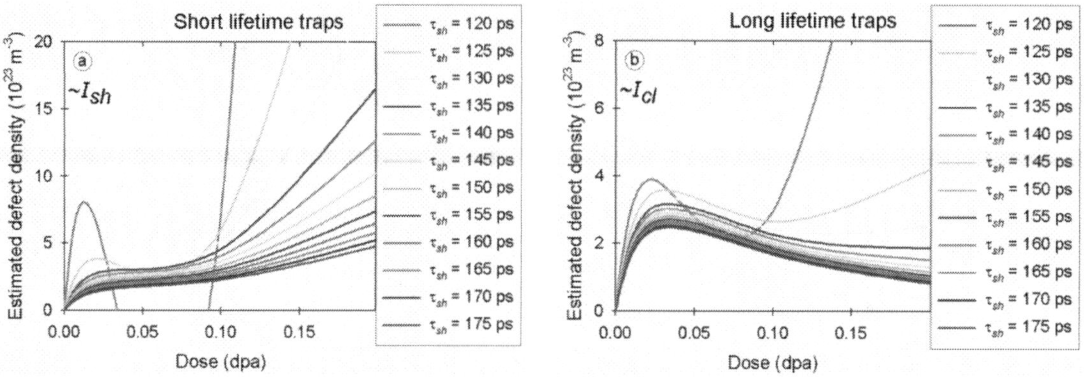

Figure 5. The estimated defect density for respectively the short lifetime traps (a) and the long lifetime traps, i.e. the clusters (b) are shown with changing assumption for the value of τ_{sh}. τ_{sh} varies from 120 ps to 175 ps for both calculations.

The estimated densities for both the short lifetime traps and the clusters show unjustified trends for very low values of τ_{sh}, while the results for high values of τ_{sh} are converting to reasonable values. In order to keep the differences as small as possible, only the high values of τ_{sh} were taken into account. More precisely, the ranges are chosen between 150 ps and 175 ps for the densities of the short lifetime traps and between 140 ps and 175 ps for the densities of the clusters.

More information is needed to select the best value for τ_{sh}. Asoka-Kumar *et al.* [27] has found that the value of the lifetime for a dislocation combined with a vacancy is in-between 142 and 165 ps. While, as mentioned earlier, Hautojärvi *et al.* [23] found a lifetime of 160 ps for a V-C pair. Furthermore, for the alloys containing copper, it has been argued in section 4.3, that the lifetime component for vacancies combined with copper is lower than the single vacancy lifetime in Fe (i.e. 175 ps). It is therefore obvious that for the short lifetime traps in the Fe-based alloys, a value of 160 ps should be assigned in our case, whatever their origin is.

Now, with a fixed τ_{sh}, it is possible to solve the differential equations given by (3) and (4) and therefore determine all other parameters. The defect density of both the short lifetime traps c_{sh} and the clusters c_{cl} can be estimated, using (8).

$$\begin{cases} \kappa_{sh} = \mu_{sh} \cdot c_{sh} \\ \kappa_{cl} = \mu_{cl} \cdot c_{cl} \end{cases} \quad (8)$$

The specific trapping rate for monovacancies μ_V has been determined by Vehanen *et al.* [9], to be equal to $(1.1 \pm 0.2) \times 10^{15}$ s^{-1}. The same specific trapping rate of single vacancies is used for the short lifetime traps ($\mu_{sh} = \mu_V$), as they are, as mentioned earlier, considered to be most probably composed of single vacancies associated with a foreign element or a dislocation.

As for the clusters, unfortunately, no experimental data are available on their specific trapping rate μ_{cl}. On top of this, Nieminen *et al.* [30] have shown that lifetime measurements exhibit strong temperature dependence for the positron trapping probability at large voids in aluminium. In addition, they showed that the specific trapping depends linearly on the temperature, in the case of large voids at low temperatures, as the trapping process is transition limited (the positron mobility is large and the capture rate is small). At temperature above 200 K, however, the process becomes diffusion limited and the dependence on temperature becomes weak. The same has been observed for an iron-based alloy containing voids with a lifetime of 500 ps by Huguenin and Moser [31], ten years later. However, in the latter work, the authors suggest that the variations under consideration are strongly dependent on void size. For small vacancy clusters (with a lifetime of 260 ps), no temperature variation could be detected in the samples. In the alloys investigated in the present work, the clusters have been observed to be small [16], with a lifetime comparable to the one given by [31]. Therefore, it can therefore be assumed that the dependency on temperature of the specific trapping rate of the clusters found here is negligible. On the other hand, theoretical calculations by Nieminen and Laakkonen [32] indicated that for these small vacancy clusters in metals, the positron trapping rate is proportional to the defect volume, i.e. the number of vacancies in the cluster (N), given by $\mu_{cl} \approx \mu_v \cdot N$.

6. Results and discussions

The results have been qualitatively explained in previous papers [16], [33]. In the present paper, an attempt is made to extract quantitative information on the detected defects. It is worth mentioning at this point, that although the lifetime results obtained show rather high reproducibility and low uncertainties (i.e. less than 5 ps), the estimated sizes and densities depicted below are to be considered in conjunction with the assumptions and the model used.

6.1. Pure Fe and Fe-C

Using the obtained lifetime results, reported in [33], the average size of the clusters and the number density of both the clusters (red curve) and the short lifetime traps (blue curves) are reported, as function of the irradiation dose, in figure 6 for pure iron and in figure 7 for the Fe-C alloy, as estimated following the methodology described above.

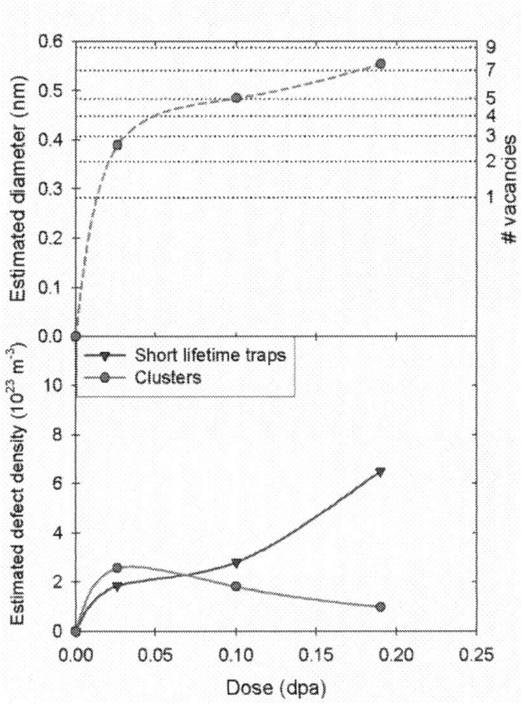

Figure 6. The estimated values for the size and density of the defects in the pure Fe alloy.

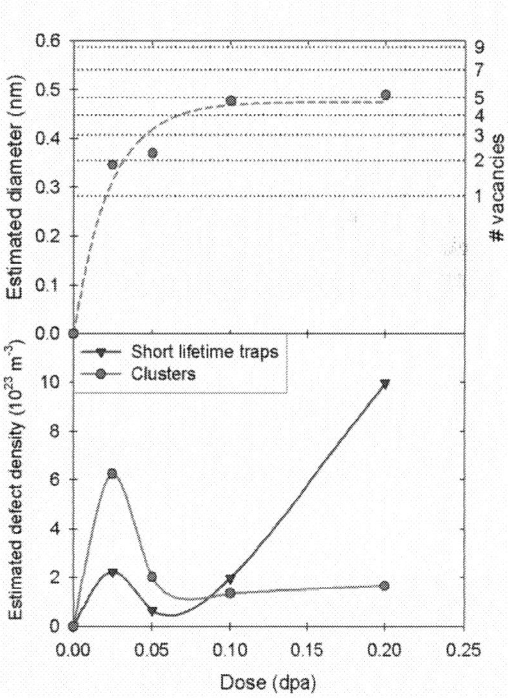

Figure 7. The estimated values for the size and density of the defects in the Fe-C alloy.

For pure iron, it is observed that the size of the clusters increases gradually with the irradiation dose, while the number density of the clusters decreases. The clusters thus grow by agglomeration of smaller clusters. With increasing the irradiation dose, the newly produced vacancies will be distributed between the clusters and the bulk. Indeed, the density of the short lifetime traps increases as well as the size of the clusters. As it was found that the density of the clusters diminish with dose, it is (as mentioned before) understandable that the amount of vacancies, which are still present in the bulk, increases.

The estimated diameter given in figure 6 agrees mainly with the one observed by Eldrup *et al.* (see figure 4 of [15]) for pure iron irradiated to approximately the same dose. The number density of the defects published by Eldrup *et al.* (see figure 5 of [15]), on the other side, is about 10 times higher than the one found here. This might be the effect of the irradiation temperature, as the alloys in the paper of Eldrup *et al.* were irradiated at 70 °C, while in the current paper the irradiation temperature was 300 °C. Indeed, at higher temperatures, more vacancies and small vacancy-clusters will be more mobile, decreasing the amount of clusters.

When carbon is added to the alloy, the size of the clusters is smaller than in pure iron. In fact, carbon hinders the growth of the clusters. This size also seems almost to saturate at higher doses. The density of both the short lifetime traps and the clusters show a peak at about 0.025 dpa. This can be caused by the presence of carbon. The vacancy-carbon pairs are rather stable [34] and the agglomeration will only start after supersaturation of vacancies. After they are formed, they grow until about 5 vacancies and then stabilize again. Newly produced vacancies will especially be present in the bulk material, combined again with the carbon, but also some new clusters will be formed.

6.2. Binary Fe-Cu alloys

As for these alloys, the results for the average size of the clusters and the number density of both the clusters and the short lifetime traps are given in figure 8 for the Fe-0.1% Cu alloy and in figure 9 for the Fe-0.3% Cu alloy. Here it is assumed that the clusters are spherical Cu-V co-clusters where the vacancies are in fact covered by Cu atoms. In red, the average size of the clusters is expressed as the average number of vacancies in the clusters, while the curve in purple is the apparent size of the clusters, including both vacancies and Cu-atoms (i.e. the degree of copper coverage is added to the values in red).

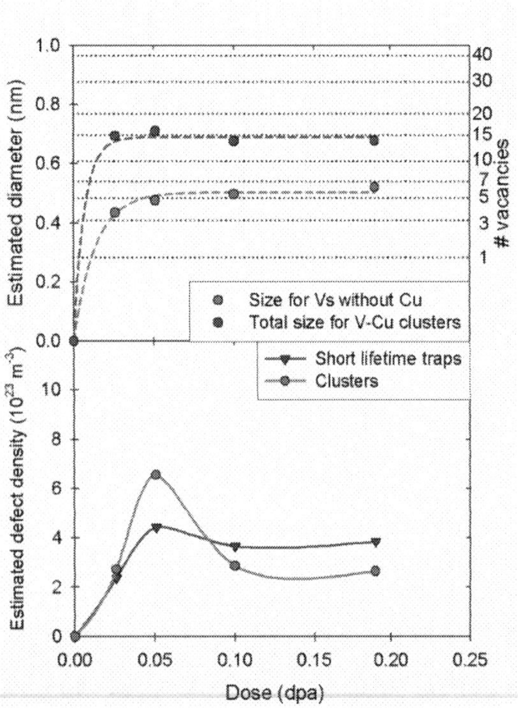

Figure 8. The estimated values for the size and density of the defects in the Fe-0.1% Cu alloy.

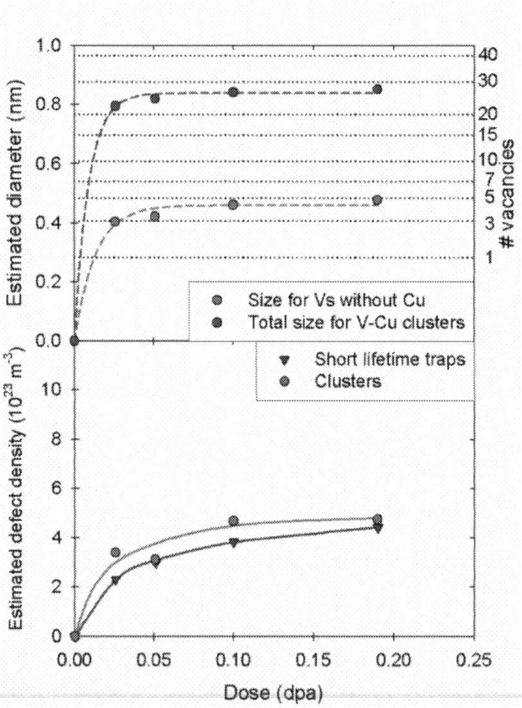

Figure 9. The estimated values for the size and density of the defects in the Fe-0.3% Cu alloy.

In the alloy containing 0.1 wt% of copper, the vacancy clusters seem to be very stable with dose. Nevertheless, a decrease of number density is found after about 0.05 dpa. This is most probably due to the agglomeration of the small clusters. Thereafter, the amount of clusters remains almost constant. Newly produced vacancies will thus be trapped by the clusters, as proposed by Verheyen *et al.* [20]. Indeed, the number of short lifetime traps is constant after 0.05 dpa, while the size of the clusters seems to grow very gently.

For the alloy containing 0.3 wt% of copper, the same observation can be made, except that the number density of both short lifetime traps and clusters keeps on growing with irradiation dose. This indicates that there is still copper available in the matrix with which new clusters and short lifetime traps are formed. As more clusters (and short lifetime traps) can be formed in this alloy, the clusters will remain smaller in term of the included amount of vacancies. Nevertheless, the total V-Cu clusters will be bigger due to a larger amount of copper atoms.

6.3. Comparison with other techniques

In figure 10, the results obtained by Meslin *et al.* [35] from four different microstructural techniques, that examine the same samples, are reported for the dose of 0.1 dpa. All measurements were performed within the European PERFECT project [36]. Each technique provides the number density and the mean size of the microstructural feature to which it is sensitive [16].

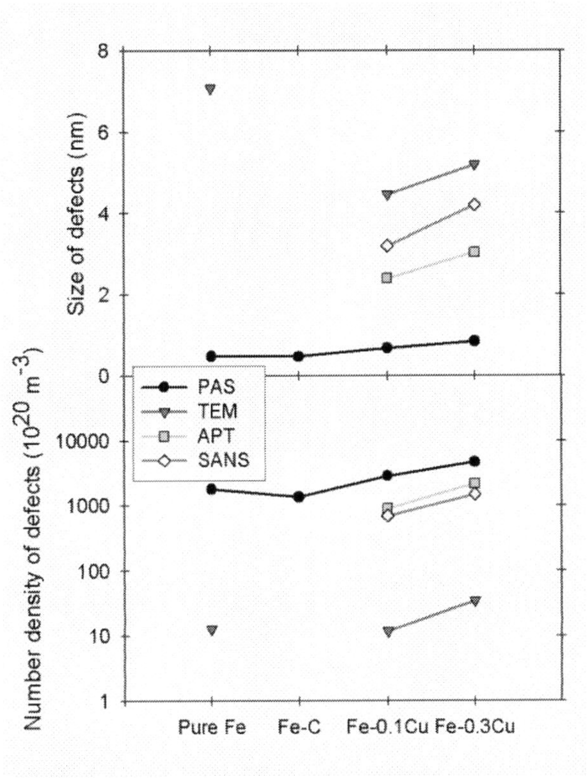

Figure 10. The estimated values for the size and density of the defects, found for 4 different techniques.

The objects detected by APT are unambiguously known to be precipitates, whose composition is reliably provided by the technique. Nevertheless, APT cannot detect the presence of vacancies associated with the observed precipitates, if they are present. The precipitate sizes obtained from the

APT are apparent and they do not include the very small vacancy-solute clusters, containing only few vacancies and solute atoms.

The SANS defects appear to give bigger mean sizes than it is observed by APT, while their number density is a little lower. This can be explained by the fact that very small defects, resolvable by APT, were not detected by SANS. Nevertheless, APT and SANS results are still in good agreement, and it can therefore be assumed that they correspond to the same defect populations.

The defects, visible by TEM, on the other hand, do not correspond to any of the previous categories. These defects are known to be mainly self-interstitial loops, limited to those of size larger than 2 nm. Indeed, defects smaller than 2 nm are coherent with the matrix, and therefore not detected by TEM. As these defects were not observed by any other technique, the TEM results are needed for a full picture of the defects.

Finally, the average size obtained from the lifetime measurement with PAS reflects the amount of vacancies and, if present, copper atoms. However, one should expect an important size distribution due to the stochastic character of neutron irradiation. It is thus obvious that PAS cannot detect all the clusters especially the ones with very low density (mostly the larger ones). At the same time, with positrons it is possible to see all kind of defects containing vacancies, even only a few vacancies close to foreign elements are detectable. These types of defects cannot be resolved by any other technique. So, the defect population detected by PAS is understood to correspond to small vacancy clusters and voids, associated or not with solute atoms.

The results obtained by Eldrup *et al.* for the density of defects in pure iron alloys irradiated up to the same dose (see figure 5 of [15]) are very comparable with ours, as they also found about 2 order of magnitude difference in-between the values given by TEM and those given by PAS. This indicates that the results that were obtained here are very reasonable.

The combination of the four experimental techniques for microstructural characterization leads to a good comprehension of all defect populations created by irradiation inside the model alloys. A sensible assessment of the strength associated with the different defect population, using the Orowan equation for each contribution, and their combination using the proper superposition law, provides a reasonable rationalization of the hardening results, as it has been done in [37].

7. Conclusions and outlook

An attempt to extract numerical information on the defects detected by PAS has been made based on the existing knowledge of the interaction between positrons and vacancy type of defects in Fe. It is thus demonstrated that using an educated guess, it is possible to access to the most useful information about the neutron irradiation induced defects, namely size and density of the trapping sites.

It was shown that in Fe, the average size of the defects observed can be estimated using a reasonable assessment of the literature data, but in binary alloys, such as Fe-Cu, a special care should be taken for the degree of copper coverage of the vacancy clusters. The latter could be estimated using the coincidence Doppler broadening results. In order to calculate the defect density, it is needed to assume at least two types of defects. A parametric study has shown that the short lifetime traps have a lifetime component of about 160 ps.

These quantitative results have then been compared with other techniques, such as APT, SANS and TEM. It has been proven that the results coming from the PAS experiments are reliable, and the combination of the four techniques leads to a good comprehension of all defect populations created during irradiation.

Acknowledgments

This work is partly financed by the European Union in the framework of the PERFECT project, under contract FI60-CT-2003-5088-40. We would like to thank all the involved partners for their contribution to this research, in particularly Dr. Mercedes Hernandez Mayoral *et al.* from CIEMAT (Spain), Dr. Estelle Meslin *et al.* from the University of Rouen (France) and Dr. Frank Bergner *et al.* from FZD (Germany).

We would also like to thank people who helped with the understanding of quantifying PAS results, such as Prof. Yasuyoshi Nagai form the Tohoku University (Japan) and Dr. Jan Kuriplach from the Charles University (Czech Republic).

References

[1] Phythian W J and English C A 1993 *J. Nucl. Mater.* **205** 162–77
[2] Malerba L 2006 *J. Nucl. Mater.* **351** 28–38
[3] Nagai Y, Tang Z, Hasegawa M, Kanai T and Saneyasu M 2001 *Phys. Rev.* B **63** 134110
[4] Eldrup M and SinghB N 1997 *J. Nucl. Mater.* **251** 132–38
[5] Puska M J, Lanki P and Nieminen R M 1989 *J. Phys. Condens. Matter.* **1** 6081–93
[6] Dai G H, Moser P and Van Duysen J C 1991 in: *Proceedings of the IXth International Conference on Positron Annihilation*, Szombathely, Hungary 941–44
[7] Hautojärvi P, Vehanen A and Mikhalenkov V S 1976 *Appl. Phys.* **11** 191–92
[8] Ohkubo H, Tang Z, Nagai Y, Hasegawa M, Tawara T and Kiritani M 2003 *Mat. Sci. Eng.* A **350** 95–101
[9] Vehanen A, Hautojärvi P, Johansson J and Yli-Kauppila J 1982 *Phys. Rev.* B **25** 762–80
[10] Hautojärvi P, Judin T, Vehanen A, Yli-Kauppila J, Johansson J, Verdone J and Moser P 1979 *Sol. Stat. Comm.* **29** 855–58
[11] Sterne P A and Kaiser J H 1991 *Phys. Rev.* B **43** 13892–98
[12] Hautojärvi P, Heiniö J, Manninen M and Nieminen R 1977 *Phil. Mag.* **35** 973–81
[13] Puska M J and Nieminen R M 1983 *J. Phys.* F: *Met. Phys.* **13** 333–46
[14] Hempel A, Saneyasu M, Tang Z, Hasegawa M, Brauer G, Plazaola F, Yamaguchi S, Kano F and Kawai A 2000 *Proc. 19th Int. Symp. on Effects of Radiation on Materials* ed M L Hamilton *et al* (West Conshohocken: American Society for Testing and Materials) pp 560–578
[15] Eldrup M, Singh B N, Zinkle S J, Byun T S and Farrell K 2002 *J. Nucl. Mater.* **307-311** 912–17
[16] Lambrecht M, Malerba L and Almazouzi A 2008 *J. Nucl. Mater.* **378** 282–90
[17] Jardin M, Lambrecht M, Rempel A A, Nagai Y, van Walle E and Almazouzi A 2006 *Nucl. Instr. and Meth.* A **568** 716–22
[18] Jardin M, Almazouzi A, Lambrecht M, Rempel A A, Nagai Y and van Walle E 2007 *Phys. Stat. Sol.* C **4** 4001–3
[19] Kansy J 1996 *Nucl. Instr. and Meth.* A **374** 235–44
[20] Verheyen K, Jardin M and Almazouzi A 2004 SCK•CEN-BLG-957.
[21] Lynn K G, MacDonald J R, Boie R A, Feldman L C, Gabbe J D, Robbins M F, Bonderup E and Golovchenko J 1977 *Phys. Rev. Lett.* **38** 241–44
[22] Hautojärvi P and Vehanen A 1981 *Rad. Eff. Lett.* **58** 77–82
Puska M J and Nieminen R M 1982 *J. Phys.* F: *Met. Phys.* **12** L211–16
Hautojärvi P, Pöllänen L, Vehanen A and Yli-Kauppila J 1983 *J. Nucl. Mater.* **114** 250–59
Dlubek G, Brümmer O, Mikhalenkov V S, Yli-Kauppila J, Vehanen A and Hautojärvi P 1984 *Cryst. Res. Tech.* **19** 627–31
Puska M J 1991 *J. Phys.: Condens. Matter* **3** 3455–69
Brauer G, Puska M J, Sob M and Korhonen T 1995 *Nucl. Eng. Des.* **158** 149–56
Barbiellini B, Puska M J, Korhonen T, Harju A, Torsti T and Nieminen R M 1996 *Phys. Rev.* B **53** 16201–13
[23] Hautojärvi P, Johansson J, Vehanen A, Yli-Kauppila J and Moser P 1980 *Phys. Rev. Lett.* **44** 1326–29
[24] Hempel A, Hasegawa M, Brauer G, Plazaola F, Saneyasu M and Tang Z 1999 *Proc. 9th Int. Symp. on Environmental Degradation of Materials in Nuclear Power Systems – Water Reactors* ed F P Ford *et al* (The Minerals, Metals & Materials Society) pp 835–844
Nagai Y, Hasegawa M, Tang Z, Hempel A, Yubuta K, Shimamura T, Kawazoe Y, Kawai A and Kano F 2000 *Phys. Rev.* B **61** 6574–78

Nagai Y, Tang Z and Hasegawa M 2000 *Rad. Phys. Chem.* **58** 737–42
Nagai Y, Hasegawa M, Tang Z, Kanai T and Saneyasu M 2001 *Mat. Sci. For.* **363–365** 91–93
Tang Z, Hasegawa M, Nagai Y and Saito M 2002 *Phys. Rev.* B **65** 195108
Nagai Y, Takadate K, Tang Z, Ohkubo H, Sunaga H, Takizawa H and Hasegawa M 2003 *Phys. Rev.* B **67** 224202
Nagai Y, Toyama T, Nishiyama Y, Suzuki M, Tang Z and Hasegawa M 2005 *Appl. Phys. Lett.* **87** 261920
Toyama T, Nagai Y, Tang Z, Hasegawa M, Almazouzi A, van walle E and Gerard R 2007 *Acta Mater.* **55** 6852–60

[25] Nagai Y, Takadate K, Tang Z, Ohkubo H and Hasegawa M 2004 *Proc. 21th Int. Symp. on The Effect of Radiation on Materials* ed M L Grossbeck *et al* (West Conshohocken: American Society for Testing and Materials) pp 590–602
[26] Kuriplach J, Melikhova O, Domain C, Becquart C S, Kulikov D, Malerba L, Hou M, Almazouzi A, Duque C A and Morales A L 2006 *Appl. Surf. Sci.* **252** 3303–08
[27] Asoka-Kumar P, Hartley J H, Howell R H, Sterne P A, Akers D, Shah V and Denison A 2002 *Acta Mater.* **50** 1761–70
[28] Glade S, Wirth B D, Odette G R, Asoka-Kumar P, Sterne P A and Howell R H 2005 *Phil. Mag.* **85** 629–39
[29] Nagai Y, Nonaka T, Hasegawa M, Kobayashi Y, Wang C L, Zheng W and Zhang C 1999 *Phys. Rev.* B **60** 11863–66
[30] Nieminen R M, Laakkonen J, Hautojärvi P and Vehanen A 1979 *Phys. Rev.* B **19** 1397–402
[31] Huguenin D and Moser P 1989 *Appl. Phys.* A **48** 583–85
[32] Nieminen R M and Laakkonen J 1979 *Appl. Phys.* **20** 181–84
[33] Lambrecht M, Jardin M and Almazouzi A 2007 *Phys. Stat. Sol.* C **4** 3477–80
Lambrecht M and Almazouzi A 2008 *OECD Handbook: Nuclear Science "Structural Materials for Innovative Nuclear Systems (SMINS)"* p 457–66
Lambrecht M and Almazouzi A 2009 *J. Nucl. Mater.* **385** 334-38
[34] Takaki S, Fuss J, Kugler H, Dedek U and Schultz H 1983 *Radiat. Eff.* **79** 87–122
[35] Meslin E, Lambrecht M, Hernández-Mayoral M, Bergner F, Malerba L, Pareige P, Radiguet B, Barbu A, Gómez-Briceño D, Ulbricht A and Almazouzi A 2010 *J. Nucl. Mater.* **406** 73-83
[36] http://www.fp6perfect.net.
[37] Lambrecht M, Meslin E, Malerba L, Hernández-Mayoral M, Bergner F, Pareige P, Radiguet B and Almazouzi A 2010 *J. Nucl. Mater.* **406** 84-89

PALS determination of defect density within friction stir welded joints of aluminium alloys

J Kansy[1], **K Mroczka**[2] **and J Dutkiewicz**[2]

[1] Institute of Material Science, Silesian University, Bankowa 12, 40-008 Katowice, Poland
[2] Institute of Technology, Pedagogical University of Cracow, Podchorążych 2, 30-084 Kraków, Poland

E-mail: kansy@us.edu.pl

Abstract. Positron annihilation spectroscopy is employed to investigate the density of defects in samples of aluminium alloys (2017 A and 6013) welded using the Friction Stir Welding method. The vacancy and dislocation densities were determined at the weld junction as a function of various parameters and conditions: Travel and rotational speed of welding tool, cooling of the surface of the welded material and the compositions of the welded alloys. The 3-state trapping model used in the computer analysis allowed to separate a vacancy component from a component related to dislocations. The determined lifetime of positron trapped by dislocation was much shorter than its experimental values referred to in literature, however, it is closer to the theoretical predictions.

1. Introduction

Conventional welding methods of the materials can be applied only to a few aluminium alloys due to many unfavourable phenomena occurring in the melt during the welding process e.g. oxidation of the weld, fracture due to contraction, structure degradation in the heat-affected zone as a result of overheating. An alternative method of joining metallic materials of flat shapes is the Friction Stir Welding (FSW) technology worked out at The Welding Institute, Cambridge. The basic principle of the method is making the joint in the solid state (materials are not melted). Friction between the welding tool and the materials causes a local increase in temperature up to the plastifying point. Simultaneously, the tool, due to its shape and movement (rotation with simultaneous movement along the welding line), causes mixing of

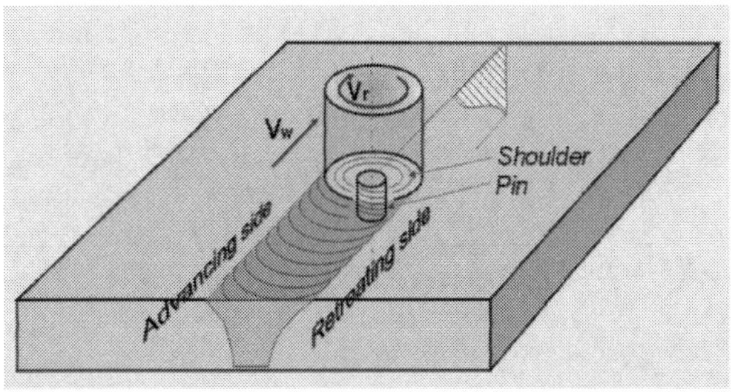

Figure 1. The scheme of the Friction Stir Welding process

the plastified materials and thus the formation of the joint. The FSW welding process is presented schematically in figure 1. The structure of the FSW welds is complicated because the welding proceeds very dynamically and the material flow around the tool is strongly oriented [1].

The processes occurring during the formation of the structure of FSW joints are not fully discovered which is clearly visible in the studies of structure and mechanical properties of welds (e.g. distribution of microhardness) [2,3].

The aim of this work was to determine the changes in the defect structure at the region of FSW joint as a function of various parameters and conditions of welding: Travel and rotational speed of the welding tool, the cooling of the surface of the welded material and the compositions of welded alloys. It is expected that the results allowed to steer the conditions of the welding process towards the optimization of the structure of the friction stir welded joints.

2. Materials

Examined aluminium alloys 2017A and 6013 of composition given in table 1 were welded in the

Table 1. Chemical composition of 2017A and 6013 aluminium alloy (Al balance).

Additional alloying elements [%]	Fe	Si	Cu	Zn	Ti	Mn	Mg	Ni	Sn	Cr
2017 A	0,23	0,42	3,94	0,07	0,03	0,55	0,57	0,01	-	0,02
6013	0,28	0,81	1,1	0,02	0,04	0,71	1,04	-	0,08	0.08

Institute of Welding, Gliwice, PL using a conventional friction stir butt welding.

In the aluminium alloys of the series 2017A a grain size of about 5 μm was found within the weld nugget. Grain refinement comparing to the original material took place due to a dynamic recrystallization. The observation performed by means of transmission electron microscopy showed higher dislocation density than in an annealed material.

A complex dislocation structure was found in the FSW welds of the 7075 alloy that contained screw dislocations, dislocation loops and dislocations interacting with precipitates [4,5]. Besides, in the structure of the weld nugget of 2017A alloy some helical dislocations were found, which indicates a process of structure formation with considerable influence of vacancies. Figure 2 shows the structure of the weld nugget of the FSW joint corresponding to the 6013 alloy with higher density of dislocations and also some dislocations interacting with the precipitates. On the basis of the TEM findings the following conclusion can be made: regardless of the type of the alloy higher density of defects in the crystal lattice for the FSW welds is observed. Fast cooling of the material during the welding should increase the density of defects.

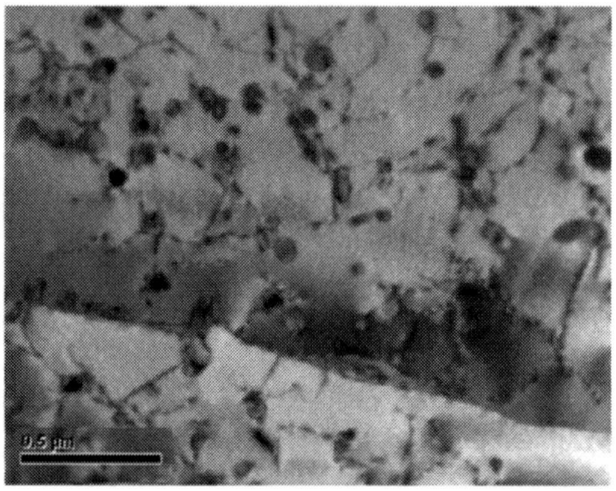

Figure 2. TEM micrograph of the dislocation structure at the FSW joint (weld nugget) of 6013 aluminium alloy.

3. The PALS experiment and its numerical analysis

The PAL spectra were measured at room temperature with a conventional fast-fast spectrometer of time resolution FWHM = 300 ps (value from fitting of Na^{22} lifetime spectra). The positron source of activity 400 kBq, covered by 5 μm Ni foil, was placed between two pieces of the investigated sample at the weld line. Thickness of the metal sheets was about 5 mm. Since the range of a positron emitted from Na^{22} in aluminium is only about 0.1 mm, PALS measurements were performed first with the positron source placed between the top sides (faces of weld) of the sample pieces and for the second time between their opposite sides. For each sample and its corresponding surface orientation a series consisting on 6 to12 lifetime spectra was recorded. A single spectrum collected 1.2 million of counts. The peak to background ratio was 15700 : 4.4 and the calibration coefficient 6.37 ps/channel.

The experimental data were analyzed with the LT-9-2 software [6], which enables to fit many spectra simultaneously during the same fitting process. The simultaneous fitting reduces the number of free parameters used in the fitting, because some of the parameters are common for various spectra.

The spectrometer time resolution function and the source contribution were estimated on the basis of 33 PAL spectra (of total statistics 140×10^6 counts) measured for pure well annealed silicon. The source correction for the Si spectra consisted of two *source components* of lifetimes 99 ps and 1.4 ns and intensities 33.6 and 0.4%, respectively.

Numerical analysis of PAL spectra for the investigated samples was based on the assumption that each spectrum contained three *sample components* (besides the *source components*) related to the positron annihilation in the bulk material and to two kinds of defects. Therefore, for theoretical description of the spectra, the well known three-state trapping model[1] [7,8] was employed. The fitting procedure searched directly for values of the model parameters, i.e. the positron lifetimes in bulk (τ_b), the lifetimes in two types of defects ($\tau_{1,2}$) and the positron trapping rates into the defects ($\kappa_{1,2}$). During the analysis the lifetimes originating from the positron source were fixed at values determined previously with help of the Si sample. However, because of possible different positron scattering coefficients for Si and Al the component contributions were free. The found contributions were similar to those for Si, i.e. 33.1% (for 99 ps) and 0.3% (for 1.4 ns).

4. Results and discussion

At the beginning, the values of the lifetimes τ_b, τ_1 and τ_2 were searched in a series of a preliminary fitting. The values that were found, turned out to be the same, within the range of error, for both investigated materials. The determined bulk lifetime 166 ± 2 ps was close to the measured value 163 ps [9] and to the calculated value 168 ps for pure aluminium [10]. The lifetime τ_1 was found to be 253 ± 3 ps. Because this is similar to the measured positron lifetime 248 ps [9] and to its theoretical value 254 ps [10] for monovacancies in Al, we also assigned τ_1 to positron annihilations in monovacancies in the investigated alloys. The determined value of τ_2 (190 ± 3 ps) was assigned to positron annihilation in dislocations. According to some theoretical calculations, the positron lifetime in dislocation lines in Al is only by 9% longer than in bulk Al [11] or even less (1.7% of τ_b) [12]. On the other hand, the measured positron lifetime in dislocations (in deformed Al and dilute Al-Si, Al-Mg and Al-Cu alloys) by Hashimoto et al. [13-15] are much longer (about 235 ps). It seems that the big discrepancies between the theoretical predictions and the experimental results can be explained by a complexity of the component assigned by the authors to positron annihilations in dislocations. The authors decomposed their analyzed spectra into two components only, of lifetimes τ_f and τ_d, related to bulk material and to dislocations. Most probably τ_d indicates the trapping of positrons at more than one type of defects. Indeed, when we attempted to employ the 2-state trapping model in our analysis, the lifetime related with defects was found to be 232 ± 2 ps, which is very close to value 235 ps determined by Hashimoto et al. In our opinion, the 3-state trapping model, used by us, allowed to separate the dislocation component from a component originated from vacancies. Maybe the separation is not yet complete so τ_2 could indicate resulting positron annihilation in the dislocation lines and in defects associated with them (e.g. jogs).

[1] The trapping model is introduced into the code of the program as its option. More details about this option and its operating one can find in the program manual.

During the main fitting, the lifetimes τ_b, τ_1 and τ_2 were fixed at the previously found values — 166, 253 and 190 ps, respectively. In this way only two model parameters were searched, i.e. κ_1 and κ_2. It is well known that the trapping rate of positron by defects is proportional to the defect concentration

$$\kappa = \mu c \tag{1}$$

where μ is the trapping coefficient characteristic for the structure of a defect. Its unit depends on the unit chosen for c. In this paper the defect concentration c will be expressed either as the number density of vacancies or as the density of dislocation lines (m^{-2}).

With equation (1) we recalculated the found values of κ_1 and κ_2 into the respective defect concentrations. In this calculations we used $\mu_V = 4.5\times10^{14}$ s^{-1} as a trapping constant for monovacancies in Al [9].

A problem arose with the estimation of the trapping constant μ_d for dislocations. In literature, there is a big scatter of the experimental values for μ_d in Al, ranging from 0.066×10^{-4} to 2×10^{-4} m^2s^{-1} [16]. Moreover, μ_d can depend on the content of impurities in Al and even on the dislocation density [13-15]. However, we supposed, that the dependence of μ_d on the dislocation density observed in those works may result from the method of lifetime spectrum analysis which did not separate the positron trapping at dislocations from the trapping to vacancies. We hope that the separation of those contributions during our analysis makes μ_d less dependent (or even independent) on the dislocation density. For a rough estimation of the density of dislocation lines, we decided to use a constant value for μ_d equal to 1×10^{-4} m^2 s^{-1}.

The estimated values of monovacancy concentrations (c_V) as well as the density of dislocation lines (c_d) in the vicinity of the weld line as a function of the travel speed (v_t) of the welding tool are shown in figure 3. Both c_V and c_d decrease when v_t increases. It turned out that, in the tested ranges of v_t, the variations of c_V and c_d can be well fitted by a logarithmical function

$$c = A\ln(v_t) + B \tag{2}$$

with fit correlation coefficients R^2 better than 0.995.

The observed dependencies can be easily explained. The motion of the welding tool releases a big amount of heat to the material

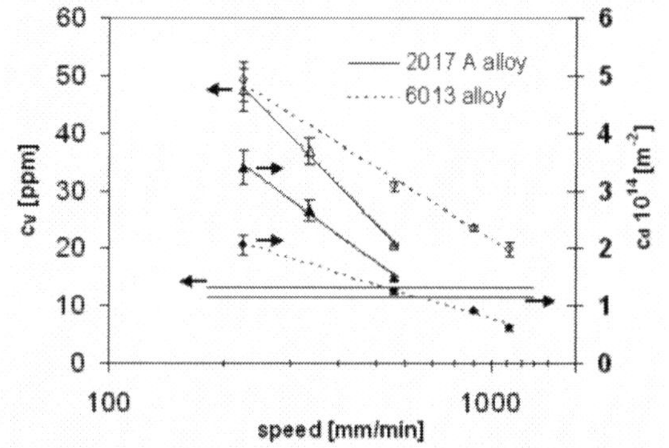

Figure 3. Variations in densities c_V of vacancies (*open symbols-left axis*) and in the density of dislocation lines (c_d) (*solid symbols-right axis*) at the FSW joint for different travel speeds of the welding tool in both investigated materials. The rotational speeds were 355 min^{-1} and 710 min^{-1} for 2017A and 6013 alloys, respectively. The values are averaged for the top (face of weld) and bottom side of sample. The horizontal lines represent c_V and c_d measured in a sample of 2017A alloy annealed at 450°C for 1 h.

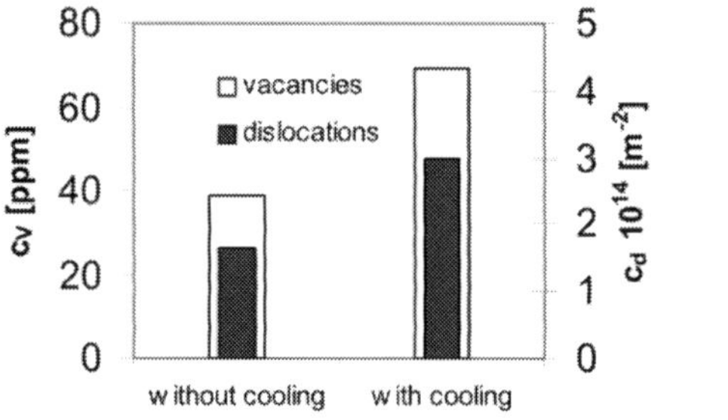

Figure 4. Densities of vacancies (left axis) and the dislocation lines (right axis) at the weld line at the welding surface in 6013 alloy for a non-cooled sample and in a sample cooled at the top side with solid carbon dioxide. In both cases the travel speed of the rotational tool were the same (450 mm/min and 710 min^{-1}).

Journal of Physics: Conference Series **265** (2011) 012010 doi:10.1088/1742-6596/265/1/012010

and its temperature increases. At high temperature the lattice defects, produced during the welding process, are partially annealed. Of course, the higher temperature, the higher recovery of the material.

In order to investigate the effect of the cooling rate on the density of defects in the welding region, a sample of the 6013 alloy was welded once with cooling its top surface with solid CO_2 and for the second time without any cooling. In both cases the travel speed and the rotational speed of the welding tool were the same. Figure 4 compares the defect densities at the weld junction on the top surface of the samples determined by PALS measurements. It is seen that the recovery of the sample is less efficient in the case of the cooled sample.

5. Conclusions

The analysis of the PAL spectra of the investigated samples with the 3-state trapping model allowed to separate the contributions of different kinds of defects (vacancies and dislocations), therefore such determined positron lifetimes at the dislocation lines are lower than the experimental lifetimes usually referred in literature, simultaneously the values are closer to the theoretical predictions.

The PALS measurements enabled to monitor the density of defects in friction stir welded materials in relation to the alloy's composition and the welding parameters such as the travel and rotational speeds of the welding tool and the efficiency for cooling the samples.

During the motion of the welding tool, the process of structural defect production competes with that of the material recovery. The densities of vacancies and dislocations at the welded part of the studied samples decrease linearly with the logarithm of the welding tool speed.

Acknowledgement
Partial financial support of the Ministry of Science and Higher Education of Poland under the grant number 3 T08A 035 30 is acknowledged. FSW joints were made in the Institute of Welding, Gliwice, Poland.

References

[1] Chan Z W, Pasang T and Qi Y 2008 *Materials Science and Engineering A* **474** 312
[2] Peel M, Steuwer A, Preuss M and Withers P J 2003 *Acta Materialia* **51** 4791
[3] Cavaliere P, Nobile R, Panella F W and Squillance A 2006 *International Journal of Machine Tools and Manufacture* **46** 588
[4] Mroczka K, Dutkiewicz J, Lityńska-Dobrzyńska L and Pietras A 2008 *Archives of Material Science and Engineering* **33** 93
[5] Mroczka K, Dutkiewicz J and Pietras A 2008 *Journal of Microscopy* **237** *521*
[6] Kansy J 1996 *Nucl. Instr. Meth. A* **374** 235
[7] Bergersen B and Stott M J, 1969 *Solid State Commun.* **7** 1023
[8] Connors D C and West R N 1969 *Phys. Lett. A* **30** 24
[9] Schaefer H-E 1987 *Phys. Stat. Sol. (a)* **102** 47
[10] Puska M J and Nieminen R M 1983 *J. Phys. F: Metal Phys.* **13** 333
[11] Shen J Q, Lung C W, and Wang K L 1986 *Phys. Stat. Sol. (b)* **134** 97
[12] Martin J W and Paetsch R 1972 *J. Phys. F: Metal Phys.* **2** 997
[13] Hashimoto E, Iwami M and Ueda Y 1993 *J. Phys.: Condens. Matter* **5** L145
[14] Hashimoto E, Iwami M and Ueda Y 1994 *J. Phys.: Condens. Matter* **6** 1611
[15] Hashimoto E, Iwami M and Ueda Y 1995 *J. Phys.: Condens. Matter* **7** 9935
[16] Jensen K O, Eldrup M, Singh B N, Linderoth S and Bentzon M D 1988 *J. Phys. F: Metal Phys.* **18** 1091

Defect studies of stainless steel via positron annihilation energy spectroscopy

L Tchelidze [1], D P Wells [1, 2] and S A Maloy [3]

[1] Department of Physics, Idaho State University, 785 S. 8th Ave, Campus Box 8106, Pocatello, Idaho 83209, USA
[2] Idaho Accelerator Center, 1500 Alvin Ricken Drive, Pocatello, Idaho 83201, USA
[3] Los Alamos National Lab, P. O. Box 1663, Los Alamos, NM 87545, USA

E-mail: tchelali@isu.edu

Abstract. High Energy proton (up to 800 MeV) and spallation neutron irradiated samples of stainless steel 316L and Mod 9Cr1Mo were studied using positron annihilation energy spectroscopy. Doses delivered to 316L were up to 10 displacements per atom (dpa) and doses to 9Cr1Mo were up to 2.5dpa. We studied the change of *T*-parameter, which is calculated as the ratio of the number of counts in the wings of the Doppler-broadened 511 keV peak to the number of counts in the center of the peak. *T*-parameter is related to the density of defects in the sample of interest. Higher defect densities induce, generally, smaller *T*-parameter, although this is complicated by additional effects that include the size, nature and other properties of defects that may lead to saturation of *T*-parameter. For the large doses studied, positron annihilation energy spectroscopy showed that the *T*-parameter dropped sharply from 0 to 3 dpa, and continued dropping up to 10 dpa. In 9Cr1Mo, similarly, *T*-parameter dropped sharply from 0 dpa to 1dpa, but from 1 dpa to 2.5 dpa it remained constant, indicating that the density of defects or *T*-parameter saturated with dose above 1 dpa in 9Cr1Mo. These results, where the change in *T*-parameter from zero dose to 1 or more dpa, is much larger than the effect that we see from one irradiated specimen to another, led us in both cases to investigate lower doses. We measured energy spectra in 316L and 9Cr1Mo that were irradiated under the similar conditions as the above samples, but with doses less than 0.1dpa. These results fill in the gap between 0 and 1 dpa and suggest that most of the change in *T*-parameter occurs below 0.05 dpa.

1. Introduction

Positron annihilation spectroscopy is currently well known as a powerful tool of microstructure investigations of condensed matter. The interaction of positrons with matter is used to study configuration and properties of materials at the atomic level. This technique, which is a non-destructive spectroscopy technique to study voids and defects in solids, was developed in the early 50s and has advanced quickly since then.

The technique is based on the fact that a positron that enters a material will die away by annihilation. In the annihilation process of a positron and an electron, 511 keV photons are released that can be detected. The 511 keV spectral line is Doppler broadened due to existing finite momentum of the annihilation positron-electron pair. Doppler broadening is proportional to the momentum of the annihilation pair.

The properties of inhomogeneities (or defects) within a material are different from that of the rest of a material. These defects in turn influence various material properties. Positron annihilation energy and lifetime spectroscopy (PAES and PALS), can be used to verify both the density and size of defects in the material. Doppler broadening of the positron-electron annihilation line changes due to presence of defects in a material. If positrons are injected into a solid body, the shape of the annihilation peak strongly depends on whether they end up in a vicinity of large electron density or in a void where electrons are absent, and positrons annihilate effectively with lower momentum valence electrons of neighbouring atoms. In the latter case the broadening of the annihilation peak is less than when annihilating in the bulk material. Accurate analysis of the Doppler broadened spectral line is required to extract information about the material.

The technique requires a source of positrons. A radioactive isotope of sodium is often used. However for acquiring information from deep inside of material, a positron source inside of it is required. Activation is one method used to achieve this. A technique of activation is described below.

2. Methods

The source of positrons was radioactivity internal to the samples. Part of the samples of interest contained internal ^{22}Na positron source as a result of daughter products of high energy proton and spallation neutron irradiation. The rest of the samples were activated by bremsstrahlung radiation from a 20MeV electron linear accelerator [1]. The photo-nuclear reaction ^{54}Fe (γ, n) ^{53}Fe was used for obtaining radioactive positron source ^{53}Fe with 8.51 min half-life.

Our experimental technique is based on the measurement of the Doppler broadening of the 511 keV annihilation gamma line. A high purity germanium detector with energy resolution of 1.2MeV at 662MeV was used for measuring energy spectra. The detector was mounted to a spectroscopy amplifier and to an analogue to digital converter in order to obtain an energy spectrum from each specimen studied. The scheme of the experimental setup used is shown in figure 1. Energy resolution was controlled by keeping ^{133}Ba and ^{137}Cs in front of the detector during measurements.

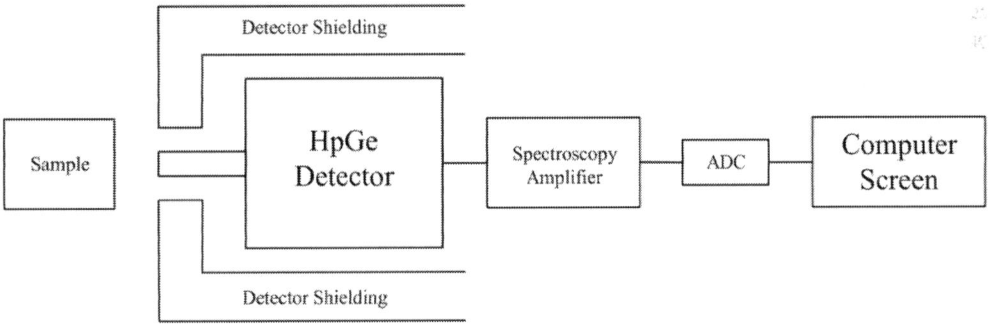

Figure 1. Schematic sketch of positron annihilation energy spectroscopy experimental setup.

STW analysis of Doppler broadened 511 keV lines was completed by computer code written at the Idaho State University. *S*, *W* and *T*-parameter values were extracted from the peak analysis. *S* and *W* have been defined as relative counts in the centre and wing of the peak with respect to total counts in the peak. *T* was defined as ratio of *W* and *S*. One should then expect to get lower *T*-parameter for highly damaged samples, due to less broadening of the annihilation peak.

3. Results

Samples studied were irradiated at Los Alamos National Lab with high energy protons (up to 800 MeV) and spallation neutrons. Materials used were stainless steel 316L and Mod 9Cr1Mo [2, 3]. A table below shows composition of these steel samples.

Journal of Physics: Conference Series **265** (2011) 012011 doi:10.1088/1742-6596/265/1/012011

Table 1. Chemical composition of 316L and 9Cr1Mo.

Material	Al	C	Cr	Cu	Fe	Mn
316L		0.019	17.3	0.26	Bal	1.75
Lot	Ni	P	S	Si	Ti	Others
E835	12.2	0.022	0.006	0.65		
Material	Al	C	Cr	Cu	Fe	Mn
Mod 9Cr-1Mo	0.002	0.089	9.24	0.08	Bal	0.47
Lot	Ni	P	S	Si	Ti	Others
10148	0.16	0.021	0.006	0.28	0.002	V-0.21; Nb-0.054; Co-0.019; N-0.035; O-0.008

Doses delivered to 316L were up to 10 displacements per atom (dpa) and doses to 9Cr1Mo were up to about 2.5dpa [4]. The high dose samples resulted in 42 – 430 mR/hr activities at contact or around 1 mR/hr at one meter after allowing to decay ~ 8 years after irradiation. The low dose samples were irradiated under the same conditions, with irradiation temperature of 74C and 127C and activities obtained after irradiation were 6 mR/hr at contact or 0.5 mR/hr at one meter.

Results from analysis of Doppler broadened energy spectra are given in table 2.

Table 2. *T*-parameter values of irradiation damaged 316L and 9Cr1Mo.

9Cr1Mo Dose (dpa)	***T*-parameter**	**Error in *T*-parameter**
0	0.52341	0.00336
0.054	0.3889	0.0028
0.086	0.37443	0.0034
1.02	0.30987	0.0012
1.58	0.30844	9.48E-4
2.43	0.30831	8.025E-4
316L Dose (dpa)	***T*-parameter**	**Error in *T*-parameter**
0	0.49683	0.0032
0.054	0.41973	0.00762
0.084	0.40503	0.0095
3.56	0.35322	9.8E-4
10.3	0.33135	0.00145

The dependences of *T*-parameter on dose for each type of material are shown on figures 2 and 3.

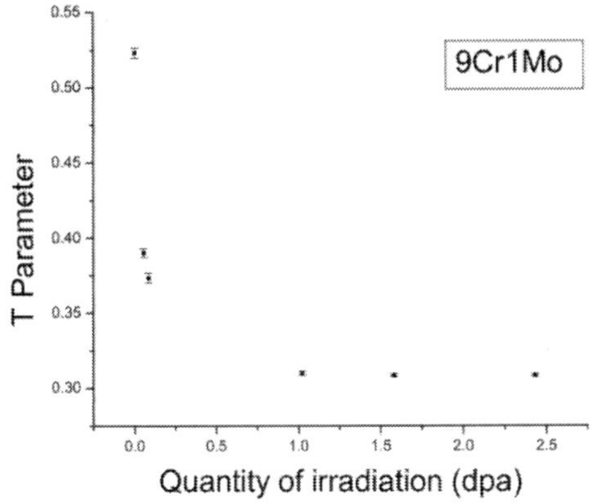

Figure 2. *T*-parameter – dose dependence for 9Cr1Mo.

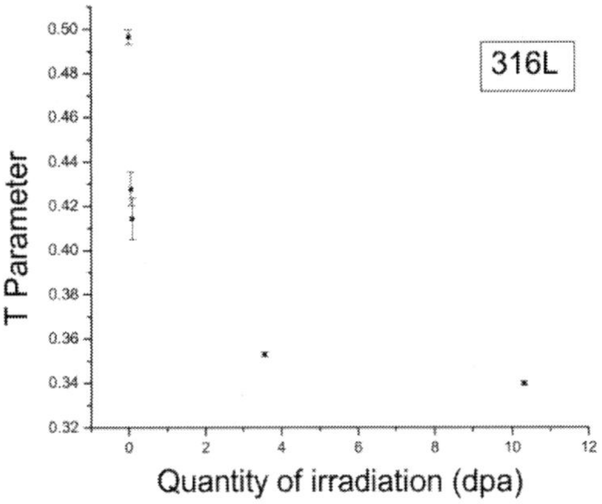

Figure 3. *T*-parameter – dose dependence for 316L.

4. Conclusions

We investigated the change of *T*-parameter, which is calculated as the ratio of the number of counts in the wings of the Doppler-broadened 511 keV peak to the number of counts in the center of the peak. *T*-parameter is related to the density of defects in the sample of interest and generally higher defect densities induce smaller *T*-parameter.

As a result we saw that for 316L, the *T*-parameter drops quickly after low dose and monotonically to higher doses up to 10 dpa. In 9Cr1Mo also drops quickly for low doses but saturates for doses above 1 dpa. In addition, for both types of steel, we see that the biggest changes are for doses less than 0.05 dpa. The reason that the 9Cr1Mo shows a saturation in defect density after low dose compared to what is observed in the 316L is that the 9Cr1Mo has a much finer microstructure which would serve to act as sinks for defects leading to the observed saturation in defect density at a lower dose.

In order to understand the early saturation in 9Cr1Mo *T*-parameter (~ 1dpa) PALS studies will be performed, to compose defect size evolution as a function of dose.

Acknowledgements

This work was supported by DOE under contract number DE – FG04 – 02AL68026 and DE – FC07 – 06ID 14780. The authors gratefully acknowledge the review of this manuscript and helpful suggestions of Dr. Bulent Sencer.

References

[1] Selim F A, Wells D P, Harmon J F, Scates W, Kwofie J, Spaulding R, Duttagupta S P, Jones J L, White T and Roney T 2002 *Nucl. Instr. Meth. B* **192** 197

[2] Sencer B H, Garner F A, Gelles D S, Bond G M and Maloy S A 2002 *J. Nucl. Mater.* **307-311** 266

[3] Sencer B H, Bond G M, Hamilton M L, Garner F A, Maloy S A and Sommer W F 2001 *J. Nucl. Mater.* **296** 112

[4] Wells D P, Hunt A W, Tchelidze L, Kumar J, Smith K, Thompson S, Selim F, Williams J, Harmon J F, Maloy S A et al 2006 *Nucl. Instr. Meth. A* **562** 688

Journal of Physics: Conference Series **265** (2011) 012012 doi:10.1088/1742-6596/265/1/012012

Presentation of the DSIP method for study of recrystallization process in subsurface zone induced by sandblasting

Jerzy Dryzek[1,2]

[1] Institute of Nuclear Physics PAN, ul. Radzikowskiego 152, 31-342 Kraków, Poland
[2] Institute of Physics, Opole University, ul. Oleska 48, 45-052 Opole, Poland

E-mail: Jerzy.Dryzek@ifj.edu.pl

Abstract. In the paper we demonstrate the application of the positron annihilation experimental technique based on the scan of the positron implantation profile for observation of the recrystallization process in the subsurface zone (SZ). The SZ was created in the copper sample whose surface was exposed to a sandblasting treatment and then annealed at different temperatures. Application of $^{68}Ge/^{68}Ga$ positron source allows us to scan the depth of c.a. 150 μm. It was found that recrystallization goes faster close to the surface than in deeper regions. The complete rebuilding of the microstructure takes place at temperature of 600 °C. Some changes in the defect structure begin at temperature of 300 °C.

1. Introduction

The scan of the defect depth profile using positron annihilation technique is usually possible applying the slow positron beam. In that case, only the close to surface regions at the depth up to a few micrometers can be observed. Nevertheless, in some cases the defect profiles are extended even to the depth of a few hundreds micrometers [1, 2]. In order to implant positrons at such a depth one may use a several MeV relativistic positron beam. Unfortunately, no beam is available at present for such studies.

To perform the detection of such depth profiles one can use the conventional positron techniques combining them with the sequenced etching method. After removing the sample material layer-by-layer deeper regions of the sample can be studied, for instance, by conventional positron lifetime spectroscopy. Numerous papers, devoted to the studies, of subsurface regions, or subsurface zone (SZ) created in the friction treatment, proved the usefulness of this method [3]. But this is a destructive technique, which cannot be applied to solution of some problems.

To solve this inconvenience the author proposed to employ the positron implantation profile which is always present when positrons from a radioactive source are injected into condensed matter. Before annihilation the positrons traverse a relatively large distance, which depends on the implant energy and material properties, and this fact can be applied to scan the defect profiles. Fortunately, positrons which have nonthermal energy are not trapped by the open volume defects, thus the implantation profile is limited only by the implant energy, density and to a minor degree by the atomic number of the implanted matter.

In general terms, the implantation profile is described by the near-exponential decay function. Thus, scanning that depth using, for instance, the HpGe detector and measuring the value of the *S*-parameter, defined as the ratio of the central part area to the total area under the annihilation line, one can get information on the defect depth distribution below the surface. The high sensitivity of the *S*-parameter to the defect type and concentration is worth noticing. The HpGe detector must be located behind the lead or tungsten shields with a narrow slit and the sample with the positron source is precisely moved perpendicular to the slit. The scheme of the setup is depicted in figure 1. We proposed to call the presented experimental method the depth scanning of the implantation profile (DSIP), and more details one can find in Ref. [4].

The positron implantation profile should more or less coincide with the defect depth profile. In the previous paper of the author that experimental technique was used for study of pure aluminium samples, whose surface was sandblasted [4]. The ^{22}Na source was applied because in the aluminium the reciprocal value of the linear absorption coefficient for positrons from that source is 93.4(2.3) μm [5]. This allowed us to detect the SZ, which was extended to the depth of c.a. 250 μm. The present studies are performed for copper. The SZ may be extended to a similar depth but the reciprocal value of the linear absorption coefficient for positrons emitted from the ^{22}Na source in copper is 26(2) μm. Thus, the ^{58}Ge/ ^{68}Ga positron emitter is employed. In that case, the end-point of positron energy spectrum is c.a. 1.9 MeV and in copper the reciprocal value of the linear absorption coefficient is equal to 133.5(8.3) μm [5]. We expected that the defect profile could be scanned to the depth of c.a. 150 μm. The aim of the performed studies was also the examination of the thermal stability of the SZ created by sandblasting.

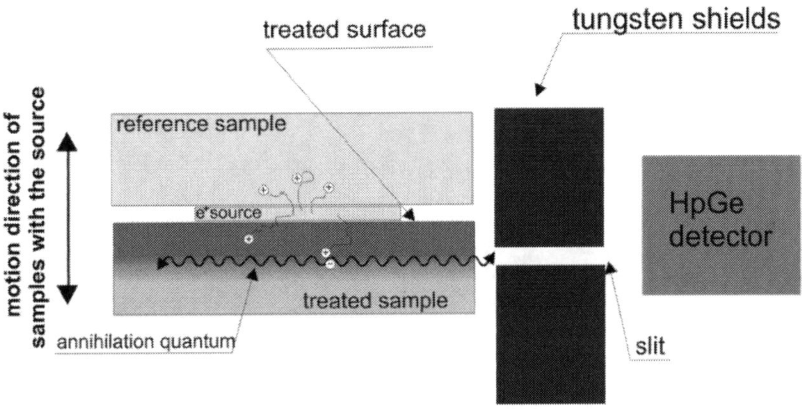

Figure 1. The scheme of the DSIP apparatus used for scanning of the defect depth distribution in the SZ. Positrons emitted from the source are implanted into reference and treated samples, which surrounded the source. The samples and the source form a sandwich, which can move perpendicularly to the slit. The HpGe detector detects only one of two annihilation quanta. From the annihilation line, the *S*-parameter, sensitive to the open volume defects, is extracted as the function of the sandwich position.

2. Sample preparation

The sample in the shape of 0.2 cm thick plate of size 1×1 cm^2 was made from the oxygen-free-high-conductivity (OFHC) copper purchased from Nilaco. Before the sandblasting treatment the sample was annealed at temperature of 900 °C under air pressure in a stream of nitrogen gas during 1 hour and then slowly cooled to room temperature. This allowed obtaining the virgin sample with residual defects. Before the treatment, the surface was etched for cleaning using 1:3 nitric acid and distilled water. In the measured positron lifetime spectrum for such a sample only the single-lifetime component equal to 120(1) ps was found. This value coincides well with the data for pure copper reported by other authors (see in Ref. [6]). This indicated that initial, virgin sample is almost defect free. Then the surface of the sample was sandblasted by silicon carbide particles of diameter less than 0.5mm. The pressure of the compressed air used for blasting was c.a. 6.5 bar.

In the experiment, the sandblasted sample was exposed to successive isochronal annealing and then the measurements of the positron lifetime spectrum and the *S*-parameter measurements in the DSIP apparatus were performed. Temperature step was 100 °C, starting from 200 °C. Samples were annealed for 1 hr at each temperature in the atmosphere of inert gas (N_2) and then they were cooled to the room temperature. All measurements were performed at this temperature.

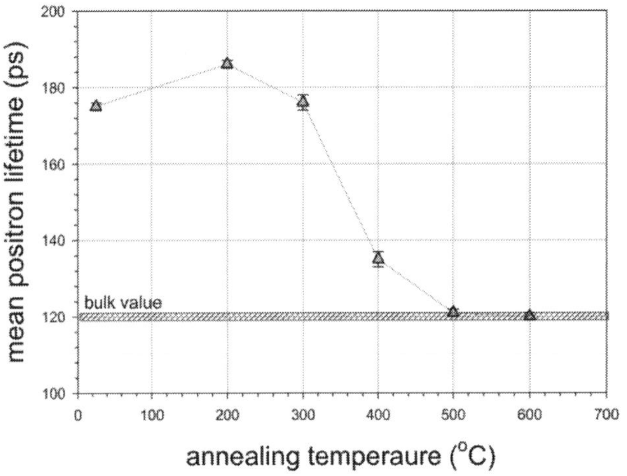

Figure 2. The dependency of the positron mean lifetime on the annealing temperature for sandblasted copper sample.

3. Results and discussion

During the sandblasting treatment the region close to the surface was highly deformed. Then the stored energy is released in processes called recovery and recrystallization initiated by temperature increase. The recovery refers to changes in the properties of a deformed material which occur prior to recrystallization; these changes are such as to partially restore the properties to their values before deformation. During recovery the stored energy of the material is lowered by dislocation movement, i.e. annihilation and rearrangement of dislocations. The recrystallization involves the formation of new strain, and defects free grains in certain parts of the sample and subsequent growth of these to consume the deformed or recovered microstructure.

In the positron lifetime spectra two lifetime components were resolved, the value of the second one was equal to 230(5) ps with intensity c.a. 35%. This indicates that in the sample after sandblasting the microvoids are present which consist of three or four vacancies [6]. The annealing process induces the increase of this lifetime up to 271(10) ps and decrease of the intensity to 28%, which at 400 °C almost disappears. This indicates the onset of the recrystallization process and growth of new grains. In figure 2 we depicted only the mean value of the positron lifetime which is a robust parameter sensitive to the open volume defects versus the annealing temperature. The increase of the mean positron lifetime, observed at 200 °C can be explained by motion of dislocations with jogs which always induces creation of point defects like vacancies and interstitial atoms [7].

The measurement of the positron lifetime using the ^{22}Na source enables us to observe only the subsurface layer of thickness of c.a. 26 μm, which was directly exposed to the treatment. The layers below can exhibit other properties, especially when specific surface processes take place. Then, we skip to the DSIP technique which allows us to observe deeper layers in the treated sample. However, we have to measure the *S*-parameter instead of the mean positron lifetime, but at this stage, only such a measurement mode is available.

In figure 3 a) there is shown the number of counts per second in the positron annihilation peak as the function of slit position related to the sandwich of two samples and the positron source (see figure1) between them. It says how many positrons from the source can reach the particular depth position in the sandwich. This dependency is symmetric because, the source emits positrons homogenously into the full solid angle and it is located between two samples. The left part corresponds to the reference well-annealed virgin copper sample and right part corresponds to the sandblasted sample. Apparent minimum in the middle of figure 3 a) can be explained by nonhomogeneous distribution of the isotope in the source region. In the samples regions the positron implantation profile can be described by the exponentially decreasing function either in left or in the right part of figure 3 a). In both regions corresponding to copper samples the positron implantation profile, not sensitive to the defects and other microstructure features, is identical, as we expected. Nevertheless, careful observation of the positron annihilation peak and extraction of the *S*-parameters reveals new feature. Figure 3 b) shows the value of the *S*-parameter as a function of the slit position.

As we expected the value of the *S*-parameter is constant and low in the part where the virgin well-annealed sample was located. The *S*-parameter value increases up to a certain maximal value in the part corresponding to source region. Positrons annihilate also in the source and one should expect a certain value of this parameter, which depends on the source design and source envelope material. Then, in the part of the sandblasted sample, the *S*-parameter value is again lower than in the source region. Generally, the detected dependency in this region is different than in the region of the virgin sample. It is due to the presence of defects generated during surface treatment. Since the number of positrons, which reach deeper layers exponentially decreases, the DSIP method allows to scan only the region to the depth of c.a. 150 μm from the treated surface. The right part (not hatched area) represents the interesting region of the treated sample. It can be concluded that the *S*-parameter and hence the defect concentration decreases with the depth. We can apply the equation (3) taken from Ref. [4] for description of the *S*-parameter as a function of the sandwich position. (In the treated sample, the exponential decrease of the *S*-parameter is assumed (see Ref. [8]), but in virgin sample and source regions, the constant values of the *S*-parameter are assumed. Certainly, this dependency is convoluted with the spatial resolution of the apparatus.) These assumptions allow describing fairly well the detected dependency. The solid line in figure 3 b) represents the best fit. From this fit for the treated sample, we extracted the parameter d_0 (see (3) in Ref. [4]) which defines how fast the *S*-parameter profile and therefore the defect profile decays when the depth or sandwich position increases. Its value is equal to 157(40) μm and corresponds well with that obtained using another experimental technique based on sequential etching and measurement of the *S*-parameter (see Ref. [3]). Then the sample was annealed and the profile of the *S*-parameter was measured again.

It can be noticed that annealing at the temperature of 200° C does not change the *S*-parameter profile in comparison with the initial profile except the region near of the surface. However, for annealing at 300 °C the change in the gradient of the *S*-parameter is noticeable. More significant change is observed after annealing at 400 °C. That observations correlate with the results presented in figure 2. Nevertheless, it can be seen that the value of the *S*-parameter drops on the surface, then increases and reaches the maximum value at the depth of c.a. 60 μm from the treated surface and then decreases. Such a behavior indicates that the defects are annealed in the recrystallization process first close to the treated surface and than in deeper regions.

This is well understood if we consider that plastic deformation energy originates at the surface and then is distributed in the SZ zone. It is well known that the recrystallization process which is responsible for the reduction of the *S*-parameter is driven by that energy. We suppose that at the depth of c.a. 60 μm the energy stored in the plastic deformed area is relatively low and recrystallization process is not so efficient as in the region close to the surface.

The increase of the temperature to 600° C almost removes all defects produced previously in the sandblasting treatment, and the *S*-parameter profile becomes constant as it is in the left part corresponding to the reference sample. The increase of the *S*-parameter below 150 μm can be explained by two reasons. First, still the defects are not completely removed at this depth. Second the number of positrons which reach this depth is low and mainly background contributes to that value.

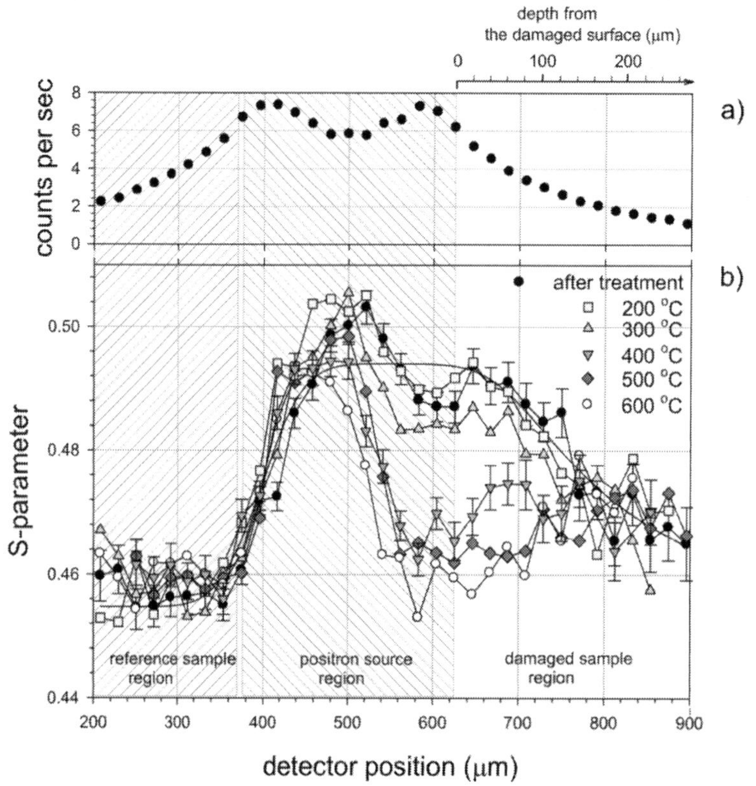

Figure 3. The positron implantation profile (a) and the profile of the *S*-parameter across the sandwich, which consists of the reference copper sample, the $^{68}Ge/^{68}Ga$ positron source and the sandblasted copper sample. The hatched area indicates the two first mentioned above parts of the sandwich. The positron implantation profile is shown (a) and the *S*-parameter profiles are presented for different annealing temperatures of the sandblasted sample (b). The solid line represents the best fit of the equation (3) in Ref. [4]. The axis on the top represents the depth from the treated surface of the sandblasted sample. (The error bars are marked only for the points of the dependency obtained for the sample after treatment and after annealing at 400 °C. In other cases the error bars are identical.)

4. Conclusions

Finally the following conclusions can be drawn:

The non-destructive technique, called DSIP applied to the studies of in the SZ in the sandblasted OFHC copper samples allows to observe microstructural changes induced by annealing at the depth of c.a. 150 μm from the treated surface. This was possible due to the application of the $^{68}Ge/^{68}Ga$ positron source.

The gradient of the *S*-parameter in the SZ caused by the gradient of defect concentration induced by sandblasting decreases after annealing at temperatures 300 °C and almost disappears at temperature higher than 400 °C. Annealing at 600 °C causes removing of all defects and recrystallization of the SZ.

The recrystallization process is faster close to the surface than in the deeper regions of the SZ. After annealing at 400 °C the *S*-parameter depth dependence exhibits different character, i.e. the *S*-parameter has a maximum value at the depth of c.a. 60 μm.

It seems that the SZ exhibits strong stability because only annealing at temperature higher than 300 °C induces some changes in the defect structure.

References

[1] Dautzenberg J H, Zaat J H 1973 *Wear* **23** 9
[2] Rigney D A and Hirth J P 1979 *Wear* **53** 345
[3] Dryzek J and Dryzek E 2003 *Trib. Lett.* **15** 309
[4] Dryzek J 2005 *Appl. Phys. A* **81** 1099
[5] Dryzek J and Singleton D 2006 *Nucl. Inst. Meth. B* **252** 197
[6] Čížek J, Procházka I, Cieslar M, Kužel R, Kuriplach J, Chmelík F, Stulíková I, Bečvář F and Melikhova O 2002 *Phys. Rev. B* **65** 094106
[7] Hull D and Bacon D J 2001 *Introduction to Dislocations* (Oxford: Butterworth-Heinemann)
[8] Dryzek J and Dryzek E 2003 *Tribol. Lett.* **15** 309

Free volume change in $Zr_{50}Cu_{40}Al_{10}$ glassy alloy by the annealing studied by positron annihilation spectroscopy

A Ishii[1], F Hori[1], Y Fukumoto[1], A Iwase[1], Y Yokoyama[2] and T J Konno[2]

[1] Department of Materials Science, Osaka Prefecture University, 1-1, Gakuen-cho, Nakaku, Sakai, Osaka 599-8531, Japan
[2] Institute for Materials Research, Tohoku University, 2-1-1, Katahira, Aoba-ku, Sendai 980-8577, Japan

E-mail: bm0801@mtr.osakafu-u.ac.jp

Abstract. It is well known that structural relaxation in a bulk metallic glass affects some properties such as viscosity, electrical conductivity, and ductility. The free volume in glassy alloy, which is open volume retained by rapidly solidification from liquid state, has a significant roll for those properties. In order to discuss the nature of free volume in a $Zr_{50}Cu_{40}Al_{10}$ bulk metallic glass, positron lifetime measurements have been performed for this metallic glass before and after annealing, and all positron lifetime spectra have been decomposed into multi components as a size distribution function by use of CONTIN-PALS II program. The positron lifetime distribution for this metallic glass has a broad spectrum comparing to crystal metal including a single vacancy, and its average lifetime corresponds to the free volume size. This width of positron lifetime (free volume size) distribution decreases by the annealing at 673 K. This change of the positron lifetime distribution can be attributed to a free volume relaxation.

1. Introduction

Bulk metallic glasses (BMGs) have advanced properties, including high specific strength, good corrosion resistance and large elastic strain limits [1]. The metastable and disordered state of metallic glass raises some questions about thermal stability. Frozen excess open volume in the metallic glass has been perceived as an important factor to describe some properties for metallic glasses under the thermal activation process in some temperature ranges: below glass transition temperature (T_g), above T_g and above crystallization temperature (T_x) according to relaxation theory in metallic glasses [2]. However, not all studies to date have revealed the nature of the free volume directly. So far, we have investigated the free volume relaxation process in $Zr_{50}Cu_{40}Al_{10}$ ternary system of bulk metallic glass below T_g using positron annihilation techniques, which can detect and estimate the size of free volume [3]. In that study, we have reported that the free volume relaxation process depending on temperature directly reflects to changes in density. Recently, C. Nagel *et al.* have demonstrated through positron annihilation measurements that the free volume relaxation process far below T_g in the Johnson glass ($Zr_{46.7}Ti_{8.3}Cu_{7.5}Ni_{10}Be_{27.5}$) does not need high activation energy to promote the evolution to more stable amorphous state [4].

In the free volume theory, free volume in amorphous materials is assumed as the hole derived from density fluctuation of element [5]. On the other hand, recently, some groups suggest that quasicrystal like local clusters exist in some metallic glasses; icosahedral type of clusters exist in Zr-based metallic glasses [6], and trigonal prism-like structural clusters exist in Fe-based metallic glasses [7]. These studies imply that it is difficult for the simple "hole type" free volume model to describe the nature of free volume. Furthermore, our results demonstrate that the long range diffusion does not occur during relaxation, suggesting the possibility of internal cluster relaxation [3]. Accordingly, development of the free volume model is becoming important.

In this work, we discuss the nature of free volume and the size distribution of free volume from the view point of positron annihilation and estimate change in the distribution after relaxation below T_g.

2. Experimental

A rod-shaped sample of $Zr_{50}Cu_{40}Al_{10}$ bulk metallic glasses (ϕ 8 mm × 60 mm) was produced by the tilt casting method in an arc furnace [8,9]. The glass transition temperature of this sample was 675 K, which was determined by differential scanning calorimetry (DSC) measurement. For positron annihilation measurements, this sample was cut into specimens of about 0.5 mm thickness. These specimens were annealed isothermally at 673 K up to 36000 s. The as-prepared and annealed specimens were characterized by X-ray diffraction (XRD) measurements.

Positron annihilation lifetime spectra were obtained by using a conventional fast-fast circuit with a time resolution of about 200 ps (FWHM) at room temperature. In order to derive fully accurate spectra, a digital stabilizer of SEIKO EG&G was used. As a positron source, we used $^{22}NaCl$ with an activity of 286 kBq, which was sandwiched by thin Kapton foils. The positron annihilation lifetime spectra consist of more than 3.0×10^6 counts. All the positron annihilation lifetime spectra were analyzed by the POSITRONFIT program [10] and the CONTIN-PALS II program [11,12] which execute the continuous lifetime distribuiton analysis. The CONTIN program provides the annihilation probability density function $f(\lambda)$. The continus probability density function $f(\lambda)$ is gained by deconvolution of the experimental lifetime spectra using a Laplace inversion technique [11,12]. Positron annihilation coincidence Doppler broadening (CDB) measurements were carried at room temperature. Each CDB spectrum consists of more than 10^9 counts.

3. Results and Discussion

XRD profiles of as-prepared and isothermally annealed $Zr_{50}Cu_{40}Al_{10}$ bulk metallic glass sample at 673 K for 10 h are shown in figure 1. This result reveals that the amorphous state of this sample was retained even after long time annealing just below T_g.

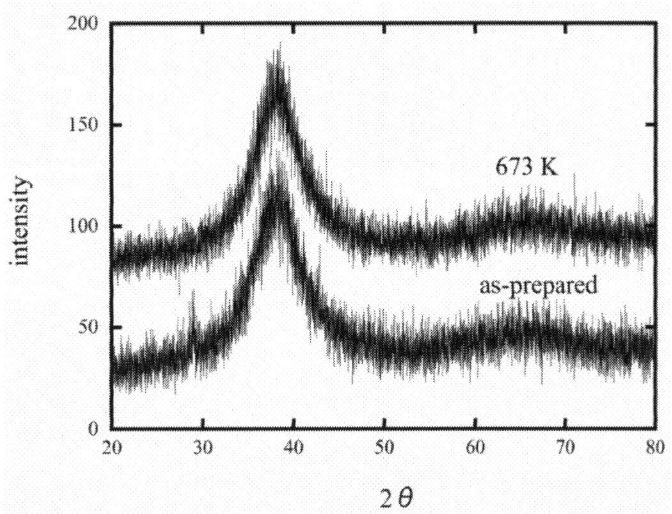

Figure 1. XRD profiles of as-prepared and 10 hours annealed at 673 K for $Zr_{50}Cu_{40}Al_{10}$ bulk metallic glass.

Figure 2 shows CDB spectrum of relaxed $Zr_{50}Cu_{40}Al_{10}$ bulk metallic glass expressed in the form of ratio to that of Al metal. For comparison, we also show the CDB ratio curves of Cu and Zr to Al. It can be recognized from this figure that the electron momentum distribution profile after relaxation is essentially the same as that of as-prepared specimen, although it was observed that low electron momentum region (0-0.006 m_0c) decreases with increases of inner part of momentum distribution (about 0.015 m_0c) only reflecting the change in free volume size. This result implies that long range diffusion or drastic rearrangement do not occur by long time annealing below T_g, while the free volume decreases and the bulk density clearly increases [13].

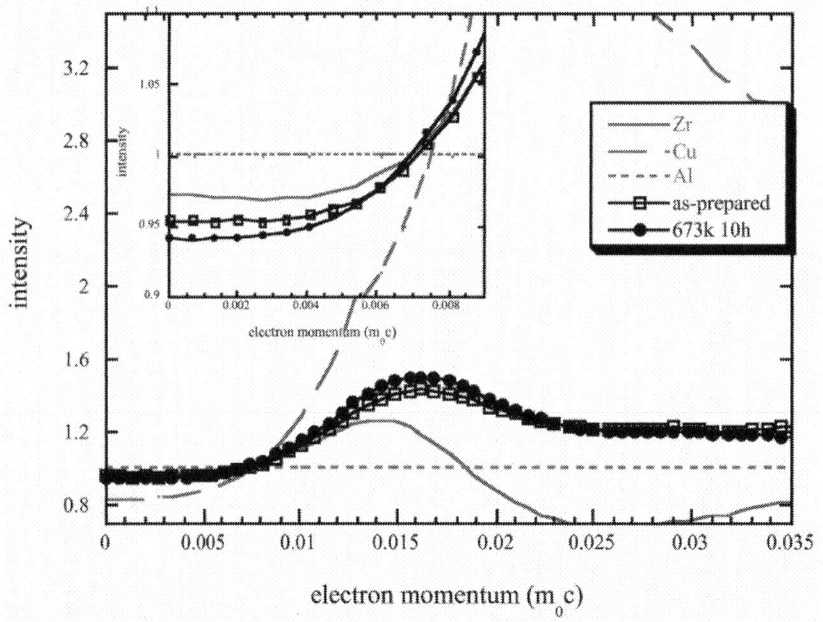

Figure 2. CDB spectra of as-prepared and annealed $Zr_{50}Cu_{40}Al_{10}$ bulk metallic glasses with those of Cu and Zr metals, expressed as a ratio of CDB intensity of Al metal. Enlarged figure in the range of low electron momentum is also shown.

On the other hand, figure 3 shows the change of the positron annihilation probability density distribution $f(t)$ between as-prepared and annealed samples of $Zr_{50}Cu_{40}Al_{10}$ bulk metallic glasses. This result indicates the size of free volume in this glassy alloy has a broad distribution. The mean positron lifetime of the as-prepared sample analyzed by POSITRONFIT program is about 163 ps. This value agrees with that of our previous work [3].

Table 1. Positron lifetimes in ZrCuAl bulk metallic glass and other experimental value of pure metals.

	$Zr_{50}Cu_{40}Al_{10}$ bulk metallic glass			Zr	Cu	Al
	as-prepared	673 K	473 K			
τ (ps)	163 [3]	147 [3]	155 [3]	165 [14]	110 [14]	163 [14]

According to the free volume model suggested by Cohen and Turnbull [5], a free volume in amorphous materials is presumed as the "hole" derived from density fluctuation of element. Assuming the existance of two or more annihilation states in amorphous, one is defect type open volume and other is defect free matrix, it is thought that two or more peaks in the positron distribution function should appear. In this case, however, only one peak was observed. This means that the large amount of free volume exists, or disorderd and broadend atomic distances have distribution in this sample.

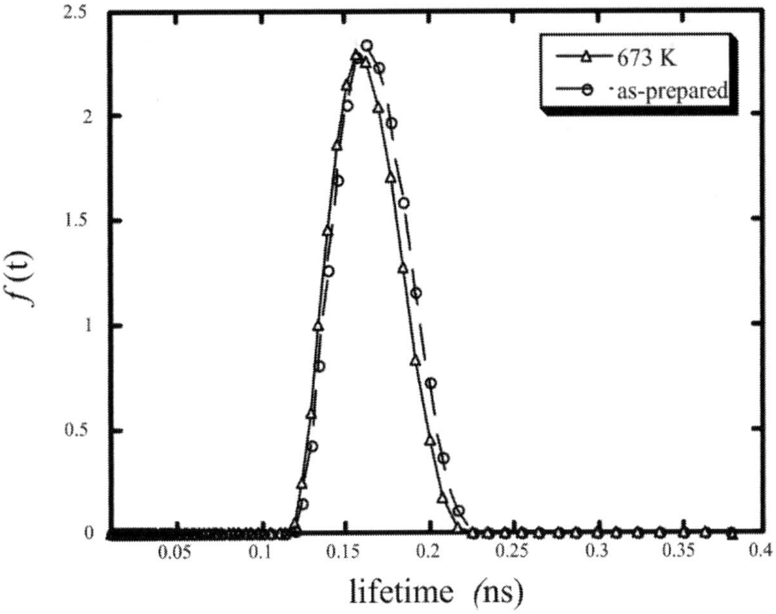

Figure 3. The probabilities of annihilation site *f*(t) are plotted as a function of positron lifetime, *t*, for as-prepared and isothermally annealed $Zr_{50}Cu_{40}Al_{10}$ bulk metallic glass.

Secondly, structural relaxation by the annealing below T_g leads the positron lifetime distribution shift to lower value. The decrease of the peak value, which is about 12 ps, was derived by the analysis of the positron lifetime distribution function. This lifetime change approximately agrees with the mean lifetime change derived by two state trapping analysis (POSITRONFIT). Moreover, the width of distribution function for relaxed sample becomes narrow, mainly on the large size part of free volumes, comparing to that for the as-quenched sample. These results suggest that annealing out of free volume of larger size will takes place preferentially.

4. Conclusions

Distribution of free volume in the $Zr_{50}Cu_{40}Al_{10}$ bulk metallic glass during relaxation below T_g has been studied by positron annihilation lifetime analyzed by positron continue lifetime function and coincidence Doppler broadening techniques, and the following conclusions have been obtained.

It can be recognized from the result of CDB spectra that the electron momentum distribution profile after relaxation is essentially the same as that of as-prepared specimen. This result implies that the long range diffusion or the drastic rearrangement around free volume do not occur during relaxation.

The size of free volume quenched in bulk metallic glass has a broad distribution with only one main peak.

It was found that the width of positron lifetime distribution function, of which reflects the free volume size distribution for relaxed sample, is narrower than that for as-quenched one. This relaxation of free volume takes place mainly in the part of large size free volume.

Acknowledgments

This work was supported by a Grant-in-Aid for Scientific Research by Ministry of Education, Culture, Sports, Science and Technology of Japan (Grant number 15074218). Authors would like to express cordial thanks to Prof. Haruyama for useful discussions on the free volume relaxation.

References

[1] Ashby M F and Greer A L 2006 *Scr. Mater.* **54** 321
[2] Van den Beukel A and Sietsma J 1990 *Acta Metall. Mater.* **38** 383
[3] Ishii A, Hori F, Iwase A, Fukumoto Y, Yokoyama Y and Konno T J 2008 *Mater. Trans.* **49** (in press)
[4] Nagel C, Rätzke K, Schmidtke E, Wolff J, Geyer U and Faupel F 1998 *Phys. Rev. B* **57** 10224-27
[5] Cohen M H and Turnbull D 1959 *J. Chem. Phys.* **31** 1164
[6] Saida J, Matsushita M and Inoue A 2002 *Intermetallics* **10** 1089-98
[7] Inoue A 2000 *Acta. Mater.* **48** 279
[8] Yokoyama Y, Inoue K and Fukaura K 2002 *Mater. Trans.* **43** 2316-19
[9] Yokoyama Y, Fukaura K and Inoue A 2002 *Intermetallics* **10** 1113-24
[10] Kirkegaard P, Eldrup M, Mogensen O E and Pedersen N J 1981 *Comp. Phys. Commun.* **23** 307–35
[11] Provencher W S 1982 *Comput. Phys. Commun.* **27** 229
[12] Gregory B R and Zhu Y 1990 *Nucl. Instrum. Methods A* **290** 172
[13] Yano T, Yorikado Y, Akeno Y, Hori F, Yokoyama Y, Iwase A, Inoue A and Konno T J 2005 *Mater. Trans.* **46** 2886-92
[14] Seeger A, Banhart F and Bauer W 1991 *Mater. Sci. Forum* **105-110** 275–7

Ion implantation induced defects in Fe-Cr alloys studied by conventional positron annihilation lifetime spectroscopy

V Kršjak[1], S Sojak[2], V Slugeň[2] and M Petriska[2]

[1] Joint Research Centre, Institute for Energy, European Commission, P.O. Box 2, 1755 ZG Petten, The Netherlands

[2] Department of Nuclear Physics and Technology, FEI, Slovak University of Technology, Ilkovičova 3, 812 19 Bratislava, Slovak Republic

E-mail: vladimir.krsjak@ec.europa.eu

Abstract. The influence of chromium on the radiation damage resistance of the iron based alloys has been studied using conventional positron annihilation lifetime spectroscopy (PALS). Experimental data evaluation has been supported by the former theoretical calculation of positron lifetimes in the studied materials and well-defined types of defects. For this purpose, density functional theory (DFT) computation method has been applied. The spectrum of used ^{22}Na positron source was decomposed into discrete fractions to better calculate efficiency of near surface layers study. For the experimental simulation of α-radiation and obtaining of defined cascade collisions in the materials, helium implantation was used. Different level of the implanted dose (6.24×10^{17} – 3.12×10^{18} cm^{-2}) corresponds to local damage up to 90 DPA acquired in thin <1 μm region. Experimental measurement has been performed using the PALS technique on the four different Fe-Cr binary alloys (2.36; 4.62; 8.39; 11.62 wt% of Cr). The results showed that chromium has a significant effect on the size and density of the implanted defects and specific Cr content should prevent the vacancy clusters formation.

1. Introduction

Conventional positron lifetime technique based on ^{22}Na positron source is usually applied for study of the bulk of the specimen. Regular surface (or near surface) study requires slow positron beam with adjustable energy to observe specific depth of the specimen. Nevertheless, considerable fraction of positrons is absorbed in the near surface region, where the implantation profile decreases very fast [1]. This paper discusses our recent positron annihilation lifetime experiments focused on the radiation treated Fe-Cr alloys. To simulate high neutron fluencies in these materials, the helium implantation has been used. The interpretation of the measured data has been based on previous theoretical calculation of Troev et al. on pure iron and chromium [2, 3].

Fe-Cr binary alloys have been chosen for our study of chromium influence on the radiation damage resistance. This alloying element is an important component of the reactor structural materials since the 1970's. However only high chromium ferritic/martensitic steels application in the nuclear industry (fast reactors) showed the importance of this element on the material properties after radiation treatment [4]. In particular the decrease of the void swelling due to chromium addition has been often discussed in the last decade [5, 6]. Although this role of chromium in swelling resistance is already quite known, the complex impact on the radiation damage resistance is not fully understood yet.

2. Experimental

2.1. Materials and sample preparation

To study the influence of chromium concentration, four Fe-Cr binary alloys with different Cr content have been selected. The detailed chemical composition of the alloys can be seen in table 1. All the alloys were heat-treated to ensure well defined martensitisation. More detailed information about these materials and the fabrication processes can be found in [7].

Table 1. Chemical composition of the studied materials.

Alloy ID	Cr*	O*	N*	C*	Mn	P	Si	Al	Ti	Ni	Cu	V
L251	**2.36**	0.035	0.012	0.008	0.009	0.013	0.002	0.003	0.004	0.044	0.005	0.001
L259	**4.62**	0.066	0.013	0.02	0.02	0.011	0.006	0.003	0.003	0.06	0.01	0.001
L252	**8.39**	0.067	0.015	0.021	0.03	0.012	7E-04	0.007	0.003	0.07	0.01	0.002
L253	**11.62**	0.031	0.024	0.028	0.03	0.05	0.006	0.003	0.004	0.09	0.01	0.002

* measured after heat treatment

The as-received materials have been cut to the desired dimensions, grinded and carefully polished to mirror like surfaces (polishing paste – grade 1 μm) before exposure to helium implantation. Subsequently, an untreated reference sample and the implanted samples were investigated with PALS techniques.

2.2. Radiation treatment

To obtain cascade collisions in the microstructure of the studied materials without neutron activation, accelerated helium ions have been used. Implantations of 250 keV helium ions at five different fluence levels have been performed at the linear accelerator of the Slovak University of Technology in Bratislava [8]. The depth profile of collision events can be seen in figure 1. The DPA (displacement per atom) parameter has been calculated for the volume of 99% positron absorption (see table 2).

Table 2. Calculations of average DPA for different level of implantation in first 100μm layer (DPA_{PALS}) of studied Fe-Cr alloys.

Dose [ions/cm^2]	6.24×10^{17}	1.25×10^{18}	1.87×10^{18}	2.5×10^{18}	3.12×10^{18}
[C/cm^2]	0.1	0.2	0.3	0.4	0.5
DPA (99% positron absorption) [a.u.]	0.15	0.30	0.45	0.60	0.74
DPA (Bragg peak) [a.u.]	18.55	37.10	55.64	74.19	92.74

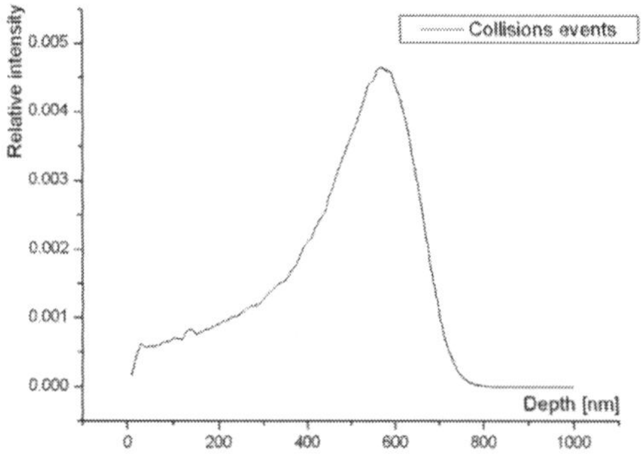

Figure 1. Depth profile of the helium implantation, E = 250 keV (SRIM simulation of 10^6 ions).

2.3. Application of conventional PALS technique

Positron lifetime measurements have been performed on the fast-fast coincidence setup of experimental equipment. Typical measurements took six hours and the time resolution of the apparatus was determined to be less than 210ps. According the long term measurements of iron based alloys, uncertainty of the results τ_2, I_2 and MLT were estimated to be ±5ps, ± 10% and ±3ps. The experimental data have been evaluated with the LT ver. 9.0 program [9] using the three component decomposition. This spectra fitting was sufficient (fit's variance < 1.1 in almost all cases) and moreover the fourth component cannot be found in any spectra. However this approach means that only one component describes the lattice defects and other two components correspond to material bulk and the long component with small intensity (< 1.5%) is typical for the conventional PALS source-sample setup. Therefore the τ_2 component may differ from known positron lifetimes of particular lattice defects and combination of several types of defects must be considered by this data interpretation.

3. Results

The PALS results show, that positron lifetime in the defects which correspond to the size of the vacancy type defects (vacancy clusters) is increasing with the implanted dose in all specimens (figure 2a). For the low Cr alloys (L251, L259) there is a clearly defined maximum for the fluency about 1.87×10^{18} ions. The highest positron lifetime (the largest clusters of vacancies) was measured for 2.56% of Cr alloy (L251) and its value 235ps should correspond to a cluster of 4 – 5 empty vacancies or slightly larger clusters containing helium. However, the absolute value of this parameter may be underestimated due to fact that here are included initial defects, typical for tempered ferritic/martensitic steels [10] and only certain number of positrons annihilated in the ion implantation damaged layer. Higher Cr content alloys (8.39 and 11.62% of Cr) are more resistant to creating of such a large vacancy clusters which can be seen via lower positron lifetime τ_2. From this aspect, Fe8.39%Cr specimen has the most optimal chemical composition for the radiation damage resistance and correspondent also with the choice of currently most perspective materials for advanced nuclear power systems (e.g. EUROFER97 – 8.96% of Cr). The intensity of the annihilation in defects (figure 2b) shows characteristic behavior for the Fe8.39%Cr as well. The significant intensity peak for this alloy can be interpreted as a higher density of the uniformly distributed small defects. This state is more acceptable from the structural point of view than the clustering of these defects into large voids, which can be seen from figure 2a for low-Cr alloys and which may lead to radiation embrittlement.

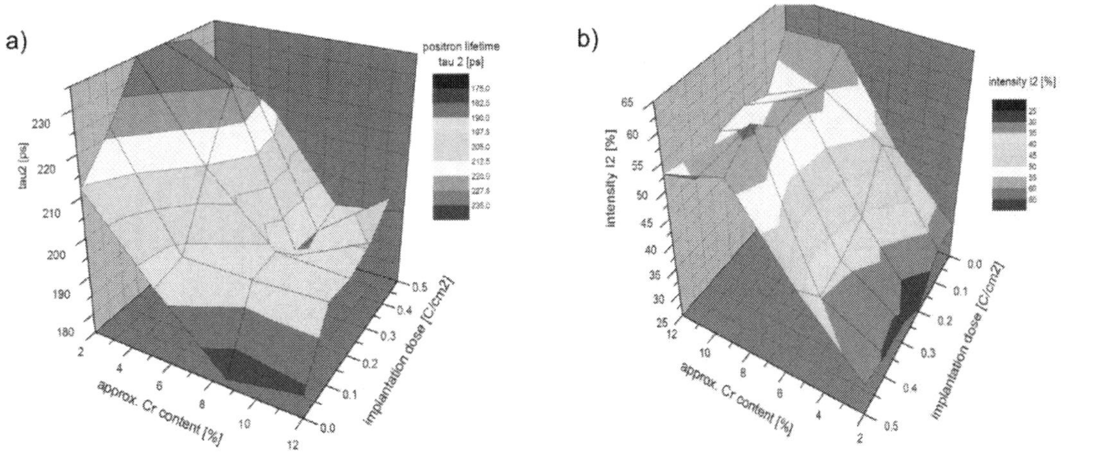

Figure 2. Positron annihilation in defects. Positron lifetime τ_2 (a) and intensity I_2 of annihilation (b).

The MLT parameter is increasing with the implantation dose in all materials (figure 3a). It confirms the fact that conventional PALS technique has been successfully applied on the near surface region study. Different slopes of the MLT curves indicate unequal behavior of the materials under radiation treatment. The plane projection of the MLT dependency (figure 3b) shows that Fe-Cr alloys with Cr content 5 – 7 wt% seems to be most resistant to the ion implantation induced defects.

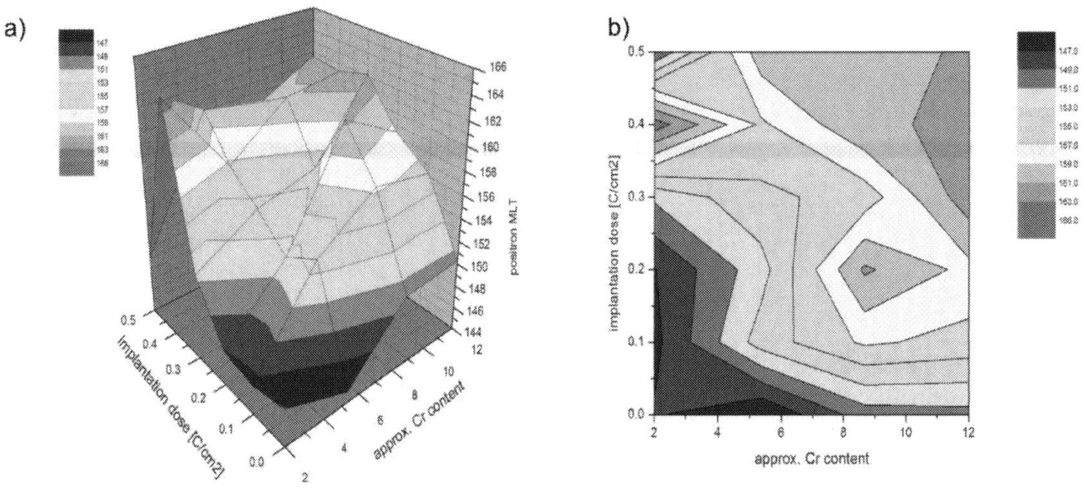

Figure 3. Positron mean lifetime (MLT) in the He implanted Fe-Cr alloys. 3D view (a) and its planar projection (b).

4. Conclusions

Our study showed that conventional positron lifetime technique can be used also with advantage for study of near surface layers damaged by light ion implantation.

Performed measurements showed that chromium plays an important role in the formation of the microstructure under radiation treatment. In particular, chromium content in Fe-Cr alloys has an influence on the vacancy cluster forming. In the low chromium alloys the small vacancy clusters have been observed. However, such clusters of the vacancies have not been observed in high chromium materials as long as presented DPA values are considered. From the studied materials, the Cr concentration on the level about 9 wt% was evaluated as a most suitable having in mind the resistance to the radiation induced defects agglomerations. Mono- and/or di-vacancies are more acceptable than their agglomerations from the radiation embrittlement resistance point of view.

Acknowledgement
This project was supported by VEGA 1/3188/06 and 7RP-Euratom/CU.

References

[1] Dryzek J and Sieracki J 2007 *Nucl. Instr. Meth. B* **258** 493
[2] Troev T, Markovski A, Peneva S and Yoshiie T 2006 *J. Nucl. Mater.* **359** 93
[3] Petrov L, Nankov N, Poppov E and Troev T 2008 *AIP Conf. Proc.* **996** 177
[4] Klueh R L 2004 *Elevated-temperature Ferritic and Martensitic Steels and their Application to Future Nuclear Reactors*, ORNL/TM-2004/176 report
[5] Porollo S I, Dvoriashin A M, Vorobyev A N and Konobeev Yu V 1998 *J. Nucl. Mater.* **256** 247
[6] Konobeev Yu V, Dvoriashin A M, Porollo S I and Garner F A 2006 *J. Nucl. Mater.* **355** 124
[7] Matijasevic M and Almazouzi A 2005 *Modellisation of Irradiation Effects: Modeling Oriented Experiments on Fe-Cr Model Alloys*, SCK•CEN-R- 4196 report
[8] Kovác P, Pavlovic M and Dobrovodský J 1994 *Nucl. Instr. Meth. B* **85** 749
[9] Kansy J 1996 *Nucl. Instr. Meth. A* **374** 235
[10] Matijasevic M and Almazouzi A 2008 *J. Nucl. Mater.* **377** 147

Study of Co-containing Ni-Mn-Ga by positron annihilation

D Mérida[1], J A García[2], F Plazaola[1], V Sanchez-Alarcos[3], J I Pérez-Landazábal[3] and V Recarte[3]

[1] Elektrizitate eta Elektronika Saila, Euskal Herriko Unibertsitatea, p.k. 644, 48080 Bilbao, Spain
[2] Fisika Aplikatua II Saila, Euskal Herriko Unibertsitatea, p.k. 644, 48080 Bilbao, Spain
[3] Departamento de Física, Universidad Pública de Navarra, Campus de Arrosadia, 31006 Pamplona, Spain

E-mail: david.merida@ehu.es

Abstract. Positron annihilation spectroscopy measurements have been carried out in the ferromagnetic shape memory alloy Ni-Mn-Ga-Co. Positron experiments have been performed at room temperature after subsequent isochronal annealing up to 600 °C. Positron lifetime results have been compared with Differential Scanning Analysis experiments performed in the same samples. Experiments show a large variation of the average positron lifetime values with the annealing temperature of the samples, indicating a clear relation between the concentration of vacancies and the properties of the martensitic transformations of these alloys.

1. Introduction

Since the giant magnetic-field-induced strain (MFIS) first reported on Ni-Mn-Ga ferromagnetic shape memory alloys (FSMA) by Ullakko et al. [1], a great amount of work has been performed to understand and improve the system in order to implement it for practical devices. The main drawbacks are the high brittleness and the low values of the martensitic transformation (MT) and Curie temperatures of Ni-Mn-Ga alloys. These limitations have stimulated the research of new ferromagnetic shape memory alloys having better mechanical properties. Indeed, several studies have been performed to improve the mechanical and thermal properties through the addition of a fourth element to the Ni-Mn-Ga alloy. Substituting Mn or Ga by Co martensitic transformation temperatures are higher than those found in Ni-Mn-Ga alloys [2]. On the other hand, Curie temperature increases substituting Ni by Co [3]. Therefore this quaternary alloy is a good candidate for high temperature applications under selected addition of Co. However, the effect of Co addition on the atomic order and on the vacancy density must be taken into account in order to better control transformation temperatures.

Different techniques have been used to study the properties of Ni-Mn-Ga-Co, but very little work has been performed to study the role of vacancies in the martensitic transformation. Positron annihilation spectroscopy is a very powerful technique to study vacancy-type defects in metals. We have used positron annihilation spectroscopy measurements to study the role of vacancy-type defects in the martensitic transformations of a Ni-Mn-Ga-Co polycrystalline alloy. The positron lifetime experiments have been performed at room temperature after subsequent isochronal annealing up to

600 °C. The positron results have been compared with calorimetric experiments performed in the same sample.

2. Experimental

Polycrystalline ingot of $(Ni_{49.5}\ Mn_{28.5}\ Ga_{22})_{96}\ Co_4$ (numbers indicate at. %) were prepared from high purity elements by arc melting under protective Ar atmosphere. The ingot was homogenized in vacuum quartz ampoules at 1000 °C during 24 hours. Small samples for calorimetric measurements were obtained from discs previously cut from the center of the ingot by slow speed diamond saw. These discs were used for positron lifetime measurements. Subsequent annealing treatment of 30 minutes at 900 °C followed by quenching into ice water was performed on the alloy in a vertical induction furnace. The composition of the sample was analyzed before and after thermal treatment by energy-dispersive X-ray spectrometry (EDX) in a JSM-5610LV scanning electron microscope and it was confirmed that no compositional change took place due to thermal treatments. Differential scanning calorimetry (DSC) measurements were carried out in a TA Q100 DSC instrument to study the thermal behavior of the alloy. The sample was polished after quenching in order to ensure a good thermal contact with the equipment. The MT temperatures of the as-quenched sample were obtained from a direct measurement up to 400 °C at a heating/cooling rate of 10 K/min. In order to observe the evolution of the MT temperatures with the heat treatment temperature, thermal cycles through the MT after heating up to different temperatures were performed at the same heating/cooling rate. All DSC measurements were performed under nitrogen atmosphere.

For the positron lifetime measurements a fast system with a resolution of 235 ps was used and a conventional ^{22}Na source on a Kapton foil was employed as a positron source. All lifetime spectra were analyzed after subtracting a constant source contribution: a long source of 1930 ps with an intensity of 1 % and a shorter one of 400 ps with an intensity of 16 %. For the isochronal annealing in positron annihilation measurements we used the same heating/cooling rate of 10 K/min. The points in each positron lifetime curve correspond to the average obtained after 6-10 measurements. The error bars correspond to the maximum deviation of the fitted positron parameters in those measurements.

3. Results and discussion

3.1. Calorimeter measurements

Figure 1 shows the results of the DSC measurements carried out up to 400 °C on samples quenched from 900 °C. An endothermic peak at 124 °C can be observed corresponding to the reverse MT (martensite-austenite). It is also worth noting the presence of an exothermic peak far above the MT. This peak could be related to processes affecting the MT [4]. The exothermic peak is observed only on the first heating up to 400 °C, and it does not appear on subsequent cooling and posterior cycling, so the peak must be linked to an irreversible process. The presence of small endothermic peaks at temperatures above 200 °C indicates that the reverse MT has not been fulfilled at this temperature. This stabilization of the martensite points out to defect pinning of the martensite interfaces, which could be a consequence of a high concentration of quenched-in defects in martensite [5].

In order to study the origin of the processes linked to the exothermic peak, as well as the effect they have on MT temperatures, consecutive thermal cycles from 150 °C to 400 °C inside the exothermic peak range were performed. These intermediate treatments only allow the partial development of the irreversible process.

Since martensitic transformations are first order phase transitions, the temperatures of their beginning and end are their characteristic parameters. The transformation of austenite into martensite (direct transformation) is characterized by the temperature Ms of the appearance of nuclei of martensitic phase in the austenitic matrix and the temperature Mp at which the transformation of the martensite ends. In figure 2, the evolution of the temperature of the MT is shown as a function of the treatment temperature. This evolution is presented in function of the parameter $\Delta Mp = Mp - Mp_0$, where Mp_0 is the temperature of the MT (position of the endothermic peak) for the sample just quenched from 900 °C and Mp correspond to the MT temperature (position of the endothermic peak) for different annealing temperature. The exothermic peak in the DSC is also showed in figure 2. It can be seen that the MT of the sample first decreases up to 275 °C and from this point increases. The maximum of the exothermic peak corresponds to the lowest MT temperature at around.

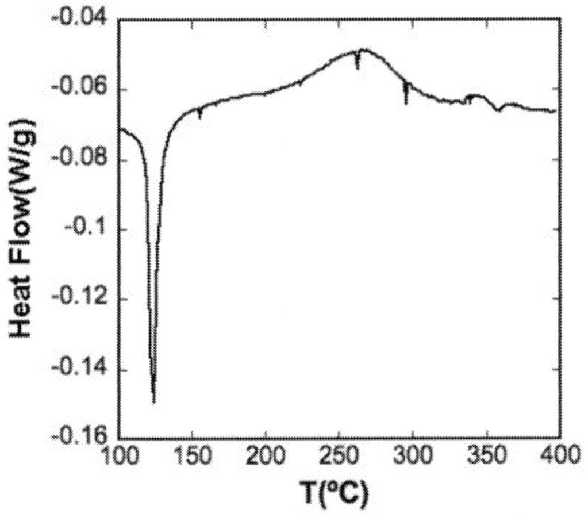

Figure 1. DSC measurements of $(Ni_{49.5}\,Mn_{28.5}\,Ga_{22})_{96}\,Co_4$.

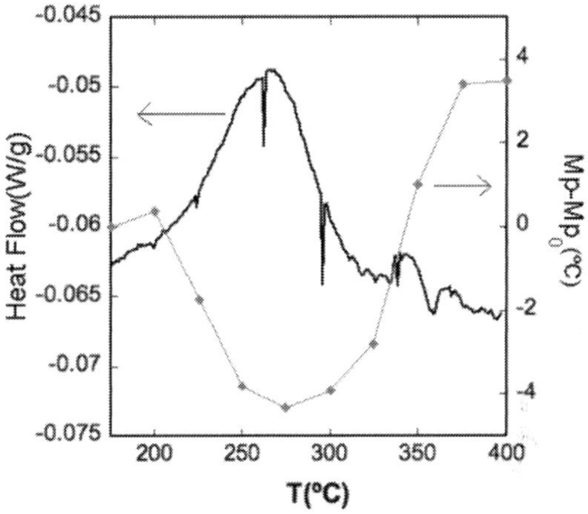

Figure 2. Exothermic peak of the DSC and martensitic transformation change, $\Delta Mp = Mp - Mp_0$ versus isochronal annealing temperature.

3.2. Positron lifetimes measurements

The one component fit of the experimental positron lifetime spectra is only possible up to isochronal annealing temperature of 275 °C. For higher temperatures the χ^2 is greater that 1.2. Therefore a two-component fit of the spectra is needed.

Figure 3 shows the fitted average lifetime (τ_{av}) as a function of the isochronal annealing temperature. The first two points are fitted with one component, the others with two components. The χ^2 of the fit is below 1.1 in all the temperature range. The average lifetime changes only a little from room temperature up to 200 °C isochronal annealing temperature. For isochronal annealing temperature above 200 °C a strong decrease of the lifetime occurs, until a minimum is reached at 450 °C, for a value of around 145 ps. The decrease rate is on average of 0.16 ps/°C. At 500 °C a sudden increase up to 155 ps takes place. From this temperature the average lifetime remains between 150 and 155 ps and we can think that the average lifetime reaches a stable situation. The total change in the average lifetime is around 30 ps in a range of 300 °C. All this indicates that below 275 °C positrons are trapped in the sample in saturation. From that temperature another lifetime is necessary for a good fit of the experimental spectra. For high temperatures the fit with only one component does not give reliable results.

Results clearly show that a process of elimination of positron traps is happening during this temperature range.

Positron lifetime parameters, τ_1, τ_2 and I_2, obtained from lifetime spectrum decomposition in two components are shown in figure 4 as a function of isochronal annealing temperature. It can be seen in the figure that below 275 °C there is saturation trapping and the lifetime after quenching from 900 °C is 183 ps, which must correspond to the lifetime value of the positron trapping defect. At 275 °C the saturation trapping disappears and a little short component appears. The long component remains quite constant up to isochronal annealing temperature below 450 °C. However, at 450 °C the long lifetime increases to around 200±5 ps.

Journal of Physics: Conference Series **265** (2011) 012015 doi:10.1088/1742-6596/265/1/012015

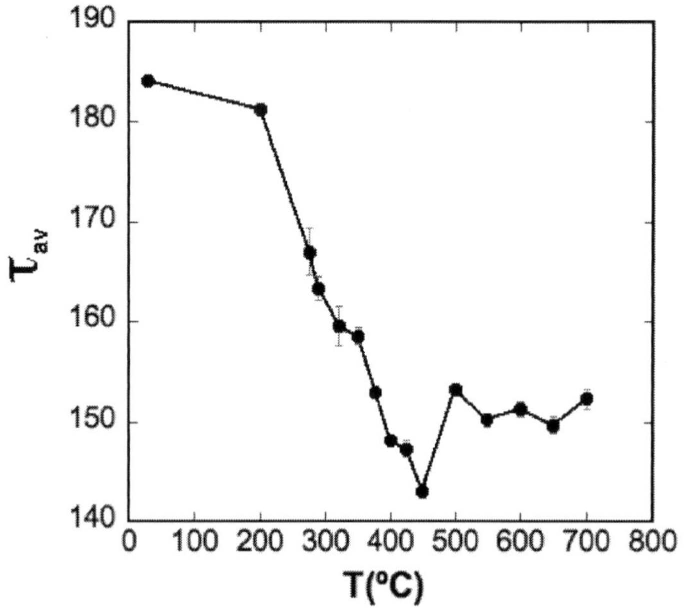

Figure 3. τ_{av} versus isochronal annealing temperature.

Figure 4 shows the behaviour of the intensity, I_2, of the second long lifetime τ_2 with isochronal annealing temperature, too. The intensity I_2 decreases from 100 % at 200 °C to 52 % at 450 °C. This behaviour indicates the reduction of the concentration of the associated trapping centres.

In figure 4 τ_1 increases until a quite constant value of about 100 ps. This behaviour is in agreement with the one trap trapping model.

Figure 5 shows bulk lifetime, τ_B, and the trapping rate K_V, calculated using one trap trapping model from a two-component fit:

$$\tau_1 = \tau_B / (1 + K_V \tau_B) \quad (1)$$

$$\tau_2 = \tau_V = \lambda_V^{-1} \quad (2)$$

$$I_2 = K_V / (\lambda_B - \lambda_V + K_V) \quad (3)$$

The value of the bulk lifetime calculated using equations (1), (2) and (3), is near constant. Indeed, it is of 136±6 ps. So, it is possible to calculate the trapping rate from 320 °C on. The trapping rate K_V, starts at a value of 0.11×10^{12} s^{-1} and it goes down rapidly to a constant value of around 0.005×10^{12} s^{-1}. The value of 136 ps for the lifetime annihilation at the bulk is quite reasonable taking into account the lifetime annihilation bulk values for the individual bulk lifetimes of the different constituent metals in the alloy [6].

Taking into account that the trapping model works for one trap in all the studied temperature range, the results would indicate that below 450 °C there is a positron trapping defect with a life time of 183±1 ps that above 450 °C evolves into another with a life time of 200±5 ps. However, as the long lifetimes measured below and above 450 °C are within 20 ps one could explained the observed behaviour assuming that below 450 °C τ_2 correspond to the contribution of at least two traps, where the defect with shorter lifetime between them is the one with the largest concentration. Above 450 °C the trap with a lifetime of around 200 ps lifetime is the one with the largest concentration. It has to be noted that the decomposition of the spectra with three components does not work.

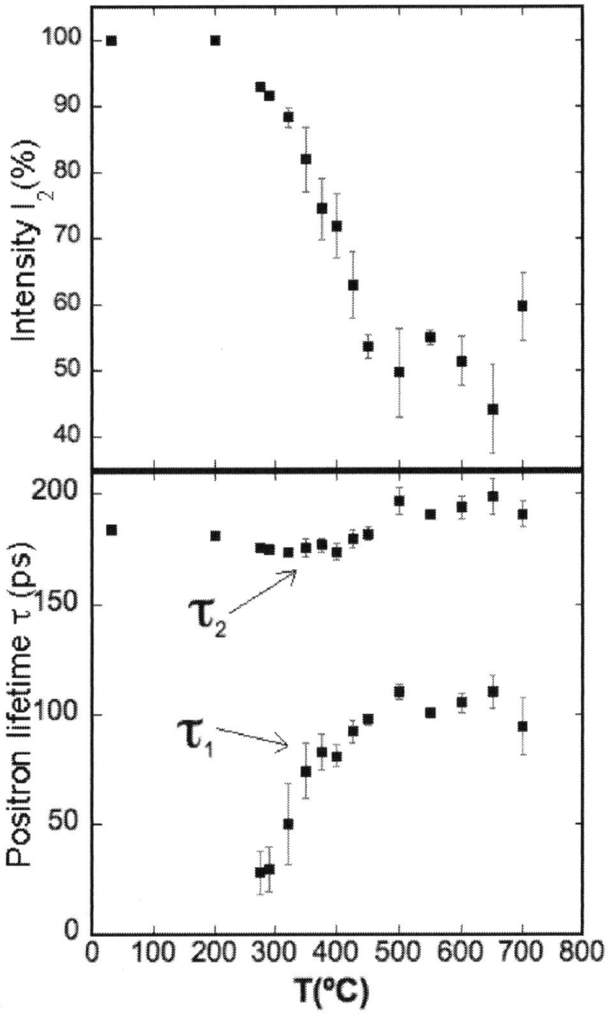

Figure 4. Lifetime parameters τ_1, τ_2 and I_2 versus isochronal annealing temperature.

The lifetime values of τ_2, 180 ps and 200 ps, associated with the trapping centres in the temperature ranges 30-450 °C and 450-700 °C respectively can be ascribed to positron annihilation lifetimes from vacancy-type defects. Indeed, taking into account that the ratio $\tau_2/\tau_B \approx 1.3$-$1.5$, we notice that it corresponds to the typical value for monovacancies in metals [6,7]. Therefore, the τ_2/τ_B ratio indicates that monovacancies are the positron trapping defects. Taking into account that in the studied sample monovacancies can be of Ni, Mn or Ga and that the covalent radius of Ni is lower than the ones of Mn and Ga encourages us to associate the shorter lifetime defects with Ni vacancies and the longer lifetime defects with Mn or Ga vacancies.

3.3. Comparison between calorimetric measurements and positron lifetime measurements

As it can be inferred from the figure 2 and 3, the temperature range in which the MT temperature (ΔMp) changes correspond to the temperature range in which the average lifetime decreases. It extends from 200 °C to 400 °C. Up to 200 °C there are no appreciable changes neither in τ_{av} nor in ΔMp. Above 200 °C, τ_{av} decreases at the same time that ΔMp begins to decrease. At 300 °C ΔMp increases while τ_{av} is still decreasing. At 400 °C the increase of ΔMp stops and τ_{av} reaches a static value of about 150 ps. This suggests that the decrease of vacancy concentration in the alloy could be related with the martensite transformation of NiMnGaCo. The decrease and subsequent increase in the

MT temperature, figure 2, suggests the presence of more than one process. One of the processes is responsible for the increase of ΔMp and the other one for the decrease of ΔMp. One of these processes can be the elimination of vacancies. The other one can be associated to another vacancy type defect or to an atomic reordering process [4].

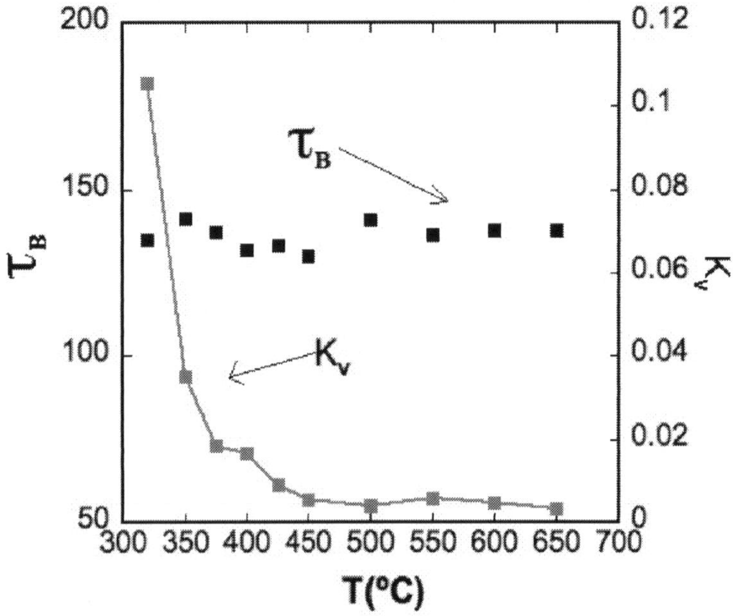

Figure 5. τ_B and K_V calculate from one trap trapping model versus isochronal annealing temperature.

Figure 2 and 3 shows that the exothermic peak in the DSC and the decrease in the lifetime happen in the same temperature range. This temperature range is also the same as the temperature of the MT (ΔMp) change. It can indicate that the elimination of vacancies is related with exothermic peak and this would influence the change in the MT temperature.

4. Conclusions

The role of vacancy-type defects in the martensitic transformation of Co containing Ni-Mn-Ga polycrystalline alloy, $(Ni_{49.5}\ Mn_{28.5}\ Ga_{22})_{96}\ Co_4$, has been studied measuring the effect of thermal treatment on positron lifetime spectra.

By means of the trapping model a bulk lifetime of 136±6 ps has been obtained for the studied alloy.

The positron trapping centres are monovacancies and at least two different types of monovacancies are needed to explain the experimental results.

The positron average lifetime decrease occurs in the same temperature range in which Mp varies. This suggests that the elimination of vacancies plays a role in the martensitic transformation. The way in which they are related and the specific mechanism that takes places in the process will be the subject of future investigations.

Acknowledgement
Financial support from the Spanish CICYT and Basque Government under Grants MAT2009-14398 and IT-443-10 is appreciated. One of us (D. Mérida) is indebted to the University of the Basque Country (UPV/EHU) for financial support.

References

[1] Ullakko K, Huang J K, Kantner C, O'Handley R C and Kokorin V V 1996 *Appl. Phys. Lett.* **69** 1966

[2] Glavatskyy I, Glavatska N, Södergerg O, Hannula S P and Hoffmann J U 2006 *Scripta Mater.* **54** 1891
[3] Sozinov A, Likhachev A A, Lanzka N and Ullakko K 2002 *Appl. Phys. Lett.* **80** 1746
[4] Sanchez-Alarcos V 2008 *Sci Ang. A* **481-482** 293-297
[5] Kustov S, Pons J, Cesari E and Van Humbeeck J 2004 *Acta Mater.* **52** 3083
[6] Campillo Robles J M, Ogando E and Plazaola F 2009 *J. Phys.: Condens. Matter* **19** 176222
[7] Campillo Robles J M and Plazaola F 2003 *Defect and Diffusion Forum* **213-215** 141-236

Quenched-in vacancies in Fe_3Al based alloys: a positron annihilation study

O Melikhova[1], J Čížek[1], J Kuriplach[1], I Procházka[1], F Lukáč[1], M Cieslar[1], W Anwand[2] and G Brauer[2]

[1] Charles University in Prague, Faculty of Mathematics and Physics, V Holešovičkách 2, 180 00 Prague 8, Czech Republic
[2] Institut für Strahlenphysik, Forschungszentrum Dresden-Rossendorf, Postfach 510119, D-01314, Dresden, Germany

E-mail: oksivmel@yahoo.com

Abstract. In the present contribution, high-resolution positron lifetime spectroscopy and slow positron implantation spectroscopy are used to characterize defects in $Fe_{75.99}Al_{24.01}$ and $Fe_{71.98}Al_{28.02}$ alloys. In order to facilitate defect identification, we also perform a theoretical study of basic vacancy-like defects in three phases of the Fe_3Al system: ordered ($D0_3$), short-range ordered (B2) and disordered (A2). Positron characteristics, i.e. positron lifetime and positron binding energy to defects are calculated from the first principles for various defect configurations. The results are discussed in the context of experimental data obtained here and available in literature.

1. Introduction

Intermetallic alloys based on the Fe_3Al system represent perspective materials for industrial applications at elevated temperatures due to a number of advantageous features such as a low density, a high strength and a good corrosion resistance. It is also their low cost, which makes these alloys attractive for industry. Very interesting property of Fe_3Al-based alloys is low formation energy of vacancies. The equilibrium concentration of thermal vacancies observed in Fe_3Al-based alloys at high temperatures appears to be as high as several at.% [1]. Several experimental works reported that vacancies have a significant impact on hardness of Fe_3Al based alloys [2-4]. By quenching from elevated temperatures, a high concentration of non-equilibrium vacancies can be retained at room temperature.

Hence it is obvious that a detailed investigation of vacancies in Fe_3Al based alloys is indispensable for a deeper understanding of physical properties of these materials. Positron annihilation spectroscopy (PAS) is a well-recognized, non-destructive method exhibiting very high sensitivity to open-volume defects like vacancies and related defects. [5,6]. The PLS method was applied to defect studies in Fe-Al alloys several times [7-10]. For instance, Schaefer et al. [7,8] performed extensive PLS measurements in situ in a wide temperature interval in order to deduce activation energies of vacancy formation and migration in Fe-Al alloys. A PLS investigation of binary Fe-Al alloys with varying Al content and quenched from 1000°C was carried out by Diego et al. [10].

Although previous papers on Fe-Al based alloys have brought a valuable knowledge, one has to state that most of the earlier PLS measurements on Fe-Al were performed using spectrometers with relatively poor time resolution (for example 320 ps in [7,8]) and low statistics. Due to these reasons, a reliable decomposition of positron lifetime spectra into individual components could not be performed and only the mean lifetimes (i.e. weighted averages of individual lifetimes) were determined in these works. It was also not possible to include more kinds of defects in the analysis of experimental spectra (e.g. mono- and di-vacancies). Theoretical calculations of positron lifetimes were performed in [10] for several types of defects, for example for different vacancy sites in Al and Fe sublattices as well as for a vacancy-antisite atom defect complex.

The aim of the present work is to characterise the nature of defects in quenched-in Fe_3Al based alloys with over- and under-stoichiometric Al content. Defect studies were performed by two complementary PAS techniques: high resolution positron lifetime (LT) spectroscopy and slow positron implantation spectroscopy (SPIS). The combination of LT and SPIS enables to determine even a very high concentration of vacancies typical for Fe_3Al alloys. PAS investigations were accompanied by Vickers microhardness testing. First principles calculations of positron lifetimes and binding energies to vacancies have been performed in order to characterise different vacancy-type defects that can be present in the studied samples.

2. Computational methods

Positron calculations of basic vacancy-like defects in different phases of the Fe_3Al system [11]: ordered ($D0_3$), short range ordered (B2) and disordered (A2), were performed employing the so-called atomic superposition (ATSUP) method [12]. The lattice parameter used in calculations was $a = 5.792$ Å [13]. We employed the supercell approach, considering 1024 atom-based supercells. Monovacancies were created by removing one Al or Fe atom. Supercells for B2 and A2 structures were prepared by randomly filling atomic positions in the corresponding sublattice. The electron-positron correlations were treated according to Boroński-Nieminen (BN) [14] and according to the gradient-correction (GC) scheme formulated by Barbiellini et al. [15].

3. Experimental details

In the present work we studied $Fe_{75.99}Al_{24.01}$ and $Fe_{71.98}Al_{28.02}$ alloys, which represent Fe_3Al based alloys with under-stoichiometric and over-stoichiometric Al content. Studied alloys were annealed at 1000°C for 1h in vacuum (10^{-3} mbar) encapsulated in silicon glass ampoules. The annealing treatment was finished by quenching of the silicon glass ampoule into water of room temperature. As the annealing was performed in the disordered A2 phase region, it is expected that A2 is at least partially retained in the quenched samples. However domains transferred into the B2 phase and even a small fraction of the ordered $D0_3$ phase should be expected in the quenched samples despite a high cooling rate.

A $^{22}Na_2CO_3$ positron source with an activity of 1.2 MBq deposited on a 2 μm thick Mylar foil was used for LT studies. The source contribution consists of two weak components with lifetimes $\tau_2 \approx 368$ ps and $\tau_3 \approx$ 1.5 ns. A high resolution digital spectrometer [16,17] was employed for LT investigations of studied alloys. The detector part of the digital LT spectrometer is equipped with two Hamamatsu H3378 photomultipliers coupled with BaF_2 scintillators. Detector pulses are sampled in real time by two ultra-fast 8-bit digitizers Acqiris DC211 with a sampling frequency of 4 GHz. The digitized pulses are acquired in a PC and worked out off-line by a software using a new algorithm for integral constant fraction timing [18]. The time resolution of the digital LT spectrometer was 150 ps (FWHM, ^{22}Na). At least 10^7 annihilation events were accumulated in each LT spectrum.

SPIS studies were performed on the magnetically guided variable energy positron beam "SPONSOR" [19]. The energy of incident positrons was varied in the range from 0.03 to 36 keV. The Doppler broadening of the annihilation line was measured by an HPGe detector with an energy resolution of (1.09 ± 0.01) keV at 511 keV and evaluated in terms of the *S* parameter.

Microhardness was examined using the Vickers method with a loading of 100 g applied for 10 s (HV0.1) using the STRUERS Duramin-2 micro-tester device.

4. Results and discussion

4.1. Theoretical calculations

Figure 1 shows the unit cell of the $D0_3$ structure. It consists of two interpenetrating simple cubic sublattices denoted A and B. The Fe and Al occupation of the sites in both sublattices differs in three phases existing in the Fe_3Al system:

1. In the case of the complete $D0_3$ order, the A sublattice is occupied exclusively by Fe atoms. The B sublattice consists of alternating sites occupied by Fe and Al atoms. All atoms in the B sublattice are surrounded by 8 nearest neighbor (NN) Fe atoms. Thus, one can distinguish two types of Fe atoms in the $D0_3$ phase: (i) Fe(A) atoms which occupy the A sublattice and are surrounded by 4 NN Fe atoms and 4 NN Al atoms, and (ii) Fe(B) atoms located in the B sublattice surrounded by 8 NN Fe atoms. All Al atoms in the $D0_3$ phase have 8 NN Fe atoms.
2. As for the B2 order in the Fe_3Al system, Fe and Al atoms occupy all sites in the B sublattice with equal probability, while the A sublattice is occupied only by Fe atoms.
3. Finally, in the disordered A2 phase all sites in both sublattices are occupied randomly by Fe or Al atoms.

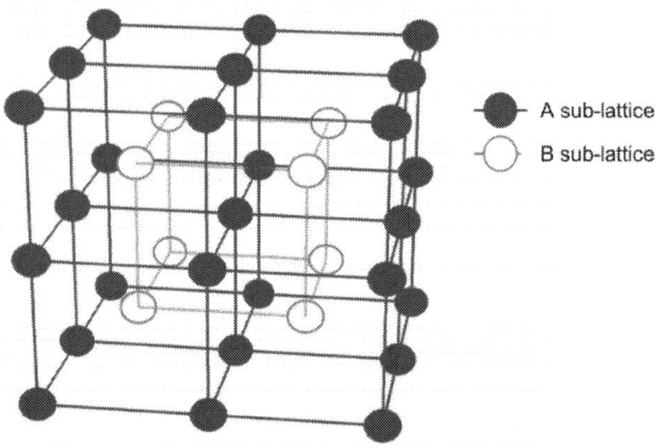

Figure 1. The unit cell of Fe_3Al based alloys showing A and B sublattices.

Positron lifetimes were first calculated using both ATSUP-BN and ATSUP-GC approaches. It is known that the GC scheme is more sensitive to details of the electronic structure than local density approximation (LDA) based approaches [20], like the BN one. Therefore, the GC approach should rather be used with self-consistent electron densities. On the other hand, within the BN approach one can observe a feedback effect, which keeps the positron lifetime unchanged independently of small transfers of the electron density [12] (see also [20]). To this end, positron lifetimes calculated by the ATSUP-BN approach are usually in a better agreement with experiment than lifetimes calculated using the ATSUP-GC approach [20,21]. We can demonstrate this for the case of the bulk lifetime for the $D0_3$ ordered Fe_3Al alloy. The bulk positron lifetime 110 ps calculated using the ATSUP-BN approach is in good agreement with an experimental value of 112 ps [7], while a bulk lifetime of 118 ps calculated with the ATSUP-GC approach is clearly too long. For these reasons, only lifetimes calculated using the ATSUP-BN approach will be considered in this paper. Bulk positron lifetimes and lifetimes of positrons trapped in vacancies calculated using the ATSUP-BN scheme are collected in table 1. Positron binding energies to vacancies (E_b) are shown in table 1 as well. The value of E_b for a given type of vacancy is calculated as a difference of the ground state energy E_0 of the delocalized positron and the energy E_V of the positron trapped in the vacancy considered, i.e.

$$E_b = E_0 - E_V \,. \tag{1}$$

In the $D0_3$ phase one can find three types of vacancies: the Al one and two different Fe ones (in A and B sublattice). Lifetimes of positrons trapped in all types of vacancies fall in the relatively narrow range from 182 to 187 ps. The longest positron lifetime was found for the Fe vacancy in the A sublattice, i.e. the only vacancy with some NN Al atoms (all the other types of vacancies have only Fe nearest neighbors). Diego et al. [10] suggested that triple defects consisting of two adjacent Fe vacancies in the A sublattice bound to a NN Al atom in the B sublattice are formed in B2 regions at high temperatures. This defect which corresponds to a Fe divacancy in the $D0_3$ phase aligned along the [100] direction was also considered in the calculations, see table 1.

Table 1. Bulk lifetimes, lifetimes of positrons trapped in vacancies, and corresponding positron binding energies (E_b) for vacancies calculated using the ATSUP-BN approach for various phases in Fe_3Al. First and second nearest neighbours of examined vacancy configurations are listed in the second and third column, respectively.

annihilation site	vacancy nearest neighbors (NN)	vacancy next nearest neighbors (2NN)	τ (ps)	E_b (eV)
$D0_3$ phase				
Bulk			110	-
Fe vacancy (A sublattice)	4 Al + 4 Fe	6 Fe	187	3.1
Fe vacancy (B sublattice)	8 Fe	6 Al	185	3.2
Al vacancy (B sublattice)	8 Fe	6 Fe	182	3.5
Fe divacancy (A sublattice) in [100] direction	6 Fe + 6 Al	10 Fe	195	4.0
B2 phase				
Bulk			109	-
Fe vacancy (A sublattice)	8 Fe	6 Fe	181	3.1
Fe vacancy (A sublattice)	8 Al	6 Fe	193	2.7
vacancy in B sublattice	8 Fe	6 Fe	182	3.5
vacancy in B sublattice	8 Fe	6 Al	185	3.1
A2 phase				
Bulk			108	-
Vacancy	8 Fe	6 Fe	181	3.1
Vacancy	8 Al	6 Al	199	3.6

The bulk positron lifetime in the partially ordered B2 phase varies due to the random occupation of the B sublattice sites by Fe and Al atoms. To estimate these variations, we calculated bulk positron lifetimes for a set of randomly generated B2 supercells. The mean value of calculated bulk lifetimes from this set is shown in table 1 and is very close to the bulk lifetime obtained for the $D0_3$ phase. The dispersion (standard deviation) of the bulk lifetime in B2 phase was found to be only 0.2 ps. Hence, the effect of the random occupation of the B sublattice sites on the positron lifetime can be considered as negligible.

In the B2 phase one can distinguish Fe vacancies in the A sublattice and vacancies in the B sublattice. The latter vacancies have always 8 NN Fe atoms. On the other hand, number of NN Al and Fe atoms surrounding Fe vacancies in the A sublattice varies randomly. Two extreme cases may be considered for Fe vacancies in the A sublattice in the B2 phase: (i) a vacancy with 8 NN Fe atoms and (ii) a vacancy with 8 NN Al atoms. The local surrounding of Fe vacancies in the A sublattice varies between these two extremes, thereby, the lifetime of trapped positrons is expected to vary between the lifetimes of positrons trapped in Fe vacancy (i) and (ii), i.e. in the range from 181 to 193 ps (see table 1). Figure 2 shows lifetimes of positrons trapped in the Fe vacancy as a function of the number of NN

Al atoms surrounding the vacancy. It is clear that the positron lifetime increases with the increasing number of NN Al atoms. It should be noted that while 2NN of Fe vacancies considered in figure 2 are always Fe atoms (sites on the A sublattice), occupation of third and higher-order NN vary due to the random occupation of the B sublattice sites. Nevertheless, we found that the dispersion of positron lifetimes due to this effect is lower than 0.1 ps for all vacancies considered in figure 2. Thus, the lifetime of trapped positrons is determined predominantly by NN and 2NN atoms surrounding vacancy. Fe vacancies in the A sublattice in the $D0_3$ phase are always surrounded by 4 NN Al atoms. Hence, the Fe-vacancy in the A sublattice surrounded by more than 4 NN Al atoms may be considered also as a vacancy in the $D0_3$ phase with corresponding number of Al antisite atoms. For example, an Fe vacancy in the A sublattice surrounded by 6 NN Al atoms can be considered also as an Fe(A) vacancy in the $D0_3$ phase with two Al antisite atoms. Lifetimes of positrons trapped in Fe vacancies in the $D0_3$ phase with various number of Al antisite atoms are shown in figure 2 as well. Clearly, these lifetimes differ only very slightly from those calculated for Fe vacancies in B2 phase surrounded by the same number of NN Al atoms.

The bulk positron lifetime in the disordered A2 phase fluctuates due to the random distribution of Fe and Al atoms. However, similarly to the B2 phase these fluctuations are rather small. The dispersion (standard deviation) of bulk positron lifetimes obtained from a set of randomly generated A2 supercells is only 0.3 ps. The mean value of bulk positron lifetimes for the A2 phase is shown in table 1.

Two extreme cases can be considered for vacancies in the A2 phase: (i) a vacancy completely surrounded by Fe atoms (i.e. 8 NN Fe + 6 2NN Fe atoms), and (ii) a vacancy completely surrounded by Al atoms (i.e. 8 NN Al + 6 2NN Al atoms). Positron lifetimes for these two extremes are shown in table 1. The lifetime of positrons trapped in a vacancy in the A2 phase varies between these extremes, i.e. in the range from 181 to 199 ps.

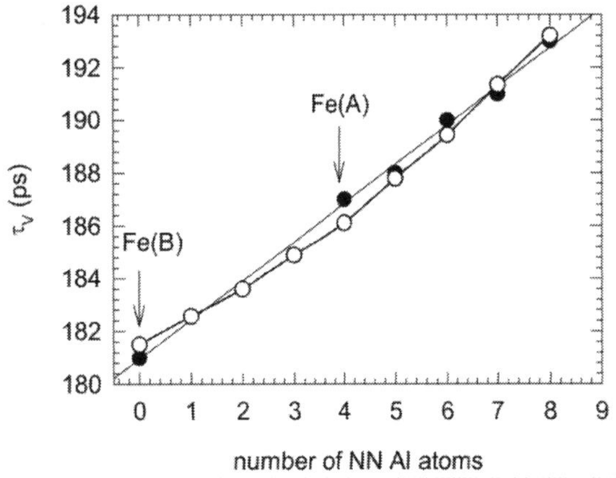

Figure 2. Calculated lifetimes of positrons trapped in Fe vacancies as a function of the number of NN Al atoms around vacancies. Open circles – B2 phase, Full circles $D0_3$ phase: Fe(A), Fe(B) vacancy and Fe(A) vacancy with various number of Al antisites.

4.2. LT and SPIS results

Both quenched alloys exhibit a single component LT spectrum with the corresponding positron lifetime τ_V shown in table 2. The lifetimes τ_V in both alloys are significantly longer than the Fe_3Al bulk positron lifetime τ_B = 112 ps [7] and fall into the range expected for vacancies. Hence, both alloys contain a high density of quenched-in vacancies, which causes saturated positron trapping. Comparison with theoretical calculations shows that experimental lifetimes τ_V are compatible with Fe vacancy in the B2 phase surrounded by 7 or 8 NN Al atoms, see table 1. Another possibility is that positrons are trapped in [001] Fe divacancies in the ordered $D0_3$ phase.

Table 2. Summarized experimental results for quenched $Fe_{75.99}Al_{24.01}$ and $Fe_{71.98}Al_{28.02}$ alloys: experimental lifetimes of positrons trapped at vacancies (τ_V), positron diffusion length measured by SPIS (L_+), concentration of quenched-in vacancies and (c_V) and microhardness (HV0.1).

sample	τ_V (ps)	L_+ (nm)	c_V (at.$^{-1}$)	HV0.1
$Fe_{75.99}Al_{24.01}$	190.8 ± 0.4	40 ± 6	(4.8 ± 0.6)×10^{-4}	412 ± 5
$Fe_{71.98}Al_{28.02}$	195.3 ± 0.5	4.0 ± 0.2	(5.0 ± 0.5)×10^{-2}	491 ± 5

The dependence of the *S* parameter on the energy of incident positrons measured by SPIS on quenched $Fe_{75.99}Al_{24.01}$ and $Fe_{71.98}Al_{28.02}$ alloys is plotted in figure 3. A local minimum of *S* at low energies 1-2 keV is due to positron annihilations in a thin oxide layer formed on the surface during annealing. One can see in figure 3 that the bulk value of the *S* parameter is higher in the $Fe_{71.98}Al_{28.02}$ alloy, i.e. in the alloy with higher Al content. The *S*(*E*) curves were fitted by the VEPFIT software package [22] assuming two layers: (i) a thin oxide layer on surface, and (ii) the bulk alloy. In both alloys we obtained a good fit, which is plotted by solid lines in figure 3. Thickness of the oxide layer is around 10 nm and 30 nm in the $Fe_{71.98}Al_{28.02}$ and $Fe_{75.99}Al_{24.01}$ alloy, respectively. The positron diffusion lengths obtained from fits of the *S*(*E*) curves in both alloys are shown in table 2. The $Fe_{71.98}Al_{28.02}$ alloy exhibits a shorter diffusion length than the $Fe_{75.99}Al_{24.01}$ alloy. This indicates that the $Fe_{71.98}Al_{28.02}$ alloy, i.e. the alloy with an enhanced Al content, contains more quenched-in vacancies compared to the other alloy.

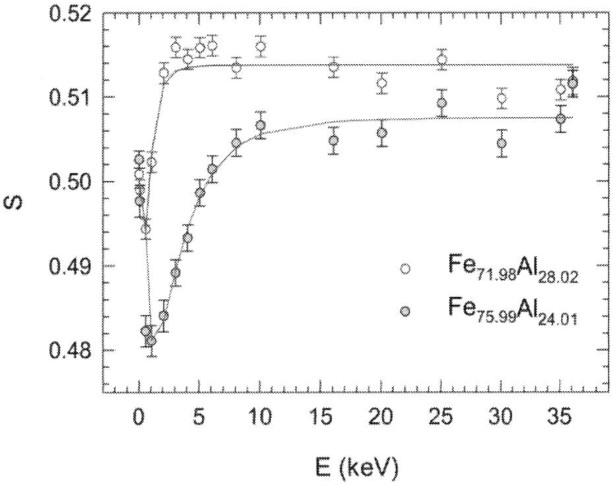

Figure 3. The dependence of the *S* parameter on the incident positron energy for $Fe_{75.99}Al_{24.01}$ and $Fe_{71.98}Al_{28.02}$ alloys measured by SPIS.

4.3. Estimation of concentration of quenched-in vacancies

The limit of vacancy concentration which causes saturated positron trapping can be estimated from the two-state trapping model. The positron trapping rate to vacancies can be calculated from the following equation

$$K = \frac{I_2}{I_1}\left(\frac{1}{\tau_B} - \frac{1}{\tau_V}\right) \tag{2}$$

where τ_B = 112 ps is the positron bulk lifetime in Fe_3Al [7]. The free positron component cannot be resolved in an LT spectrum if its relative intensity I_1 falls below ≈ 5%. This corresponds to the positron trapping rate $K_{\max,LT} \approx 7 \times 10^{10}$ at.s^{-1}. If the trapping rate $K \geq K_{\max,LT}$, then saturated positron trapping in vacancies takes place. The lowest concentration of vacancies which causes saturated positron trapping is

$$c_{V\,\max,\mathrm{LT}} = \frac{K_{\max,\mathrm{LT}}}{\nu_V} \approx 2.0\times10^{-4}\,\mathrm{at.}^{-1} \tag{3}$$

where we used the specific positron trapping rate $\nu_V = 4 \times 10^{14}$ at.s^{-1}, reported for vacancies in Fe_3Al in Ref. [7]. Clearly, the concentration of quenched-in vacancies in both alloys is higher than $c_{V\,\max,\mathrm{LT}}$.

The positron trapping rate K to vacancies can be also calculated from SPIS results

$$K = \frac{1}{\tau_B}\left(\frac{L_{+,B}^2}{L_+^2} - 1\right) \tag{4}$$

Here $L_{+,B}$ is the positron diffusion length in the defect-free material and L_+ is the positron diffusion length measured in the studied sample. The diffusion length $L_{+,B}$ is given by the equation

$$L_{+,B} = \sqrt{D_+\tau_B} \tag{5}$$

The positron diffusion coefficient D_+ in Fe_3Al was estimated as a weighted average of positron diffusion coefficients in Fe and Al

$$D_+ \approx \frac{3}{4}D_{+,Fe} + \frac{1}{4}D_{+,Al} = 3\ \mathrm{cm^2 s^{-1}} \tag{6}$$

The positron diffusion coefficients for pure Fe and Al were calculated using the deformation potential approximation [23].

The shortest positron diffusion length which can be determined by SPIS is $L_+ \approx 1$ nm. This corresponds to a limiting positron trapping rate $K_{\max,\mathrm{SPIS}} \approx 3 \times 10^{14}$ at.s^{-1}. Hence, using SPIS it is possible to determine even very high concentrations of vacancies. The upper limit of vacancy concentration measurable by SPIS is

$$c_{V\,\max,\mathrm{SPIS}} = \frac{K_{\max,\mathrm{SPIS}}}{\nu_V} \approx 0.7\ \mathrm{at}^{-1}. \tag{7}$$

Obviously the upper limit given by equation (7) is extremely high concentration, which cannot be reached in any real material. Thus, we can conclude that SPIS is able to determine even very high concentrations of vacancies and there is actually no upper limit of vacancy concentration measurable by this technique.

The concentration of quenched-in vacancies in the studied alloys determined from SPIS results is shown in table 2. Both studied alloys exhibit significant concentrations of vacancies, in particular $Fe_{71.98}Al_{28.02}$ contains extremely high concentration of vacancies, which is two orders of magnitude higher than in $Fe_{75.99}Al_{24.01}$. A higher concentration of vacancies in $Fe_{71.98}Al_{28.02}$ alloy is in concordance with higher hardness measured on this alloy. Hence, the concentration of vacancies and the hardness of the alloy increase with the increasing Al content. Coincidence Doppler broadening studies of quenched $Fe_{75.99}Al_{24.01}$ and $Fe_{71.98}Al_{28.02}$ alloy revealed that quenched-in vacancies are surrounded predominantly by Al atoms in the NN sites [24]. This finding is in agreement with lifetimes τ_v measured in these alloys, which are comparable with the lifetimes calculated for Fe vacancy surrounded by 7 or 8 NN Al atoms.

5. Conclusions

In the present work we studied $Fe_{75.99}Al_{24.01}$ and $Fe_{71.98}Al_{28.02}$ alloys quenched from 1000°C. Both alloys contain a high concentration of quenched-in Fe vacancies, which are surrounded predominantly by nearest neighbor Al atoms. The concentration of quenched-in vacancies was determined from positron diffusion length measured by SPIS. The $Fe_{71.98}Al_{28.02}$ alloy exhibits significantly higher concentration of vacancies. Thus, the vacancy concentration increases with the increasing Al content. A higher concentration of vacancies in $Fe_{71.98}Al_{28.02}$ alloy is in concordance with higher hardness measured on this alloy. This testifies a strong influence of vacancies on mechanical properties of Fe_3Al based alloys.

Acknowledgement
We are grateful to M.J. Puska for his ATSUP code that served as a basis for further developments. This work was supported by the Czech Scientific Foundation (contract № GA106/08/P133) and by the Ministry of Education, Youths and Sports of the Czech Republic through the research plan № MSM 0021620834.

References

[1] Ho K and Dodd R A 1978 *Scr. Metall.* **12** 1055
[2] Liu T, Lee E H and McKammy C G 1989 *Scr. Metall.* **23** 875
[3] Jordan J L and Derbi S C 2003 *Intermetallics* **11** 507
[4] Sassi O, Aride J, Bernardini J and Moya G 1998 *Ann. Chim. Sci. Mat.* **23** 421
[5] Hautojärvi P 1979 *Positrons in Solids*, ed Hautojärvi P (Springer-Verlag, Berlin)
[6] Hautojärvi P and Corbel C 1995 *Positron Spectroscopy of Solids, Proceedings of The International School of Physics "Enrico Fermi"*, ed Dupasquier A and Mills A P (IOS Press, Amsterdam) p 491
[7] Schaefer H E, Würschum R, Šob M, Žák T, Yu W Z, Eckert W and Banhart F 1990 *Phys. Rev. B* **41** 11869
[8] Schaefer H E, Damson B, Weller M, Arzt E and George E P 1997 *Phys. Stat. Sol. (a)* **160** 531
[9] Jirásková Y, Schneeweiss O, Šob M, Novotný I, Procházka I, Bečvář F, Sedlák B, Šebesta F and Puska M J 1995 *J. Physique* **5** C1-157
[10] de Diego N, Plazaola F, Jiménez J A, Serna J and del Río J 2005 *Acta Mater.* **53** 163
[11] *Binary Alloy Phase Diagrams*, ed Massalski T B (Americal Society for Metals, Metals Park, Ohio, 1986)
[12] Puska M J and Nieminen R M 1983 *J. Phys. F* **13** 333; Seitsonen A P, Puska M J and Nieminen R M 1995 *Phys. Rev. B* **51** 14057
[13] Wakiyama T 1972 *J. Phys. Soc. Jpn.* **32** 1222
[14] Boroński E and Nieminen R M 1986 *Phys. Rev. B* **34** 3820
[15] Barbiellini B, Puska M J, Torsti T and Nieminen R M 1995 *Phys. Rev. B* **51** 7341
[16] Bečvář F, Čížek J, Procházka I and Janotová J 2005 *Nucl. Instr. Meth. A* **539** 372
[17] Bečvář F, Čížek J and Procházka I 2008 *Acta Physica Polonica A* **113** 1279
[18] Bečvář F 2007 *Nucl. Instr. Meth. B* **261** 871
[19] Anwand W, Kissener H R and Brauer G 1995 *Acta Physica Polonica A* **88** 7
[20] Barbiellini B, Puska M J, Korhonen T, Harju A, Torsti T and Nieminen R M 1996 *Phys. Rev. B* **53** 16201
[21] Campillo Robles J M, Odo E and Plazaola F 2007 *J. Phys.: Condens. Matter* **19** 176222
[22] van Veen A, Schut H, Clement M, de Nijs J, Kruseman A and Ijpma M 1995 *Appl. Surf. Sci.* **85** 216
[23] Bardeen J and Shockley W 1950 *Phys. Rev.* **80** 72
[24] Melikhova O, Čížek J, Kuriplach J, Procházka I, Cieslar M, Anwand W and Brauer G 2010 *Intermetallics* **18** 592

Nano-structures in Al-based alloys

Paola Folegati[1]**, Martti J Puska**[2] **and Torsten E M Staab**[3]

[1] Dipartimento de Fisica e Centro L-NESS, Politecnico di Milano, Via Anzani 42, I-22100 Como, Italy
[2] Laboratory of Physics, Helsinki University of Technology, P.O. Box 1100, FI-02015 TKK, Finland
[3] Helmholtz Institut für Strahlen- und Kernphysik, Universität Bonn, Nussallee 14-16, D-53129 Bonn

E-mail: paola.folegati@como.polimi.it

Abstract. The usefulness of first-principles calculations for studying solute-atom clustering in metal alloys is discussed. This usefulness stems directly from the properties predicted by the calculations or via a related interpretation of experimental results. In this paper we review the results of our computational studies on small solute clusters in Al-based alloys. The predicted coincidence Doppler broadening spectra of the positron annihilation method are used to analyse experimental results. The calculated binding energies of small solute atom clusters explain why Cu atoms form two-dimensional platelets on the (100) planes in Al whereas Zn forms three-dimensional clusters.

1. Introduction

First-principles calculations have become an important tool to study various properties of molecules, small aggregates of atoms, and bulk solids. The availability of powerful supercomputers and of several different simulation methods has given an enormous momentum to this subfield in computational physics.

The different simulation approaches can be characterized according to their level of sophistication necessary for describing the structures and phenomena in question. E.g., one can distinguish between nanoscale and subnanoscale descriptions or between static or dynamic descriptions.

In this paper we survey possibilities of interpreting measured positron annihilation data on the basis of computer simulations, showing results for pure metals and for some Al-based alloys. Moreover, we show how first-principles calculations can explain details of solute-atom clustering in these alloys.

2. Why is solute-atom clustering in Al-alloys interesting?

Aluminium alloys are extensively employed in vehicle industry. The process of ageing, discovered by Alfred Wilm [1] already at the beginning of the 20th century, has been recognized as a tool for strengthening these alloys.

Age hardening can be ascribed as precipitation of small particles blocking the crystallographic slip, increasing the resistance to deformation, and hindering the motion of dislocations. It has been recognized that to maximize the hardening one should optimize the size and the distance of the precipitation particles.

Modern experimental techniques have established that the ageing processes in most aluminium alloys are complex and may involve several stages. Typically coherent Guinier-Preston (GP) zones and semicoherent intermediate precipitates may precede the formation of the normally incoherent equilibrium phase. For early decomposition stages, i.e. small agglomerates, the distinction between cluster and GP-zones is still controversially discussed for some alloys.

Atomic clustering, during and immediately after the quenching from the solution-treatment temperature, may also influence the succeeding precipitation process. From this point of view, it is of great interest in the materials science to understand the early stages of clustering.

In order to describe the early stages of clustering only a finite number of host atoms around few solute atoms have to be considered and the modelling can be carried out by means of first-principles electronic-structure calculations, using well-established codes such as VASP and SIESTA [2,3]. In the next section, we will give a brief review of the main computational tools available for the study of clustering and precipitation in alloys.

3. Role of simulations in interpreting measured positron annihilation data (for the solute-atom clustering in metal matrices)

A deeper understanding of the clustering phenomena at the microscopic level is a valuable objective, besides in the area of basic Solid State Physics, also in the field of Materials Science, as it provides guidelines for different thermal and mechanical treatments aimed for improving the materials properties. Clustering and precipitation phenomena in metal alloys are exploited for most applications of these materials. Many desired mechanical and thermal properties of these materials arise from suitable treatments that give rise to precipitation and clustering. Also point defects and impurities play a key role as far as mechanical properties are concerned.

The simulation of clustering and precipitation combines the attention to microscopic aspects (transport and aggregation of atoms in the bulk of a solid matrix) with the interest to industrial applications. The expected results will contribute quantitative data potentially useful for designing compositions and thermal treatments aimed at controlling the size, spatial distribution, and morphology of the aggregates at the nanometer scale. These are all important factors affecting the mechanical and chemical properties of alloys.

At the same time, the comparison of simulations with the existing experimental data gives the possibility to test the accuracy of different schemes and approximations, thus to "sharpen the tool" to be used for computational analyses of atomic rearrangement phenomena. Aluminium alloys are the systems better known from this point of view, and therefore constitute ideal test cases for theories of phase stability. The existence of a rich variety of ordered compounds, solid solutions and metastable phases in aluminium-based systems is an additional point of interest. Understanding phase stability in these systems is a prerequisite for a systematic design and control of alloy properties.

Presently, ab-initio calculations are available on Al-based alloys [4] and some results also on Mg-based alloys [5]. Extensive results are available in particular for:

- vacancy formation energies in bulk metals, both in the local density approximation (LDA) and in the generalized gradient approximation (GGA) of the density functional theory (DFT) and vacancy formation volumes in metals [4,6,7];
- formation enthalpies of ordered compounds in Al-based alloys [4];
- formation energies, enthalpies, and volumes for impurities, vacancy clusters, impurity clusters, and impurity-vacancy complexes in metals [6,10].

However, very few preliminary results are available, for instance, for divacancy formation energies, impurity-divacancy formation energy [5], and, what is most interesting in the present context, for the binding energies of impurity atoms to clusters at different cluster sizes. The last point is a crucial one, if one wants to understand clustering in metal alloys. Ab-initio studies of the energetics should justify the formation and dissolution of clusters, explain the experimental results already obtained from different techniques in Al-based alloys and provide useful input for Kinetic Monte Carlo (KMC) calculations, in which bigger atom systems can be treated (with respect to first-principle calculations). The existing experimental data on Al-based alloys to be compared with calculated properties range from positron annihilation data to differential scanning calorimetry (DSC), X-ray scattering (XRS), atom probe field-ion microscopy (APFIM), transmission electron microscopy (TEM) and high resolution transition electron microscopy (HRTEM).

The computational tools that have been used until now in this field are:

- Phase-diagram calculation methods (CALPHAD)
- First-principles electronic structure calculations based on DFT
- Kinetic Monte Carlo methods
- Molecular dymamics simulations

A brief description of the main characteristics and limits of each of these methods follows.

3.1. Computational tools

- Methods for calculating phase-diagrams, CALPHAD, use databases of thermodynamic functions to produce accurate phase diagrams and phase equilibria. The advantage of the methods is that they can be applied even to complex multi-component alloys, and for many years they have been used to study many commercial alloys like Al-based, Mg-based and Fe-based alloys. Databases have been generated on the basis of these methods. The limit of this approach is that it cannot predict new thermodynamic data (like existence of new phases) and it is then limited to the analysis of materials for which thermodynamic data are already available.
- First-principles methods have been revealed as a useful complement to CALPHAD approaches by providing energetics and other thermodynamic functions in cases where experimental data are unavailable. An example is the technologically important case of metastable precipitate phases. As far as point defects are concerned, these were studied in the past with empirical models, but recently ab-initio calculations have become the most popular scheme for the calculation of defect formation energies in metals. Although first-principles calculations are slower and far fewer atoms can be handled than in empirical methods, their advantage with respect to empirical methods is their transferability: in multi-component systems, empirical potentials are designed to work in a particular environment, while first-principles calculations are much more flexible: for example, LDA has proved to be a very powerful and sufficiently accurate approach for multicomponent systems in drastically different chemical environments. Although DFT studies are currently limited to a few hundred atoms, many important results about the energetics of defect formation can already be obtained. In addition, first-principles results can generate a large amount of information about the interatomic interaction on a microscopic scale that can be used to fit, or at least put constraints on, simple interatomic potentials. It is difficult, if not impossible, to extract such information from experiments.
- Kinetic Monte Carlo simulations [6] have been revealed to be very well suited for the study of precipitation kinetics in alloys. They allow the treatment of much bigger systems, that is, many more atoms with respect to first-principles calculations. The first simulations were based on a very simple diffusion mechanism: the direct exchange of two nearest-neighbour atoms. Recently, more realistic diffusion mechanisms by vacancy jumps have been introduced in the Monte Carlo method and special attention has been paid to the influence of such diffusion mechanisms on the kinetic pathway during nucleation and growth or phase ordering. Besides the different diffusion mechanisms, these studies have been relying on different models to compute the activation barriers and attempt frequencies of the diffusion events and on different Monte Carlo algorithms. However, in all these studies, the configurational energies and the activation barriers are computed using very simple potentials consisting of short-range pair interactions on a rigid lattice. More accurate potentials exist, but properly accounting the atomic relaxations decreases dramatically the computer efficiency. Monte Carlo studies of precipitation in alloys are based on severe assumptions, when computing both

the configurational energies and the diffusion properties. These assumptions and their consequences on the kinetics of phase transformations have to be questioned and their validity should be tested by comparison with more accurate methods. At the atomic scale, KMC methods are well suited to study diffusion-controlled phenomena, like precipitation kinetics in alloys [7]. These simulations are also useful to test and improve other large scale modelling techniques.

- Another approach to larger systems is molecular dynamics (MD) simulations based on semiempirical models for interatomic interactions. Especially they can be used to study the structures and energetics of solute atom clusters and those of precipitate-host matrix interfaces. An example of the use of MD simulations to the study of clustering in Al-Cu based alloys can be found in ref. [8].

3.2. What has been done and what has to be done

It is not a long time since powerful supercomputers have led to the possibility of studying by ab-initio methods many aspects of the energetics and clustering of defects in metal alloys. As far as Al-based alloys are concerned, some work has already been done. Vacancy formation energy in Al has been calculated by many authors, in different schemes of DFT and with different degrees of precision [9]. The later point is related to the supercell size and the related **k**-point sampling, to the consideration of relaxations in the lattice and/or in the volume of the supercell itself, and to the type of pseudopotential employed to compute the valence electron wave functions. Results for vacancy formation energies in transition and noble metals, within the LDA, with the order-N Green's function method [11] and for vacancy and substitutional impurities in Al bulk and in different small cluster environments have been computed [10]. Ref. [10] presents also a detailed analysis of the dependence of the calculated quantities on the lattice relaxation, the number of **k**-points, and on the supercell size. Other results for vacancy and divacancy formation energies in Al in the LDA and GGA approximations are compared to experimental data in ref. [6] and calculations have been performed also with the KKR Green's function method [9,12]. Impurity diffusion activation energies have also been calculated for Mg, Si, Sc, Zn in Al [14,15]. Recently, a database has been constructed for Al-rich ordered compounds, containing energetics of impurities, formation enthalpies of the ordered compounds, and solubility enthalpies of solute atoms [4]. Ab-initio calculations on vacancy and divacancy formation and binding energies in Al and Mg, together with formation volumes have also been computed [5,6].

First-principles calculations for the energetics of ordered solute - host atom structures or ordered alloys can been used to construct models, such as the mixed-space cluster expansion (MSCE) approach [16], which are applicable to systems of thousands of atoms. MSCE has been used to study, e.g., Cu [17] and Zn [18] clusters in Al alloys. However, at the moment of writing this article, to the best of our knowledge, direct first-principles calculations to study even the smallest solute atom clusters in alloys have not yet been reported in the literature. The reason is the large computer resources they demand.

3.3. Results on Al and Al-based alloys

The target of our first-principles calculations of energetics in Al-based alloys is to give an explanation for the observed clustering phenomena in Al-based alloys, both binary (Al-Cu), ternary (Al-Cu-Mg) and, possibly, also quaternary (Al-Cu-Mg-Ag) alloys. Many experimental data are available from Positron Annihilation Spectroscopy (PAS), both lifetime (LS) and coincidence Doppler broadening spectroscopy (CDB), and also from APFIM, TEM, HRTEM and DSC. The existing data seem to give a clear evidence of the tendency of Cu atoms to form platelets along the {100} planes in Al [19], while Mg addition indicates that the clusters formed have a more spherical shape. This can be justified on the basis of the calculated binding energies of different atomic species to clusters of different morphology and size. Moreover, the vacancy binding energies to different clusters have to be analysed to deeply understand positron spectroscopy data in these alloys.

4. Why positrons are useful for studying the clustering

The experimental techniques for the analysis of precipitates in alloys can be classified according to "what we want to know". Table 1 shows the techniques mainly used for different kinds of analysis. Essentially, electron microscopy and TAP (Tomographic Atom Probe) techniques are useful to see precipitates of more than 100-1000 atoms, i.e., clusters of the nanometric size. At the subnanometric level, when there is less than 10 atoms in a cluster, only positron annihilation and XAFS (X-Ray Absorption Fine Structure) can give some information [20,21]. On the other hand at this scale, first-principles calculation can be employed in the simulations. Theoretical approaches for the analysis of bigger (>10 atoms) cluster can only be realized through Kinetic KMC or CALPHAD methods. However, KMC methods need as input data some parameters that have to be computed from first-principles results and others that have to be taken from experimental data.

Table 1. Different method to investigate precipitates in Al-alloys. As a footnote the abbreviations are explained.

Physical information required on precipitates	Experimental technique
Crystal structure	XRD[a], Electron Diffraction
Internal order	Electron Diffraction, 3D-APFIM[b], XAFS
Composition	3D-APFIM, CDB-PAS[c], TEM[d], XAFS
Spatial distribution	TEM
Nanosize precipitates	TEM, SAXS[e], SANS[f], 3D-APFIM
Sub-nanosize precipitates	PAS, XAFS
Volume fraction	TEM, SAXS, SANS, 3D-APFIM
Formation temperature and time	DSC[g], PALS[h]

[a] XRD: X-Ray Diffraction
[b] 3D-APFIM: Three-dimensional Atom-Probe Field Ion Microscopy
[c] CDB-PAS: Coincidence Doppler Broadening-Positron Annihilation Spectroscopy
[d] TEM: Transmission Electron Microscopy
[e] SAXS: Small Angle X-Ray Scattering
[f] SANS: Small Angle Neutron Scattering
[g] DSC: Differential Scanning Calorimetry
[h] PALS: Positron Annihilation Lifetime Spectroscopy

The role of positrons in material science has been recognized since long time ago: they are indeed the only non-destructive method that can give information about open-volume defects, and they are sensitive to defect concentrations of the order of 1 ppm. Moreover, since the migration of solute atoms is possible thanks to presence of vacancies, the role of vacancies is a fundamental one. Therefore it is of great importance to have a good probe, like a positron, for the detection of vacancy-like defects. Three kinds of positron traps can be distinguished:

a) large clusters with high positron affinity,
b) solute-vacancy clusters,
c) misfit regions.

The traps of type a) can be found for instance in Fe-Cu alloys (see ref. [22,23]). These are usually big solute clusters (>10 atoms) that tend to trap positrons because the affinity of positrons for the solute is higher than that for the matrix.

Traps of type b) are believed to be the most common in Al-based alloys, where a few atoms usually bind to a vacancy during the migration process. The determination of the percentages of solute and matrix atoms surrounding open-volume defects can be obtained from the analysis of Doppler or Coincidence Doppler Spectra, combined with information from positron lifetime, as will be shown in the next sections [24,25,26].

Traps of type c) form because of the presence of big precipitates, semicoherent or incoherent with the matrix. The presence of these objects causes the formation of a misfit region on the precipitate edge, where open volumes can be formed, and these open volumes act as positron traps.

5. Results from PAS: experimental data and simulations

The interpretation of Doppler spectra of metal alloys requires a careful analysis to extract quantitative information about the chemical environments of annihilation sites. However, this is very difficult to perform only on the basis of experimental data. Therefore, the comparison with simulated curves for which one knows exactly the material composition is a fundamental procedure to support the experimental analysis. Currently, a linear fitting procedure based on the comparison of the alloy spectrum with the spectra for positrons trapped at vacancies in pure constituent metals has been used by the experimentalists in their data analysis [24,27]. The outcome of the fit is the coefficients of the linear combination of the spectra for pure metals. Whereas this procedure tries to attribute the fraction of solute atoms surrounding the vacancy, it cannot account for the effects due to relaxations in- or outwards around the open-volume defects. On the contrary, simulations can account for the different overlaps of the electron and positron densities at the annihilation sites and of the relaxation of atoms around the vacancies. This has revealed to be very important, for instance, in the case of Cu in Al. One of the aims of the simulations is then to verify the adequacy of the fitting procedure used by the experimentalists or to give the recipe for the correct interpretation and for the correction of the linear combination coefficients.

5.1. Electron-positron momentum spectra in Al

The comparison of simulations with experiments can give information also on the size of the open volume defects. Figure 1 (from ref. [28]) shows the results of a study of thermally generated vacancies performed on 5N purity Al single crystals, heated at 417 °C and then quenched in water. Please note that no full trapping of the positrons to vacancies is achieved that way. The momentum density distribution Δ as a function of the momentum p_x is presented in figure 1a without smoothing or symmetrising, in the form of relative difference to a reference spectrum ρ_{bulk} taken for the same single crystals after careful annealing, as given by the equation

$$\Delta = \frac{\rho - \rho_{bulk}}{\rho_{bulk}} . \qquad (1)$$

In this presentation, the effect of positron trapping is immediately seen as the deviation of the Δ-curve from the zero line.

Since the annihilation spectrum ρ is related to the trapping fraction F to vacancies, i.e., to the fraction of positrons annihilating at vacancies by

$$\rho = (1 - F)\rho_{bulk} + F\rho_{vac} \quad . \qquad (2)$$

Equation (1) can be rewritten as

$$\Delta = F\frac{\rho_{vac} - \rho_{bulk}}{\rho_{bulk}} . \qquad (3)$$

This equation tells that the shape of the Δ signal depends only on the morphology of the trap, whereas the intensity is proportional to F. For the experimental details see ref. [28].

The experiment with deformed samples was carried out by using 5N purity polycrystalline Al with compression deformation applied at room temperature up to 50% thickness reduction. The measurements were performed at liquid nitrogen temperature with the same procedure as used for the quenched specimens. The results are shown in figure 1b, once more in terms of the relative difference to bulk data (equation (1)).

The comparison of figures 1a and 1b shows the following features of similarity in the Δ curves: a) a broad maximum centered at p_x=0; b) a well-defined peak near $p_x=8\times10^{-3}\ m_0c$; c) a negative tail in the high momentum region. The same features can be found in previous CDB measurements for defected Al [29,30]. On the other hand, it may be noted that the curves in figures 1a and 1b cannot be scaled to a common master curve, as it would be predicted by equation (3) if the morphology of the traps were the same in quenched and in deformed Al.

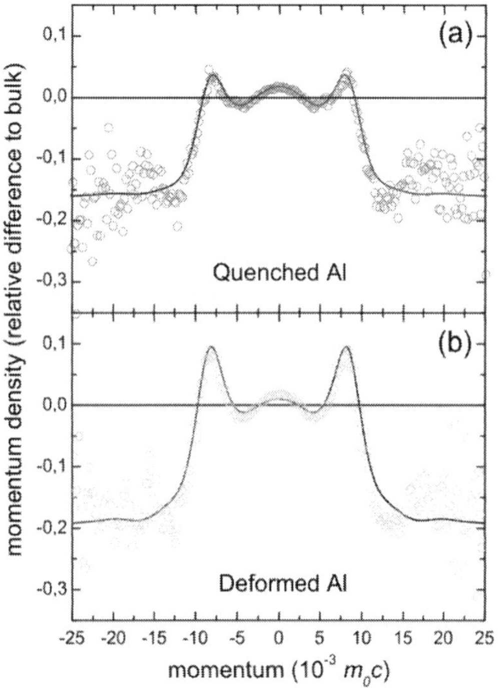

Figure 1. CDB spectrum for defected Al (relative difference to annealed Al). The solid line is the result of an ab-initio LDA calculation (see text).

Ab-initio calculations for the e^+-e^- momentum densities were performed for bulk Al and for various geometries of open-volume positron traps, with the aim of testing the possibility to reproduce the similarities and the differences in the curves in figures 1a and 1b, and to isolate the physical mechanisms to which they are to be ascribed.

A general introduction to ab-initio calculations of positron annihilation characteristics in solids can be found in ref. [31]. The procedure adopted for the simulation of curves in figures 1a and 1b was as follows. Valence electron densities were computed self-consistently in the LDA framework. The computation was carried out by employing the VASP code [35] with the projector augmented-wave (PAW) method [34] to account for the electron – ion core interaction. The PAW method enables plane-wave expansions for pseudo valence wavefunctions used in self-consistent calculations and the post-construction of all-electron valence wavefunctions resulting in accurate electron momentum densities [32]. In the calculations the Brillouin zone was sampled using a 3×3×3 Monkhorst-Pack **k**-point mesh [33]. A cutoff of 241 eV was used when calculating the pseudo valence wavefunctions and a cutoff equivalent to $70\times10^{-3}\ m_0c$ was used when forming all-electron valence wavefunctions.

Using the total charge density from the VASP-PAW calculations and the LDA [36] for electron-positron correlation potential, positron states were calculated on a three-dimensional real-space point grid [37]. The so-called "conventional scheme", in which the localized positron density does not affect the electron density, was used to describe trapped positrons. The annihilation rates of self-consistent all-electron valence states and atomic core electron states were calculated within the LDA [38] for the electron-positron correlation. These partial annihilation rates were used as weighting factors when calculating momentum densities of annihilating electron-positron pairs within the so-called state-dependent enhancement scheme [39,41].

The momentum distributions corresponding to valence electrons were obtained by the three-dimensional Fourier transform on a cubic grid with the spacing of 0.67×10^{-3} m_0c and those for the core electrons on a dense radial grid using parameterized forms of the positron wavefunction [40].

A cubic supercell of 108 atomic sites in a periodic superlattice was used for the simulation of bulk and of defected Al. Vacancy-like defects were simulated by empty atomic sites. The nearest neighbours of the missing atoms were moved isotropicly from their normal lattice sites in order to reproduce the local relaxation in the proximity of the empty site. The explored range of possible relaxations was from 2% outward to 6% inward. Larger inward displacements are expected for positron traps associated to dislocations, which can be essentially described as distorted vacancies with a reduced free volume [41]. One-dimensional momentum distributions were computed for bulk and for defected Al and convoluted with a Gaussian (FWHM=3.5×10^{-3} m_0c) to simulate the experimental resolution. Then they were used to calculate the relative difference Δ by equation (2) (with F=1).

The Δ curves in figure 2 correspond to different values of the linear relaxation at the empty site. The theoretical curves for the 0 and 4% inward linear relaxation curves are also reported in figure 1 after vertical scaling to account for a trapping fraction F<1. The best fit is for F=0.34 for the quenched sample and F=0.64 for the deformed sample.

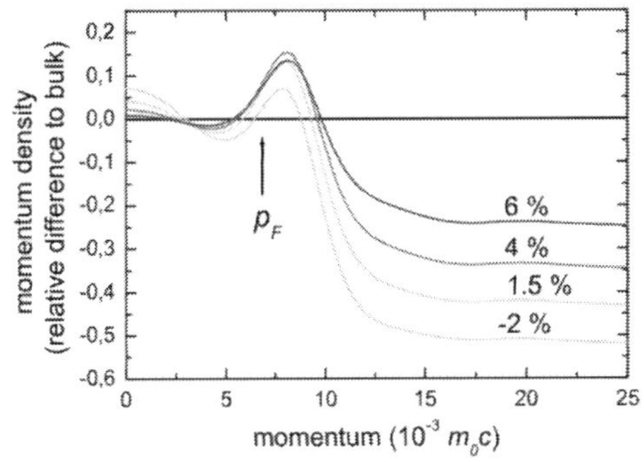

Figure 2. Calculated Δ-curves for positron trapping at an Al vacancy. Labels on the curves show the isotropic inward relaxation of the nearest-neighbour positions (here, and in the following figures, negative values are conventionally attributed to outward relaxations). The Fermi momentum p_F is denoted by the upward arrow.

The simulation offers the possibility to isolate the different factors that contribute to the formation of the characteristic shapes of the Δ curves. The following considerations can be made.

- The broad maximum at $p_x=0$ comes from the narrowing of the valence electron momentum distribution associated with the reduction of electron density at the defect. This effect is well known since the early studies of positron trapping at defects in metals [44], but is clearly visible also in semiconductors [29,30,42,43].
- The negative value taken by Δ at high momentum is simply due to the reduction of the core electron contribution to annihilation, which always occurs when the positron wavefunction becomes localized in an open-volume defect.
- The peak seen in figures 1 and 2 near to 8×10^{-3} m_0c, i.e., just beyond the Fermi cutoff of the valence electron distribution, is due to the quantum confinement of the positron wavefunction in a region of atomic dimensions. The motion of the confined particle implies a non-negligible contribution to the total momentum carried away by the annihilation radiation. Figure 3 shows

computed momentum distributions for positrons trapped at vacancy-like defects with different degrees of relaxation. A visual gauge of the momentum scale of figure 5 is provided by the horizontal double-head arrow, which shows the minimum momentum indetermination that, by the uncertainty principle, corresponds to the quantum confinement over a distance equal to the Al lattice constant of 4.0 Å.

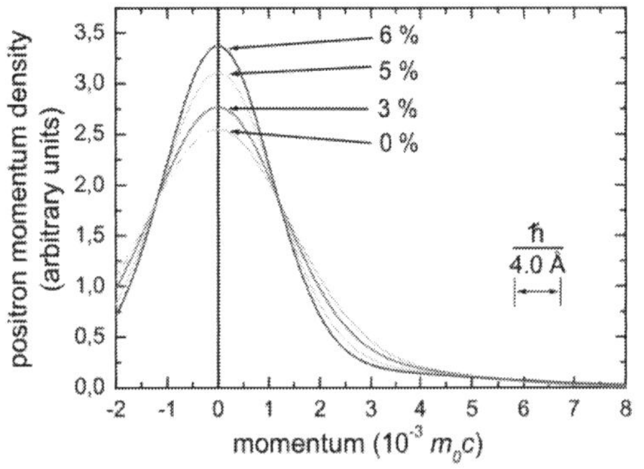

Figure 3. Positron momentum distributions for positrons trapped in vacancy-like defects. The labels on the curves indicate the inward displacement of the nearest neighbours of the empty atomic site. The horizontal bar shows the expected momentum width for quantum confinement within a box with a dimension of 4.0 Å.

The most relevant effect of positron motion is to smear the Fermi cutoff. Since the smearing enhances the counting rate beyond the Fermi cutoff, additional counts fall in a region where the counting rate in the bulk metal reference spectrum is low, thus producing a peak in the ratio curves used for presenting the CDB data. The momentum contribution of a confined positron can be added to the electron momentum distribution without including any other effect of trapping (namely, reduction of the density of valence electrons at the vacancy and changes in the annihilation rates with different electron states). This is done by convoluting the momentum distribution calculated for bulk Al, where there is no effect of positron confinement, with the positron momentum distribution that is expected in the case of trapping at a vacancy-like defect (as depicted in figure 3). In any real experiment, the visibility of the effect would depend on the resolution of the apparatus, which determines the sharpness of the Fermi cut-off.

In summary, the CDB technique shows a sensitivity to the local geometry, since the distributions taken for positrons trapped at quenched-in thermal vacancies or at dislocation-associated vacancy-like defects are clearly distinguishable.

5.2. Electron-positron momentum spectra in Al-based alloys

In the case of alloys, when CDB measurements are applied to the study of precipitations (for a review, see ref. [45]), the confinement peak can be a disturbance, as it is partially superimposed to the additional structures that are due to the decoration of vacancy-like defects by minority alloy components. This makes it impossible to reproduce the CDB distribution for an alloy as a linear combination of the distributions measured for annealed samples of pure elements. An empirical attempt to circumvent the problem has been proposed by Somoza et al. [27], who have obtained good fits of their experimental data for Al-Cu-based alloys by using linear combinations of distributions measured for defected samples of the alloy components.

First-principles calculations have been performed also for Al-based alloys, to study the possibility of quantitative chemical analysis of vacancy-solute complexes [28,29]. Figure 4 shows a comparison of simulated curves with CDB data for an Al-Cu alloy, as-quenched (a) and after 1 min ageing at 150°C (b), in order to find the most probable complexes present in the samples. The details of the procedure are explained in ref. [28].

The annihilation spectrum can be given in this case by equation (2) modified to account for the different types of traps (index k):

$$\rho = (1-F)\rho_{bulk} + F\sum_k C_k \rho_{vac,k} \quad . \tag{2'}$$

The fitting procedure by Somoza et al. [27] uses this form with $\rho_{vac,k}$'s obtained by measuring the spectra for elemental deformed metals.

In the case of the as-quenched sample the procedure by Somoza et al. gives 23 ± 1.5 % as the average number of solute atoms surrounding the vacancy, i.e., roughly three Cu atoms among the 12 nearest neighbours of the vacancy. This is in agreement with the analysis based on the first-principles calculations and shown in Fig. 4a. The figure shows best one-component fits (the trapping fraction is fitted) obtained with spectra calculated for vacancy-Cu defects with two, three and four Cu atoms as the dash-dotted (blue), solid (red) and dotted (black) lines, respectively.

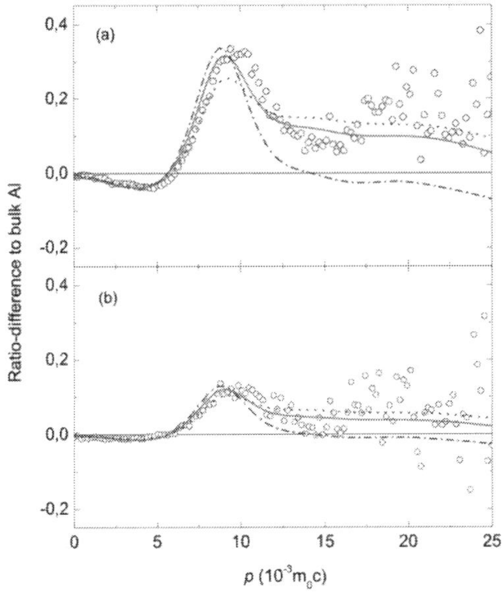

Figure 4. Comparison of simulated curves with CDB data for an Al-Cu alloy, as-quenched (a) and after 1 min ageing at 150°C (b). Dash-dotted (blue), solid (red) and dotted (black) lines show the best one-component fits obtained for vacancy-Cu defects with two, three and four Cu atoms, respectively.

In the case of the aged sample in Fig. 4b, we get a reasonable fit with three or four Cu atoms, resulting in trapping fractions of 24 and 17%, respectively. In this case the procedure by Somoza et al. gives 33 ± 5 % for the Cu contribution, that is, about four Cu atoms, and a trapping fraction of 18 ± 1 % is found.

In the case of ternary (AlCuMg) and quaternary (AlCuMgAg) alloys, the first-principles simulations of CDB spectra show that the procedure by Somoza et al. can give good estimates for the Cu and Ag concentrations and for the sum the Al and Mg concentrations but not for the Al and Mg concentrations separately. However, the conclusion about determining the Mg concentration is unclear because the calculated and measured CDB spectra for Mg are clearly different.

5.3. What is the degree of precision of the calculations

When studying defects in solids by positrons we observe changes in annihilation characteristics relative to well-known systems, typically perfect bulk lattices of elemental solids or monovacancies in them. Experimentally, the former are obtained by a careful annealing treatment and the latter can be prepared in a controlled manner by electron irradiation of the perfect bulk samples. The conclusions drawn from the changes about the defect structures and their development are then on a firm basis. This philosophy is applied further when analysing the positron data with the help of theoretical predictions. Calculated absolute values, for example the momentum distributions of annihilating electron-positron pairs, depend strongly on the treatment of the many-body effects between electrons and between electrons and the positron even for delocalized positron states in perfect lattices (For a recent work, see ref. [49]). DFT within the LDA or GGA is at the moment the only practical and reliable enough method to simulate defects in solids. However, as a mean-field theory DFT cannot accurately describe the electron-electron or electron-positron correlation effects. But when taking differences between different systems errors cancel. For example, the predicted ratio-difference curves between transition metals and Al show a good agreement with experiment [37].

Typically, the LDA underestimates slightly, usually by about 1 – 2 % the lattice constants and there may be small uncertainties in the ionic structures of the defects. The predicted positron annihilation characteristics are very sensitive to the atomic positions. In theory-experiment comparisons one approach is therefore to adjust slightly the lattice constant or in the case of vacancies in elemental solids the symmetric relaxation of atoms neighbouring the vacancy.
For example, the results for the Al-Cu alloy discussed above are obtained by adjusting the Al lattice constant to reproduce the experimental lifetime for trapped positrons in the alloy. In this case, the lifetime is about 20 ps shorter than that for Al indicating a strained lattice around the defect. The reduction of the Al lattice constant by 4 % reproduces the decrease of the lifetime. The substitution of the host atoms by solute atoms around a vacancy causes changes in the annihilation characteristics which depend on the solute atom in question. The changes are due to the changes in the electronic structure and depend also on the specific relaxation of the solute atom. In the case of Cu the changes are relatively large and therefore one can conclude that the analysis of the experimental data by the simulation results can distinguish the number solute atoms around the vacancy with the accuracy of about one atom. This is certainly enough in order to draw conclusions about the details of the clustering process.

6. Preliminary first-principles results for clustering in Al-based alloys

First-principles calculations have been performed for Al-based alloys [46] to understand, on the basis of energetics, the early stages of solute-atom clustering. Calculations of vacancy and divacancy formation energies and the impurity solubility enthalpies have also been performed in different schemes. Comparisons with existing literature values show that for these energies accurate results converged with respect to the different computation parameters can be achieved.

All the calculations of the formation and binding energies are carried out with the VASP code. It allows a comparison between LDA and GGA, and precision tests as a function of the supercell size and the number of **k**-points. Moreover, it allows the relaxation of both the atomic positions and the configurations are shown. The two-dimensional clusters are compact configurations on the (100) plane.

Although our main goal is to study the early stages of precipitation in Al-Cu-Mg(-Ag) alloys we have already performed calculations, in addition to Cu clusters in Al, also for Zn clusters in Al. This is because small Cu and Zn precipitates behave remarkably differently in Al and we want to put our first-principles approach in a stringent test. Specifically, small Cu clusters are two-dimensional platelets on (100) planes whereas small Zn clusters are three-dimensional (see, e.g., refs. [17,18,48] and the references therein).

Figures 5 and 6 show the binding energy per cluster atom for Zn and Cu clusters in Al, respectively. The clusters are build up by filling adjacent lattice sites. The two-dimensional clusters form platelets on the (100) plane. The increase of the average binding energy as the size of the cluster increases reflects the stability of the crystal growth.

The favoured configurations are two- and three-dimensional for Cu for Zn clusters, respectively. This result, which is in agreement with experimental APFIM [19,47] and HRTEM [19] data, can be interpreted on the basis of the following argument. The Wigner-Seitz radius of Zn is very close to that of Al, and therefore the elastic stress created in the matrix is minimized by the three-dimensional configuration. The Wigner-Seitz radius of Cu is smaller than that of Al. Therefore, Cu tends to create a quite a huge stress in the matrix that is minimized if the clusters form on the (100) plane of Al and the relaxation of Al atoms takes place essentially only along the [100] direction perpendicular to that plane. This is the soft direction in the fcc-Al for atomic displacements, and according to our calculations the relaxation of Al atoms below and above the (100) Cu platelet is substantial towards the platelet.

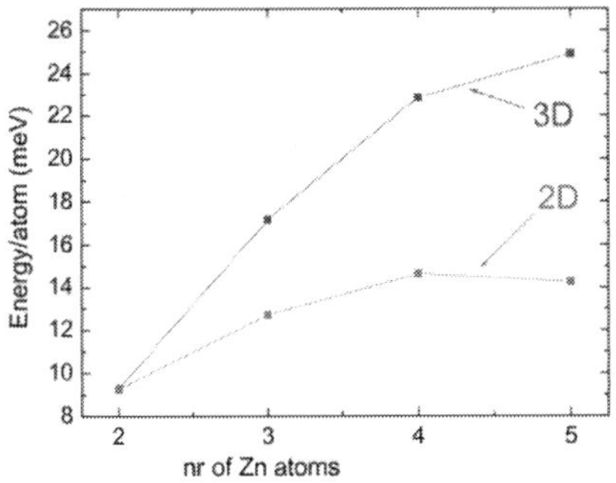

Figure 5. Binding energy per Zn atom in Zn clusters in Al. Results for two- and three-dimensional configurations are shown. The two-dimensional clusters are compact configurations on the (100) plane.

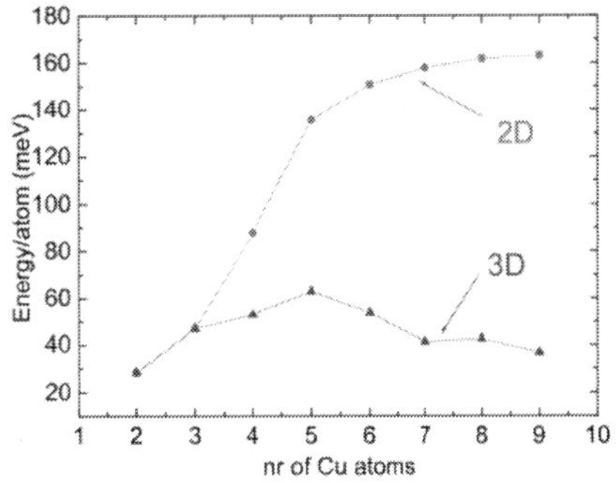

Figure 6. Binding energy per Cu atom in Cu clusters in Al.

7. Summary

In this paper we have first reviewed methods to simulate the properties of metallic alloys and the results obtained especially by the first-principles electronic structure calculations. We have advocated also another route of using the ab-initio results. Namely, in the case of positron annihilation methods we discussed how the theoretical predictions can be used to support the interpretation of measured data. The analysis of the formation of vacancy-solute-atom complexes in Al-Cu alloys was used as an example. Finally, we have presented preliminary results for the energetics of Cu and Zn clusters in Al. They give microscopic information about the early stages of the precipitation process and they can be used as the starting point for Monte-Carlo simulations for the precipitation kinetics.

References

[1] Wilm Alfred 1911 *Metallurgie* **8** 255 (1911)
[2] http://cms.mpi.univie.ac.at/vasp/, Sun G, Kürti Rajczy J P, Kertesz M, Hafner J and Kresse G 2003 THEOCHEM - *J. Molec. Struct.* **624** 37
[3] Soler J M, Artacho E, Gale J D, Garcia A, Junquera J, Ordejón P and Sánchez-Portal D 2002 *J. Phys. Cond. Matter* **14** 2745
[4] Wolverton C and Ozoliņš V 2006 *Phys. Rev.* B **73** 144104
[5] Uesugi T, Kohyama M and Higashi K 2003 *Phys. Rev.* B **68** 184103
[6] Carling K, Wahnström G, Mattson T, Mattson A, Sandberg N and Grimvall G 2000 *Phys. Rev. Lett.* **85** 3862
[7] Martin G and Soisson F 2006 *Adv. Eng. Mater.* **8** 1210
[8] Hu S Y, Baskes M I, Stan M and Chen L Q 2006 *Acta Mater.* **54** 4699
[9] Hoshino T, Papanikolau N, Zeller R, Dedrichs P H and Asato M 1999 *Comp. Mater. Sci.* **14** 56
[10] Turner D E , Zhu Z Z, Chan C T and Ho K M 1997 *Phys. Rev.* B **55** 13842
[11] Korzhavyi P A, Abrikosov I A, Ruban A V and Skriver H L 1999 *Phys. Rev.* B **59** 11693
[12] Hoshino T, Asato M, Zeller R and Dederichs P H 2004 *Phys. Rev.* B **70** 094118
[13] Clouet E, Nastar M and Sigli C 2004 *Phys. Rev.* B **69** 064109
[14] Sandberg N and Holmestad R 2006 *Phys. Rev.* B **73** 014108

[15] Turchi P E A, Drchal V, Colinet C, Kaufman L and Liu Z K 2005 *Phys. Rev.* B **71** 094206
[16] Lax D B, Ferreira L G, Froyen S and Zunger A 1992 *Phys. Rev.* B **46** 075111
[17] Wang J, Wolverton C, Müller S, Liu Z-K and Chen L-Q 2005 *Acta Mater.* **53** 2759
[18] Müller S, Wolverton C, Wang L-W and Zunger A 2001 *Europhys. Lett.* **55** 33
[19] Ringer S P and Hono K 2000 *Materials Characterization* **44** 101
[20] Staab T E M, Zamponi Z, Haaks M, Modrow H and Maier K 2007 *Phys.Stat. sol.* (rrl) **1** 172
[21] Klobes B, Staab T E M and Dudzik E 2008 *phys. stat. sol.* (rrl) **3** doi: 10.1002/pssr.200802067
[22] Nagai Y, Takadatc K, Tang Z, Ohkubo H, Sunaga H, Takizawa H and Hasegawa M 2003 *Phys. Rev.* B **67** 224202
[23] Eldrup M, Singh B N, Edwards D J, Nagai Y, Ohkubo H and Hasegawa M 2004 *Mater. Sci. Forum* **445-446** 21
[24] Somoza A, Petkov M P, Lynn K G and Dupasquier A 2002 *Phys. Rev.* B **65** 094107
[25] Folegati P, Makkonen I, Ferragut R and Puska M J 2007 *Phys. Rev.* **B 75** 054201
[26] Folegati P, Dupasquier A, Ferragut R, Iglesias M M, Makkonen I and Puska M J 2007 *phys. stat. sol.* (c) 3493
[27] Somoza A, Dupasquier A, Polmear I J, Folegati P and Ferragut R 2000 *Phys. Rev.* B **61** 014454
[28] Calloni A, Dupasquier A, Ferragut R, Folegati P, Iglesias M, Makkonen I and Puska M J 2005 *Phys. Rev.* B **72** 054112
[29] Ambigapathy R, Corbel C, Hautojärvi P, Manuel A A and Saarinen K 1995 *J. Phys.: Condens. Matter* **7** L683
[30] Ambigapathy R, Corbel C, Hautojärvi P, Manuel A A and Saarinen K 1996 *Appl. Phys.* A **62** 529
[31] Puska M J and Nieminen R N 1994 *Rev. Mod. Phys.* **66** 841
[32] Makkonen I, Hakala M and Puska M J 2008 *J. Phys. Chem Phys.* **128** 244101
[33] Monkhorst H J and Pack J D 1976 *Phys. Rev.* B **23** 5048
[34] Blöchl P E 1994 *Phys. Rev.* B **50** 17953
[35] Kresse G and Joubert D 1999 *Phys. Rev.* B **59** 1758
[36] Boronski E and Nieminen R M 1986 *Phys. Rev.* B **34** 3820
[37] Makkonen I, Hakala M and Puska M J 2006 *Phys. Rev.* B **73** 035103
[38] Alatalo M, Barbiellini B, Hakala M, Kauppinen H, Puska M J, Saarinen K, Hautojärvi P and Nieminen R M 1996 *Phys. Rev.* B **54** 2397
[39] Barbiellini B, Hakala M, Puska M J, Nieminen R M and Manuel A A 1997 *Phys. Rev.* B **56** 136
[40] Alatalo M, Kauppinen H, Saarinen K, Puska M J, Mäkinen J, Hautojärvi P and Nieminen R M 1995 *Phys. Rev.* B **51** 4176
[41] Häkkinen H, Mäkinen S and Manninen M 1990 *Phys. Rev.* B **41** 12441
[42] Ambigapathy R, Manuel A A, Hautojärvi P, Saarinen K and Corbel C 1994 *Phys. Rev.* B **50** 2188
[43] Hasegawa M, Tang Z, Chiba T, Saito M, Kawasuso A, Akahane T, Li Z Q, Kawazoe Y and Yamaguchi S 1997 *Mater. Sci. Forum* **255** 414
[44] Marceau R K W, Ferragut R, Dupasquier A, Iglesias M M and Ringer S P 2006 *Mater. Sci. Forum* **519-521** 197
[45] Dupasquier A, Kögel G and Somoza A 2004 *Acta Mater.* **52** 4707
[46] Folegati P, Kohlbach I, Makkonen I, Puska M J and Staab T E M 2008 to be published
[47] Hono K 1999 *Acta Mater.* **47** 3127
[48] Müller S, Wang L W, Zunger A and Wolverton C 1999 *Phys Rev* B **60** 16448
[49] Sormann H and Kontrym-Sznajd G 2006, *Phys. Rev.* B **73** 075111

Atomic structure of pre-Guinier-Preston and Guinier-Preston-Bagaryatsky zones in Al-alloys

Torsten E M Staab[1,4], Benedikt Klobes[1], Iris Kohlbach[1], Björn Korff[1], Matz Haaks[1], Esther Dudzik[2] and Karl Maier[1]

[1]HISKP, Universität Bonn, Nußallee 14-16, D-53115 Bonn, German
[2]Hahn-Meitner-Institut Berlin/Bessy, Albert-Einstein-Str. 15, D-12489 Berlin, Germany

E-mail: torsten.staab@isc.fraunhofer.de

Abstract. We present results on the structure of nano-sized particles (Guinier-Preston (GP) and Guinier-Preston-Bagaryatsky (GPB) zones) in Aluminum alloys. Precipitates of alloying elements like Cu, Mg, or Si hinder the motion of dislocations and, thus, are responsible for the strength of AlCuMg- and AlMgSi-alloys - used *e.g.* as AA2024 (old aircrafts) and AA6013 for the fuselage of the new Airbus A380, respectively. We will discuss the role of quenched-in vacancies for diffusive motion at room temperature (RT) enabling the growth of the precipitates. Using positron annihilation spectroscopy (PAS) – both lifetime and Doppler broadening – gives information on the local atomic environment in the vicinity of vacancies. On the other hand X-ray absorption fine structure (XAFS) spectroscopy is capable of characterizing the local atomic environment around selected elements (Cu, Mg). We will interpret the measured data by comparing them to numerical calculations of PAS and XAFS spectra. However, reliable numerical calculations of spectroscopic quantities are only possible provided that relaxed atomic positions are used as an input. We calculate those employing the ab-initio code SIESTA. Thus, considering decomposition of Al-alloys, we obtain extremely valuable information on the earliest stages, forming immediately after solution heat treatment and quenching, *i.e.* during the first few minutes of storage at RT.

1. Introduction

The hardening process in Al-alloys was discovered about 100 years ago by Alfred Wilm [1] in 1906. Wilm tried to harden different Al-alloys like steel: performing a solution heat treatment followed by quenching in water at room temperature (RT). But while steel is hardened immediately by this process it did not work for the Al-alloys Wilm was investigating. One day he had done the quenching process he left the samples for hardness testing in laboratory over a weekend. When he returned after a couple of days and did the hardness testing, the strength of the samples had increased significantly. This way Wilm found the AlCuMg alloy, which is still used in a similar composition as AA2024 for the fuselage of airplanes [2].

However, Wilm could not explain why some samples showed a higher strength than others. It took another 30 years until Guinier and Preston discovered by the new method of X-ray diffraction, independently from each other, that in pure AlCu alloys Cu precipitates as mono-atomic platelets on the (100)-directions in fcc Al – named nowadays Guinier-Preston (GP) zones [3, 4]. However, the Cu platelets became visible in the X-ray spectra first after some hours

[4] present address: Fraunhofer Institut für Silicatforschung, Neunerplatz 2, D-97082 Würzburg, Germany

Journal of Physics: Conference Series **265** (2011) 012018 doi:10.1088/1742-6596/265/1/012018

of room temperature storage following the solution heat treatment and quenching. Once the metastable phases (precipitates) have reached a certain size of some nanometers they can be imaged nowadays – at least for atoms with a significantly larger atomic charge Z than Al – by high resolution transmission electron microscopy (HRTEM) [5]. However, it is still extremely difficult to get atomic resolution pictures or selected area diffraction patterns of the earliest stages, where precipitates are smaller than 1 nm, and especially for alloying elements like Mg or Si neighboring Al in the periodic table.

Even after 100 years of technical application of these materials, it is still unclear when during the first minutes following the quenching process the first few-atom agglomerates grow to small sub-nanometer sized clusters of solute atoms, how their atomic structure looks like, and what the exact role of quenched-in vacancies is. Even though the first clusters formed consist of just a few atoms, they have a significant influence on the final strength of some aluminium-alloys. Hence, AlCuMg and AlMgSi alloys are still subject of numerous investigations (for a review see *e.g.* [6]).

We know that the precipitation process starts immediately after the heat treatment and quenching forming a super saturated solution (SSS) of the alloying elements. The formation of the agglomerates is believed to happen by diffusion via quenched-in vacancies. This is supported by the observation that once very heavy elements like lead (Pb) are added to the alloy, the precipitation process is delayed, since these large atoms capture vacancies.

Since positron annihilation spectroscopy (PAS) is a unique tool to detect vacancies and vacancy-like defect in metals, aluminum alloy were systematically investigated – first by Dlubek an co-worker at the end of the 1970s [7] (for recent reviews see [8, 9]).

However, it was discovered recently that one has to freeze the decomposition process immediately after quenching to prevent the sample from changing during the course of taking one single spectrum within positron annihilation lifetime spectroscopy (PALS) or measuring the momentum distribution with Doppler broadening of the annihilation radiation (DBAR) [10].

Even if freezing the stage of early decomposition is possible, the problem remains to identify the atomic surrounding of vacancies and solute atoms to characterize the kind of cluster they have been forming. We have to note here that the relative concentration of quenched-in vacancies and solute atoms is about 1/1000. Hence, we need an additional method characterizing early precipitates of few atoms *without* a vacancy attached, and thus, possible not able to trap positrons. Recently, the method of X-ray absorption fine structure (XAFS) spectroscopy has been re-invented after more than 25 years for the characterization of the surrounding of selected elements in Al-alloys [11, 12].

In this paper we will show how the experimental methods PAS and XAFS can be combined with numerical simulations of obtain a characterization of early decomposition stages in Al-alloys.

2. Review of existing methods

While the measurement of pure well annealed metals has been proposed as a way obtaining a fingerprint in the momentum distribution measured by positron annihilation [13], we are convinced that these attempts did show the correct route but turned out to be insufficient.

To our mind the main problem is that in well annealed metals the positrons 'see' a totally different electronic density and momentum distribution as if the positron is localized in a vacancy or a vacancy-like defect, *e.g.* a vacancy solute-atom pair or precipitates containing structural vacancies. However, taking this route only a rough qualitative analysis is possible. We can detect certainly the influence of several alloying element in *e.g.* the momentum distribution of a AlMgSiCu alloy, as displayed in figure 1 involving a single Ge-detector with the setup and data analysis as described in [14, 15]. One can clearly see the influence of Cu-additions in AA6013. However, any qualitative analysis is impossible, *i.e.* identifying which solute atom is attached to

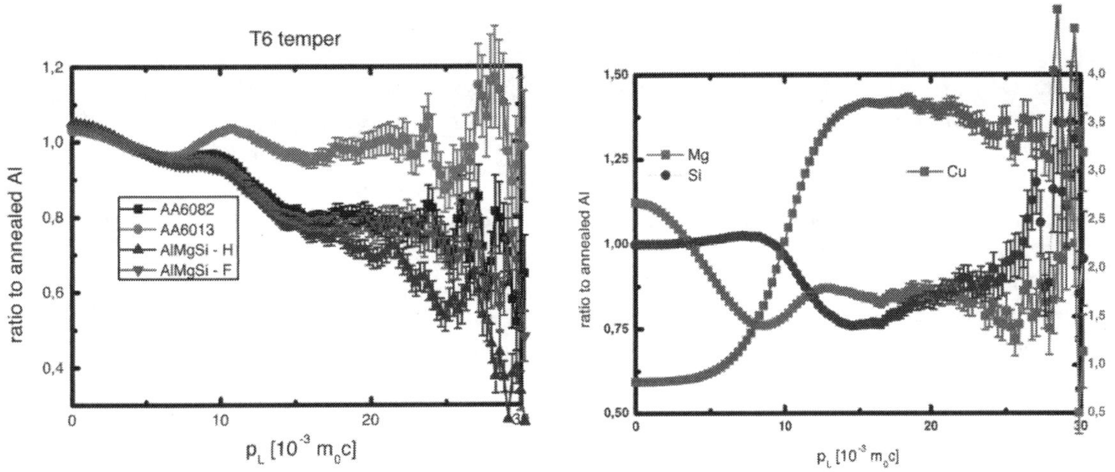

Figure 1. Pure elements – well-annealed samples as ratio to Al bulk (right) and AlMgSi(Cu)-alloys at the so called T6-state of maximum hardness (left side). The temperature treatment for T6-state is 180°C for 4h after solution heat treatment and quenching. In comparison to the pure elements (right) we can see clearly the influence of Cu-additions in the alloy AA6013 – the only one containing copper. Please, note the different scale for copper (right).

the vacancies where the positron is localized, what would the atomic structure of precipitates be.

It has been tried to overcome this problem by using samples of the alloying elements (Cu, Mg, Si) where the positron is localized in vacancy defects, in order to mimic the density and momentum of electrons around vacancy-solute atom pairs in aluminum alloys. However, the reference samples should be prepared in such a way that they show single component spectra. Otherwise it will be very hard to extract exactly the correct momentum distribution for the vacancy defects - even if a parallel measurements of the momentum distribution and the positron lifetime are performed.

However, the main problem remains the sample preparation. We have to ensure that only one type of defect – namely the mono-vacancy – will be created, and additionally, at a concentration leading to complete positron trapping. To the authors knowledge there are only three methods of sample preparation existing to achieve this goal: in-situ measurements at high temperature of thermally generated vacancies (see for a review [16, 17]), low temperature electron irradiation (see *e.g.* [18]), and quenching from a temperature just below the melting point.

For the first case the main experimental challenge is due to the problem of sealing the positron source and mounting the detectors close enough to the sample – especially for positron lifetime measurements. However, PALS measurements have been done – to the authors knowledge – only for noble metals like Cu, Ag, Au, while for *e.g.* refractory metals like W, Mo or Ta only measurements of the momentum distribution exist [19].

For the other two case the main experimental problem is that in most metals, *e.g.* the noble series (Cu, Ag, Au) but also in Al, Fe, Ni, Zn, Mg and for many others, vacancies are highly mobile at room temperature. Exceptions are found for refractory metals with a high melting point (W, Mo, Ta) but also for Pt. This means a low-temperature electron or proton irradiation experiment is necessary, where the samples have to be kept cold and transferred to the PAS measurement without warming them up. However, in rare cases even the liquid nitrogen temperature of 77 K would be insufficient for preventing vacancies to become mobile.

Quenching experiment exhibit the additional difficulty that a certain amount of vacancies is lost during the quench or may agglomerate to form vacancy clusters.

On the other hand, for In or Ga their melting point is so low that even at room temperature a measurable amount of thermal vacancies exists. Hence, for determining the positron annihilation properties for the bulk of these materials one would have to cool the samples.

However, even if it is possible to prepare all reference samples in an accurate way and perform all the measurements at a sufficient accuracy, the main problem cannot be circumvented by experiments alone. The electron density (measured by PALS) and the electron momentum density (measured by DBAR) cannot be exactly interpolated from *e.g.* mono-vacancies in pure copper (12 nearest neighbors) to a vacancy solute atom pair like Cu in Al. This is due to the fact that certain relaxations of atoms around a vacancy occur, which can only be estimate accurately by ab-initio calculations [20, 21]. Additionally, the electron density, and thus also the electron momentum density, changes by an energetically favorable redistribution of charge in the solid.

Nevertheless, those experiments, as described above, are indispensable for testing the calculation schemes involved in producing spectra to compare to experiment. Even nowadays the calculation schemes suffer from many approximations resulting in deficiencies describing the properties of solids exactly.

3. Experimentals

Two high purity Al alloy has been investigated: one containing 1.7at% of Cu (named AlCu in the following) and one with 1.87at% of Cu and additionally 1.78at% Mg (named AlCuMg in the following). The latter has a similar composition as AA2024. The alloys were solution heat treated at 560°C for 30 minutes and then rapidly quenched into water at room temperature. For the PAS measurement the samples was taken out of the water, mounted to a holder, and immediately cooled to −40°C. The time delay until the samples reached a temperature of 0°C was less than 2 minutes. The samples for XAFS measurement were taken out of the water and immediately cooled in liquid nitrogen. They were kept at temperatures below −70°C throughout transportation to BESSY and mounting to a cryostat for the synchrotron measurements (for details see [10, 12]). All XAFS measurements were taken at the 7T-MPW-mags synchrotron beamline at BESSY in Berlin. The XAFS spectra were recorded at the Cu K-edge in fluorescence mode. For the energy range from 8900 to 9400 eV each one of the 1000 points per spectrum contains about 10^5 counts.

During all measurements the samples were kept at −40°C during the measurements to assure that all decomposition kinetics was frozen [22]. In order to simulate natural aging, the samples were then exposed to room temperature (RT) for increasing time intervals and cooled again to −40°C for the measurement as shown in figure 2.

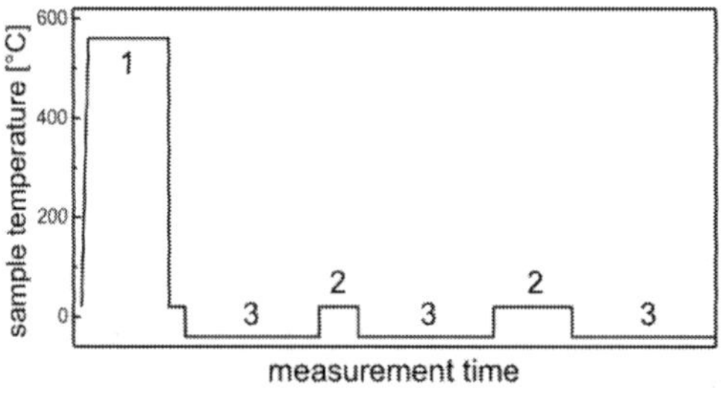

Figure 2. Measurement sequence: after solution heat treatment (1) the samples were quenched into water at RT. Measurements (3) were taken while the samples were cooled to −40°C. In between different measurements the sample were exposed successively to increasing time intervals at room temperature (natural aging).

The positron lifetime spectra were taken with a conventional fast-fast coincidence setup employing BaF_2 detectors with a FWHM time resolution of about 230 ps. Collecting about 17 million counts per spectrum guaranteed a reliable data analysis. Each sample was arranged in sandwich geometry embedding a ^{22}Na Al-foil source (30 μCi) between two identical alloy disks. The Doppler Broadening of the annihilation radiation (DBAR) has been measured at a temperature of $-40°$C with a ^{68}Ge source employing a single pure Ge-detector as described in [14, 15].

4. Numerical methods

4.1. SIESTA method

For the calculation of the total energies and the structural relaxations we used the ab-initio code SIESTA [23]. This code implements the density functional theory (DFT) in the local density approximation (LDA). The Brillouin-zone is sampled by a Monkhorst-Pack (MP) mesh. In contrast to other established ab-initio DFT-methods, instead of a plane wave basis, a local Gaussian basis set is employed within the split-valence method [24].

For our calculations we used a DZP basis set for the atomic valence orbitals of Al and Cu. The corresponding pseudo potentials were generated using the Troullier-Martins flavor without any core-corrections (for details see [25]).

Usually, testing the employed pseudo potentials involves Al in Al bulk crystals and Cu in Cu bulk crystals. However, we also tested the generated pseudo-potentials for the Al-Cu interactions by calculating the coherent and semi-coherent precipitation phases Θ and Θ' (see details in [25]). Additionally, we calculated vacancy formation energies for mono-vacancies in bulk Al and bulk Cu (108 atom supercells), leading to values close to experimental data or other ab-initio calculations (see [25] for details and references).

As atomic structure of the equilibrium S-phase we assumed an orthorhombic crystal structure (symmetry group Cmcm) with lattice parameters $a = 400$ pm, $b = 923$ pm, and $c = 714$ pm for Al_2CuMg [6].

4.2. Positron lifetime and momentum distribution calculations

We utilize here structures of Cu-platelets relaxed with the first-principles method SIESTA (see section 4.1). Then theoretical calculations of positron annihilation parameters are performed employing free atomic wave functions using the method introduced in refs. [26, 27] in the improved version of [28]. The momentum distribution is obtained by summing up the contributions from each core electron state weighted with its partial annihilation rate. The partial annihilation rate is calculated within the generalized gradient approximation (GGA) of electron-positron correlation [29]. The calculation of positron lifetimes is performed using the local-density approximation (LDA) scheme employing Boronski-Nieminen enhancement [30].

The supercells for positron calculations were created by embedding the SIESTA-relaxed atoms (3 unit cell containing 108 atoms) in larger cells extended with bulk lattice atoms to ensure the localization of the positron wave function. Thus, one avoids interactions of the simulated positron with its periodic replica. The localization of the positron inside the calculation cell is ensured by cutting the supercell with relaxed atoms at a plane with the greatest possible distance to the defect and inserting a number of ideal unit cells. A sufficient size of the supercell has been found by increasing the size until the positron lifetime did not longer change by a significant value – here chosen with 0.2 ps. A similar scheme has been used to find the right number of real space grid points for the calculation. For the mini-Guinier-Preston zones (GPZ) $5 \times 5 \times 5$ unit cells with 500 atoms have been employed (16 real space grid points per unit cell) (for more detail see [25]).

4.3. XAFS-calculations with FEFF-8.2
By employing the FEFF-8.2 code [31] we obtained the simulated XAFS-spectra presented here. The full multiple scattering calculation involved a cluster of at least 87 atoms, corresponding in fcc-Al to about six shells of atoms around the scattering center. A Debye-Waller factor ($T = 233\,\mathrm{K}$) has been employed to take into account the atomic vibrations due to the finite sample temperature (the typical measurement temperature to freeze in the decomposition [12]). A discussion of typical deviations between measured and calculated XAFS-spectra has been presented in [32].

5. Results
The relaxations around an isolated Cu atom in the Al matrix shows up as a decrease in the distances of nearest neighbor atoms from 286.34 pm to 280.5 pm, which is 2.0% less than the ideal bond length in fcc Al [25]. This result is in good agreement with other ab-initio calculations [33] and XAFS experiments for dilute AlCu-alloys giving 280 pm [34].

5.1. Pre-Guinier-Preston zones in AlCu
It is well established experimentally that Cu precipitates in Al as mono-atomic platelets on the (100)-planes in fcc Al (see *e.g.* the textbook by Haasen [35]). Hence, we consider different structures of small Cu-disks up to 5 Cu atoms arranged as closest neighbors on one face of an fcc cube (see Figure 3). However, it has been shown recently that 3D-clusters are less stable than 2D-platelets of Cu [21].

After the first stage of decomposition when the first atom pairs have formed, the precipitates have to grow under Cu-poor conditions – compared to the stoichiometry of the the stable Al_2Cu-phase Θ. Hence, we calculated some structures of Cu-disks containing vacancies. Additionally, it is experimentally well established that one detects in as-quenched pure AlCu- AlCuMg-alloys positron lifetimes which should be related to an open volume of about one mono-vacancy in Al. Hence, it may be that at least a fraction of GP zones contains structural vacancies as shown in figure 4.

For mini-GPZ with or without vacancies we estimated the positron binding energy E_b to the precipitates as the difference of the total energy of the positron in Al-bulk. As we can see from figure 3 it is increasing for the GPZ without structural vacancies with the size of the precipitate reaching some 100 meV for infinitely extended GPZ. If vacancies are involved, the binding energy is decreasing with the number of Cu-atoms in the GPZ (figure 4). This may be related to the enhanced relaxation of Al-atoms towards the Cu disk with increasing number of Cu atom.

A more detailed view on the relaxation features is given in the following figures (Figure 5 and 6). Generally, we find relaxations of Cu atoms away from each other within the (100) habit plane and strong relaxation of Al-atoms above and below the habit-plane towards the Cu-platelet. For a more detailed discussion see [25].

When one Al-atom is embedded in 4 Cu atoms on a (001)-plane in Al we did not observe any pronounced relaxation. This is in contrast to the other cases where Cu atoms are on NN positions. However, this structure is more than 0.5 eV less favorable in energy than 4 Cu atoms in closest positions on an (001)-plane in Al as in figure 6 [25].

Concerning the comparison to experimental methods we find that the positron wave function is generally localized around Cu atoms for ideal, *i.e.* not vacancy containing, precipitates (see Figure 3). The relaxation features creating more open space in between the Cu atoms enhance this effect (see Figure 3 – right). However, the positron lifetime is only slightly changed in the range from 165 to 168 ps. Thus, ideal or relaxed GPZ without structural vacancies cannot explain the experimental findings of 220 ps after quenching and storage at room temperature [36].

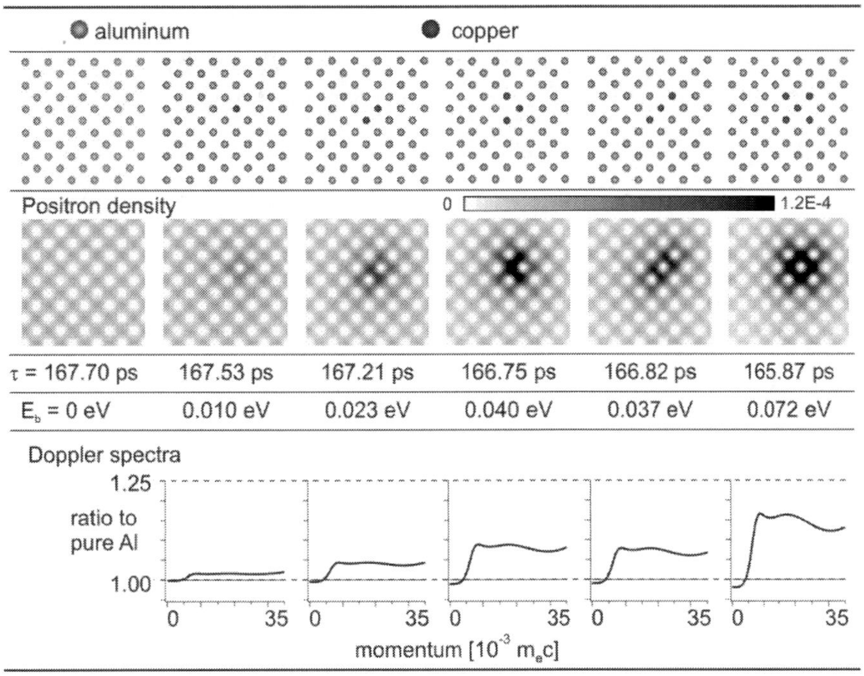

Figure 3. Early stages of GP-zones without vacancies in the (001)-plane of the Al-lattice: shown are atomic positions, positron densities, lifetimes, and binding energy E_b of the positron as well as Doppler spectra for the atomic configurations (from top to bottom). The momentum distribution is obviously changed by substitutional Cu-atoms – especially when the size is above three Cu-atoms, while the positron lifetime is largely unaffected. It stays very similar to that in Al-bulk.

Precipitates containing vacancies lead to calculated positron lifetimes of about 220 ps – in agreement with the experimental data (see Figure 4). When comparing the two configurations of four Cu, one with a vacancy inside and besides the Cu-atoms, one can see that the location of the vacancy besides (220 . . . 230 ps) is more compatible with experiment than inside the disk (205 ps). Indeed the former structure is also more stable [33].

Figure 7 shows a XAFS spectrum in terms of the fine-structure function $\mu(E)$ which is defined in the usual way after subtraction of a smooth background function.

Concerning XAFS measurements and calculations we can see from figure 7 that there is a good correspondence[5] between measured and calculated XAFS spectra for a pure AlCu sample where the applied heat treatment ensured the formation of the equilibrium Θ-phase (Al_2Cu). Other phases like the metastable Θ' or Θ" did not fit. For the atomic structure of the Θ or Θ' phase we refer to [35] or [25].

Measurements of as-quenched samples of the pure AlCu alloy showed that even during the quenching process Cu-pairs and possibly also triples must have formed, which becomes evident simply by statistical considerations. Figure 8 shows a superposition of XAFS spectra for isolated Cu atoms and for a pre-GPZ displayed in figure 5.

[5] This usually means a coincidence in the positions and height the maxima and minima of the XAFS spectra.

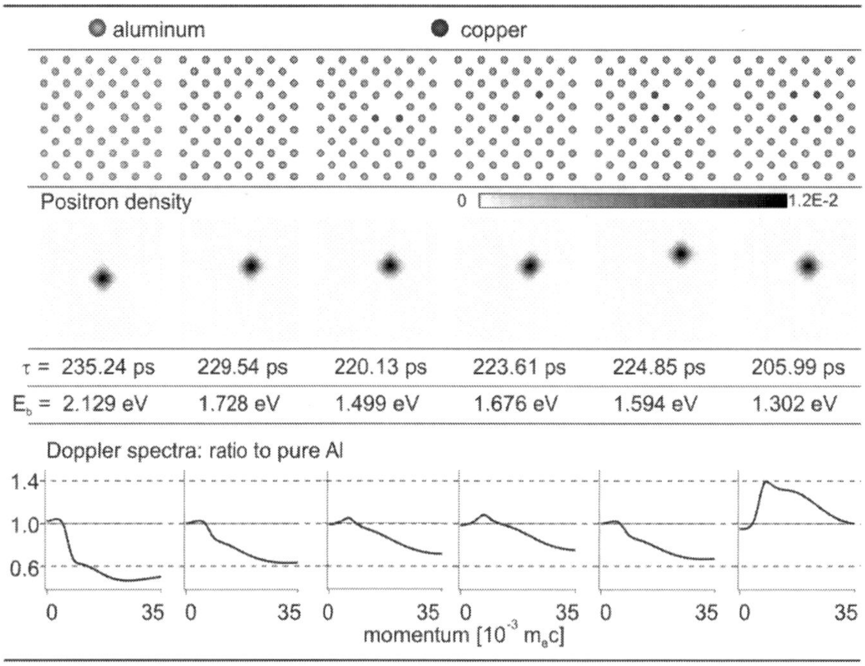

Figure 4. Early stages of GP-zones with one vacancy in the (001)-plane of the Al-lattice: Atomic positions, positron densities, lifetimes, and binding energy E_b of the positron as well as Doppler spectra of the displayed atomic configurations (from top to bottom). As expected the momentum distribution is changing from the left to the right towards more Cu-rich environment. The positron lifetime, on the other hand, is only affected significantly, if the vacancy is either inside the or outside the Cu-disk. Here the binding energy is decreasing with the number of Cu atoms

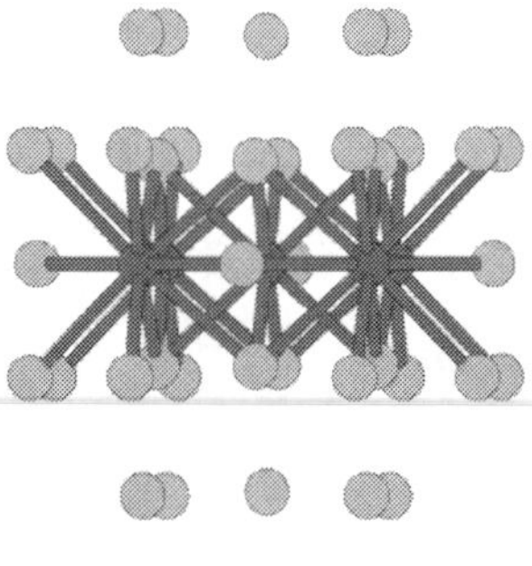

Figure 5. Mini-GPZ consisting of 5 Cu atoms (red): atomic structure as calculated with SIESTA. The strong relaxation of Al-planes (grey) above and below the Cu-platelet towards it results in some Al-Cu bond length shortened to less than 270 pm.

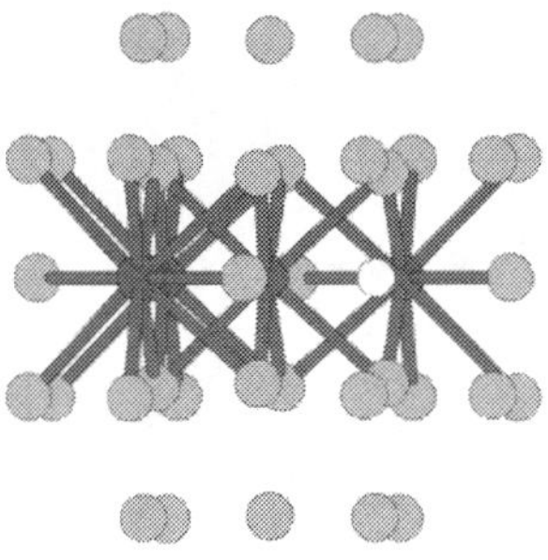

Figure 6. Mini-GPZ consisting of 4 Cu atoms (red) and a vacancy (open circle): atomic structure as calculated with SIESTA. The strong relaxation of Al-planes (gray) above and below the Cu-platelet towards it is not changed much by the presence of the vacancy.

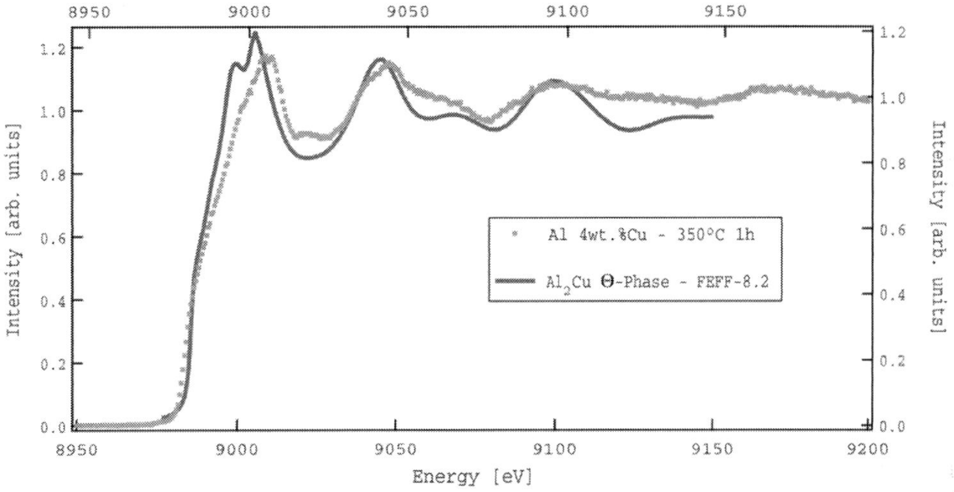

Figure 7. XAFS spectrum from the Al 1.7at.% Cu alloy heat treated for 1 hour at 350°C (green dots) in comparison to FEFF-8.2 calculations of the Θ-phase (straight line).

5.2. Pre-Guinier-Preston-Bagaryatsky zones in AlCuMg - PAS

Guinier-Preston-Bagaryatsky (GPB) zones are named after Yu. A. Bagaryatsky who investigated AlCuMg alloys by X-ray diffraction both after natural aging [37] and artificial aging [38]. Bagaryatsky proposed a certain structure for the arrangement of Cu and Mg atoms in small nanometer-sized cluster, which has been later questioned by many authors according to their own X-ray diffraction results.

In a recent paper it has been shown by both XAFS and PAS that the atomic configuration in the technical AlCuMg alloy AA2024 showed after 10 − 80 hours aging at room temperature that the atomic environment seen by positrons must be very similar to a vacancy on the copper sublattice of the equilibrium S-phase (Al_2CuMg) [39]. But the equilibrium S-phase is formed first after heat treating the AlCuMg alloys for some hours at temperature above 200°C, so it had not yet formed in our samples. However, it has been shown by Nagai *et al* that the momentum distribution 1 hour after quenching is very similar to that when the S-phase must have been formed according to their temperature treatment [40].

Now, very recently it could be shown that changes during the first minutes after quenching are detectable by PAS [10]. The samples measured by PALS and DBAR in this work were prepared according to the procedure described in section 3. For all papers published previously

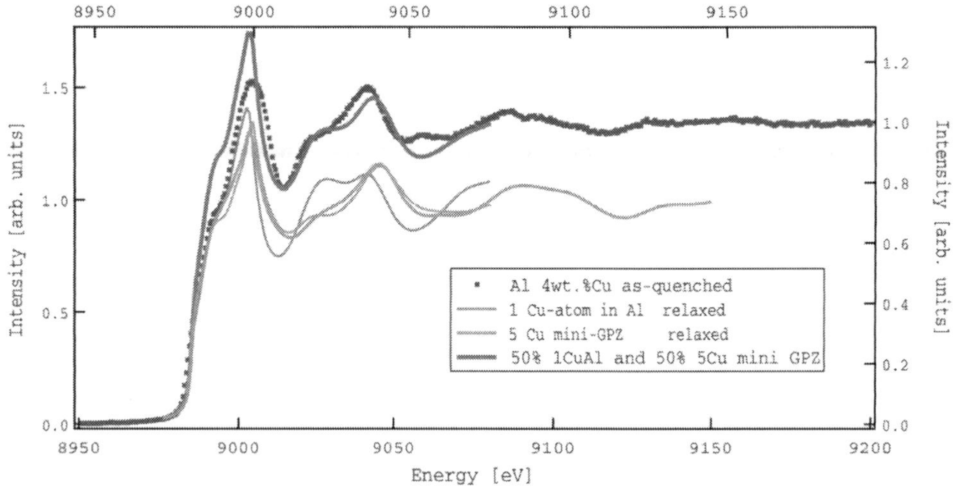

Figure 8. XAFS spectrum from the pure Al 1.7at.%Cu alloy in the as-quenched state (blue dots) in comparison to FEFF-8.2 calculations (straight lines) of an isolated Cu atom and a pre-GPZ as displayed in figure 5 (middle Cu taken as scattering center). The straight red line is a superposition of the spectra for the single Cu atoms and the pre-GPZ (50% each).

the authors reported single component lifetime spectra for AlCuMg-based alloys [41, 42, 43] when measurements started after quenching. However, taking one spectrum usually takes about 30 – 60 minutes at least. Measuring at room temperature we agree with the other authors finding an average positron lifetime of about 220 ps from single-component lifetime spectra in AlCuMg and AlMgSi alloys. But applying the experimental procedure described in section 3 (The time/temperature cycle applied has been displayed in figure 2) we find a completely different behaviour of the positron lifetime.

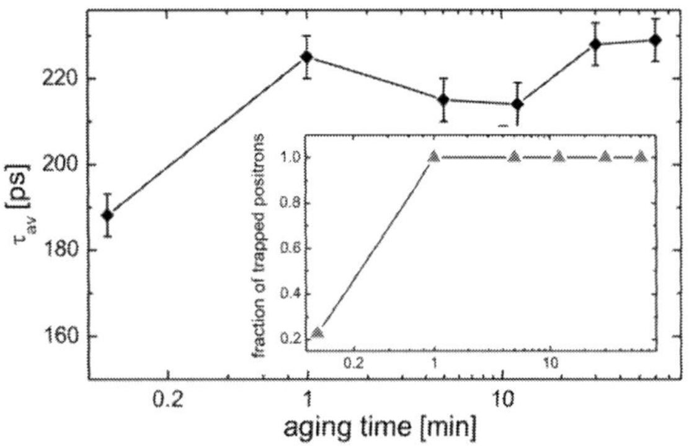

Figure 9. AlCuMg alloy: Mean positron lifetime during RT storage. Only the as-quenched state did not show a single-component lifetime spectrum and, *i.e.* complete trapping.

Figure 9 shows the changes during the first minutes of aging at room temperature. The lifetime spectrum is composed of two components – the longer one close to the known value for vacancies in Al: about 240 ps. After one minute of room temperature storage we detect a jump in the average lifetime from 188 to more than 220 ps[6], which can hardly be explained by the creation of defects. Rather the trapping potential may change as described below.

[6] 220 ps is typical for vacancy solute atom pairs or structural vacancies inside precipitates (see figure 4)

Decomposing the spectrum according to a simple two-component trapping model, where the trapping fraction amounts to 0.30, and a trapping coefficient of $10^{14}s^{-1}$ [44], we obtain a quenched-in concentration of mono-vacancies in the order of 2×10^{-5} per atom. This is a reasonable fraction of the original concentration of thermal vacancies at the temperature for the solution heat treatment. As similar value has been found for AlMgSi alloys [10].

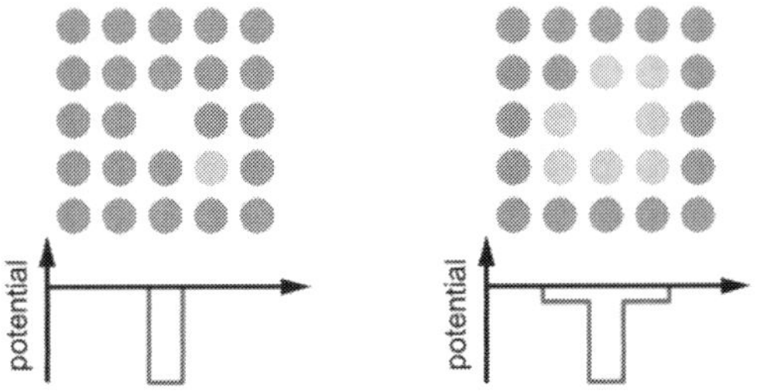

Figure 10. Before complete trapping is observed, the positrons are trapped by vacancy-solute complexes (left). After decomposition started structural vacancies inside agglomerate (right side) may be a better candidate. Gray circles: Al atoms; yellow circles: solute atoms.

However, it is difficult to identify the atomic structure of the defects detected by PALS alone. We can just assume that the signal must arise from vacancy-solute-atom pairs since isolated vacancies become mobile in pure Al at about 220 K. In principal, there are several possibilities to interpret the observed incomplete trapping of positrons during the first minutes of RT aging: The positrons could be scattered by the solute atoms [45], meaning that the diffusion length is effectively reduced so that not all positrons can reach a vacancy-solute pair during their lifetime.

During the ongoing decomposition the solute atoms aggregate. Thus, the number of scattering centers would decrease, the positrons could diffuse farther and the fraction of trapped positrons would increase, leading to the complete trapping observed (figure 10 – left).

Another possibility would be that at the beginning of RT aging the trapping is transition limited [45]. The agglomeration of solute atoms around the vacancies would then change the attractive potential sensed by the positron, leading to an increasing trapping probability (right side of figure 10). A decision may be possible according to DBAR measurements.

In figure 11 we present measurements of the same AlCuMg sample material by DBAR, as described in section 3 employing the tool of high-momentum analysis (HMA) [14, 15]. In comparison to numerical calculations we can reveal a possible atomic structure of the defects responsible for positron trapping.

We tried many different planar configurations of Cu or Mg atoms attached to the vacancy. However, none of these simple vacancy-solute atom pairs – even with two Cu or Mg atoms attached on both sides of the vacancy – would fit the experimental data; neither did any considered structure model for the metastable S'-phase do the job.

Thus, we propose here – inspired by the old work of Bagaryatsky [37, 38] – the first attempt of a 3-dimensional structural model for the initial stage of decomposition, *i.e.* an atomistic model of possible Cu-Mg-clusters formed, which may be the pre-cursor of the growing GPB-zones (see figure 12). However, up to now we could only employ ideal coordinates from the fcc Al lattice, while there is work in progress determining relaxed positions by ab-initio calculation.

Considering that the bump in the displayed momentum distribution between 7 and $10 \times 10^{-3} m_0 c$ is an artifact of the employed calculation scheme, this model could be a first building block of the long seeked atomistic structure of Guinier-Preston-Bagaryatsky zones. We will propose a probable structure in a forthcoming paper, once we find support by experimental work.

5.3. Pre-Guinier-Preston-Bagaryatsky zones in AlCuMg - XAFS

XAFS measurement were analyzed in terms of the fine-structure function $\chi(k)$ which is defined in the usual way as the difference between the measured absorption coefficient and a smooth

Journal of Physics: Conference Series **265** (2011) 012018 doi:10.1088/1742-6596/265/1/012018

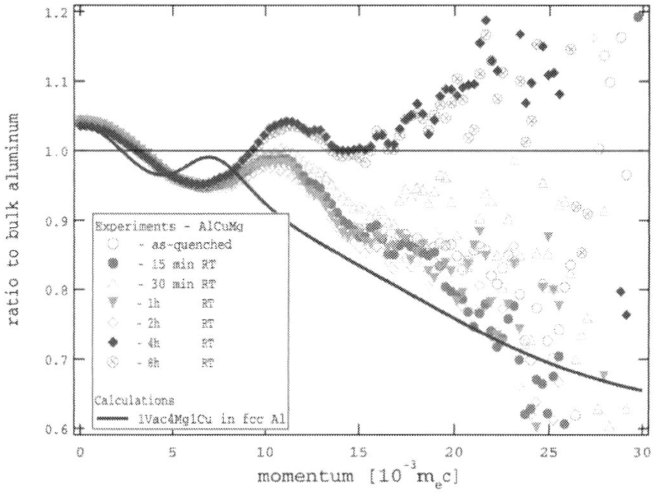

Figure 11. AlCuMg alloy (symbols) in comparison to calculations of the momentum distribution (straight line) involving a small cluster of 1 Cu and 4 Mg atoms around a vacancy as shown in figure 12.

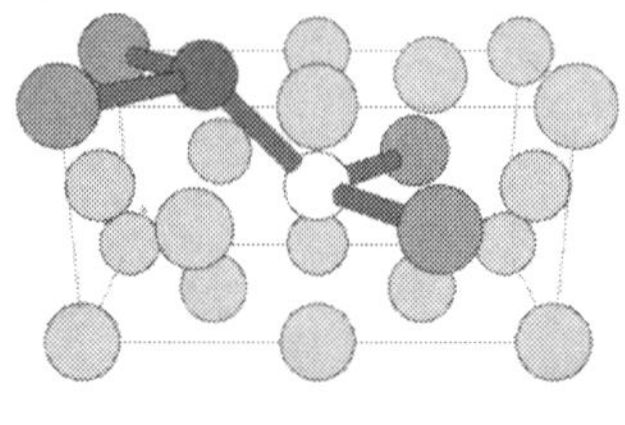

Figure 12. Atomistic model of a small cluster of 1 Cu (red) and 4 Mg atoms (green) around a vacancy (open circle) employing unrelaxed coordinates in fcc Al (Al atoms in gray). For this structure the DBAR calculation in figure 11 has been performed.

background function representing the absorption of an isolated atoms divided by the jump of the absorption coefficient at the threshold energy of the Cu K-edge energy. The applied method of data analysis is described in details elsewhere [46]. Furthermore the Fourier transform of the fine-structure function yields a pseudo radial distribution function (PRDF) of the next neighbors of the absorbing atoms.

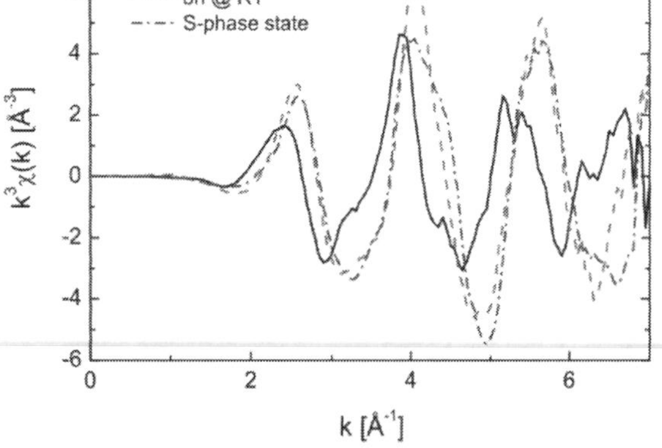

Figure 13. X-ray absorption fine-structure (XAFS) data after processing: For the ternary AlCuMg three different stages of decomposition are shown. During the first eight hours of room temperature storage the XAFS spectra change drastically (see figure 14). Relaxation processes around Cu-atoms are revealed by the pseudo radial distribution function (figure 14). They start after eight minutes of RT storage.

Figure 13 shows the spectra for three different stages of decomposition of an AlCuMg sample: the as-quenched state, after 8 hours room temperature aging, and Cu embedded in S-phase precipitates (sample treatment: 1 hour at 350°C).

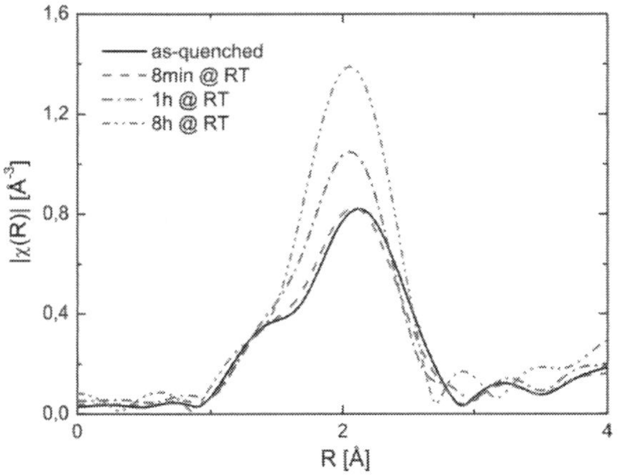

Figure 14. AlCuMg alloy: Fourier transformation of the XAFS spectra of the early stage of decomposition. Relaxation processes around Cu-atoms are revealed by the pseudo radial distribution function. They start after eight minutes of RT storage.

Note that the as-quenched state differs significantly from the other two. Zero-crossing points are shifted towards higher wavevectors and the amplitudes change as well. These systematic changes are also reflected in the Fourier transform (see figure 14). The sample exhibits an increasing relaxation of the first shell of NN atoms towards the absorbing Cu atom with increasing storage time. This can be seen from a shift of the center of the first peak of the respective PRDF towards smaller distances. After eight minutes of RT storage the onset of the relaxation process is detectable for the AlCuMg alloy continuing during the following hours of RT storage.

This is in good agreement with PAS measurements described in section 5.2 showing changes of the atomic surrounding of vacancies during the first few minutes. Additionally, results of Starink *et al* point to the same feature, who detected Cu-rich cluster in an AlCuMg alloy after 15 minutes of RT storage by atom probe investigations [47].

Finally, we present a comparison of the sample (1 hour 350°C) resembling the stable S-phase with FEFF-calculations.

Figure 15 shows the XAFS spectrum for the S-phase state as in figure 13 without the division of the jump of the absorption coefficient at the threshold energy of the Cu K-edge energy for better showing the features at the absorption edge. We can see a good agreement with the FEFF-calculation for the S-phase confirming the reliability of the calculations. Unfortunately, since so far no atomistic model for pre-GPG zones is existing, we cannot present any calculations here.

6. Summary

For aluminum alloys containing Cu and Mg we could demonstrate the sensitivity of both XAFS and PAS to the very early stages of decomposition – just a few minutes after quenching.

Besides possibly atom probe methods XAFS is one of the few tools which can give reasonable information about the first stages during the ongoing decomposition process. In conjunction with numerical simulation methods even the atomic structure of few atom clusters can be revealed.

PAS on the other hand is unique in showing the atomic surrounding of quenched-in vacancies and their fate during the ongoing decomposition process. However, as soon as PAS detects complete trapping information on the density of defects is lost. Nevertheless, their atomic structure may be revealed by combining positron lifetime spectroscopy (PALS) with Doppler broadening of the annihilation radiation (DBAR) in comparison to numerical simulation tools.

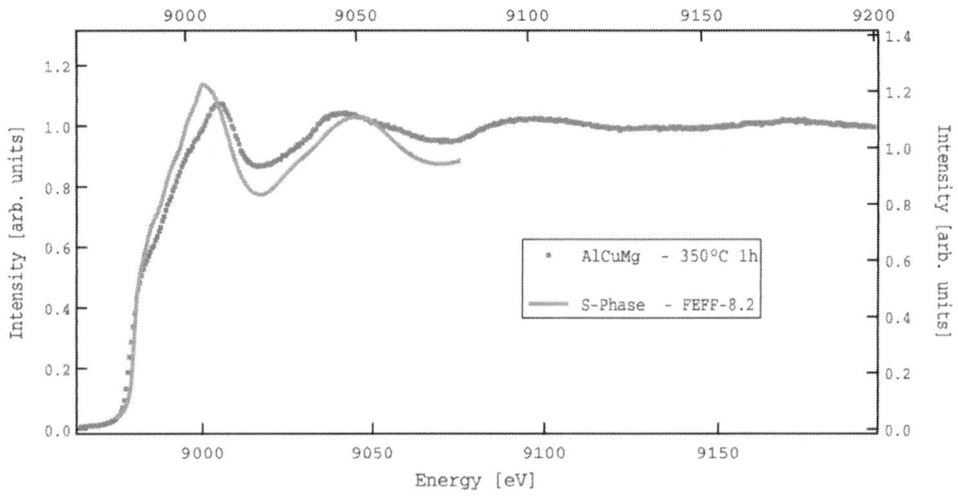

Figure 15. XAFS spectrum from the AlCuMg alloy heat treated for 1 hour at 350°C – S-phase state (red dots) – in comparison to FEFF-8.2 calculations of the S-phase (straight line).

PAS measurements of samples immediately cooled after quenching showed that the distribution of solute atoms is altering significantly during the first few minutes after quenching.

This is confirmed by XAFS showing in the case of the AlCuMg alloy that the first relaxation process involving the Cu atoms occurs just after a few minutes of RT storage following the quenching. Interpreting this as the formation of the first agglomerates is supported by recent atom probe investigations [47].

We may have found the first building block of the mysterious Guinier-Preston-Bagaryatsky zone – named pre-Guinier-Preston-Bagaryatsky zone – which has been constructed as a 3-dimensional arrangement of Cu and Mg atoms according to the guidelines suggested by Bagaryatsky more than 50 year ago. We hope that our structural model the GPB zones (to be presented elsewhere) will find support by further measurement.

Acknowledgments

Acknowledgments We have to thank the John-von-Neumann Institute for Computing Jülich for providing CPU-time on the JUMP supercomputer where most of the heavier calculation have been performed (Grant hbn14) and BESSY for providing synchrotron beam time and support in general as well as the DFG (Sta 527/3-1) for funding.

Last not least we have to thank Angela Vasilescu very much for translating the original work of Bagaryatsky from Russian to English during the PSPA-37 meeting last year.

References

[1] Wilm A 1911 *Metallurgie* **8** 255–
[2] Zschech E 1996 *HTM* **51** 137–144
[3] Guinier A 1938 *Ann. Phys.* **12** 161–
[4] Preston G 1938 *Phil. Mag.* **26** 855–
[5] Konno T, Kawasaki M and Hiragi K 2001 *JEOL News* **36E** 14–17
[6] Wang S and Starink M 2005 *Int. Mater. Rev.* **50** 193–215
[7] Dlubek G 1987 *Mater. Sci. Forum* **13-14** 15
[8] Dupasquier A, Folegati P, de Diego N and Somoza A 1998 *J. Phys.: Condens. Matter* **10** 10409–10422

[9] Dupasquier A, Kögel G and Somoza A 2004 *Acta Materialia* **52** 4707–4726
[10] Klobes B, Staab T, Haaks M, Maier K and Wieler I 2008 *phys. stat. sol. (RRL)* **2** 224–226
[11] Staab T, Zamponi C, Haaks M, Modrow H and Maier K 2007 *phys. stat. sol. (RRL)* **1** 172–174
[12] Klobes B, Staab T and Dudzik E 2008 *phys. stat. sol. (RRL)* **2** 182–184
[13] Somoza A, Petkov M P, Lynn K and Dupasquier A 2002 *Phys. Rev.* **B65** 094107
[14] Haaks M, Staab T, Saarinen K and Maier K 2005 *phys. stat. sol. (a)* **202** R38–R40
[15] Haaks M, Staab T and Maier K 2006 *Nucl. Instr. Meth.* **A569** 829–
[16] Hehenkamp T 1994 *J. Phys. Chem. Solids* **55** 907–915
[17] Kluin J E 1992 *Phil. Mag.* **A65** 1263–1268
[18] Mantl S and Triftshäuser W 1978 *Phys. Rev.* **B17** 1645–1652
[19] Maier K, Peo M, Saile B, Schaefer H E and Seeger A 1979 *Phil. Mag.* **A40** 701–728
[20] Folegati P, Makkonen I, Ferragut R and Puska M 2007 *Phys. Rev.* **B75** 054201
[21] Folegati P, Makkonen I, Puska M and Staab T 2008 Nanostructures in Al-based alloys this conference: in preparation for J.Phys.: Conf. Series
[22] Royset J, Stene T, Saeter J and Reiso O 2006 *Mater. Sci. Forum* **519-521** 239
[23] Soler J, Artacho E, Gale J, García A, Junquera J, Ordejón P and Sánchez-Portal D 2002 *J. Phys.: Condens. Matter* **14** 2745–2779
[24] Junquera J, Paz O and Sánchez-Portal D 2001 *Phys. Rev.* **64** 235111
[25] Kohlbach I, Korff B and Staab T 2008 Pre-Guinier-Preston zones in AlCu-alloys constructed from ab-initio relaxed atomic positions – comparison to experimental methods submitted to phys. stat. sol. (a)
[26] Alatalo M, Kauppinen H, Saarinen K, Puska M, Mäkinen J, Hautojärvi P and Nieminen R 1995 *Phys. Rev.* **B51** 4176–4185
[27] Alatalo M, Barbiellini B, Hakala M, Kauppinen H, Korhonen T, Puska M, Saarinen K, Hautojärvi P and Nieminen R 1996 *Phys. Rev.* **B54** 2397–2409
[28] Makkonen I, Hakala M and Puska M 2005 *J. Phys. Chem. Solids* **66** 1128–1135
[29] Barbiellini B, Hakala M, Puska M and Nieminen R 1997 *Phys. Rev.* **B56** 7136
[30] Puska M and Nieminen R 1983 *J. Phys. F: Metal Phys.* **13** 333–346
[31] Ankudinov A L, Ravel B, Rehr J and Conradson S 1998 *Phys. Rev.* **B58** 7565–7576
[32] Modrow H, Bucher S, Rehr J and Ankudinov A 2003 *Phys. Rev.* **B67** 035123
[33] Folegati P, Puska M, Staab T and Kohlbach I 2008 Ab-initio calculation of Guinier-Preston-zones in AlCu-alloys in preparation
[34] Lengeler B and Eisenberger P 1980 *Phys. Rev.* **B21** 4507–4520
[35] Haasen P 1984 *Physikalische Metallkunde* (Berlin: Springer Verlag)
[36] Gläser U, Dlubek G and Krause R 1992 *Mater. Sci. Forum* **105-110** 1025–1028
[37] Bagaryatsky Y A 1952 *Dokl. Akad. Nauk CCCP* **Tom 87** 559562
[38] Bagaryatsky Y A 1952 *Dokl. Akad. Nauk CCCP* **Tom 87** 397 – 401
[39] Staab T, Haaks M and Modrow H 2008 Early precipitations stages in aluminium alloys - the role of quenched-in vacancies appl. Surf. Sci. in press
[40] Nagai Y, Murayama M, Tang Z, Nonaka T, Hono K and Hasegawa M 2001 *Acta Mater.* **49** 913920
[41] Kratzer T, Gable B, Neirn K, Bastow J, Muddle B and Hill A 2006 *Mater. Sci. Forum* **519-521** 597
[42] Marceau R, Ferragut R, Dupasquier A, Iglesias M and Ringer S 2006 *Mater. Sci. Forum* **519-521** 197
[43] Staab T, Zschech E and Krause-Rehberg R 2000 *J. Mater. Sci.* **35** 4667–4672
[44] Dlubek G 1987 *Mater. Sci. Forum* **13-14** 11
[45] Puska M and Nieminen R 1994 *Rev. Mod. Phys.* **66** 841–897
[46] Ravel B and Newville M 2005 *J. Synchrotron Rad.* **12** 537
[47] Starink M, Gao N, Davin L, Yan J and Cerezo A 2005 *Philos. Mag.* **85** 1395

Journal of Physics: Conference Series **265** (2011) 012019 doi:10.1088/1742-6596/265/1/012019

Probing the defects in nano-semiconductors using positrons

P M G Nambissan

Nuclear and Atomic Physics Division, Saha Institute of Nuclear Physics, 1/AF Bidhannagar, Kolkata 700064, India

E-mail: pmg.nambissan@saha.ac.in

Abstract. Positron annihilation spectroscopy (PAS) is a very useful tool to study the defect properties of nanoscale materials. The ability of thermalized positrons to diffuse over to the surfaces of nanocrystallites prior to annihilation helps to explore the disordered atomic arrangement over there and is very useful in understanding the structure and properties of nanomaterials. As examples, the results of studies on FeS_2 nanorods and ZnS nanoparticles are presented. In semiconductor nanoparticles, there are positron trapping sites within the grains also and these are characterised by using appropriate models on the measured positron lifetimes. We have observed vivid changes in the measured positron lifetimes and Doppler broadened gamma ray spectral lineshapes during structural transformations prompted by substitutional effects in Mn^{2+}-doped ZnS nanorods. Interestingly, the nanoparticles did not exhibit the transformation, implying the morphologies of the nanosystems playing a decisive role. Quantum confinement effect in CdS nanoparticles was another phenomenon that could be seen through positron annihilation experiments. Coincidence Doppler broadening measurements have been useful to identify the elemental environment around the vacancy clusters that trap positrons. Recent studies on nanocrystalline oxide and sulphide semiconductors are also discussed.

1. Introduction

Positrons are extensively used as probes to explore the defect-related properties and processes in semiconductors. Since a plethora of literature on positron annihilation studies of bulk semiconductors already exists [1], the interpretation of the results of positron annihilation experiments on semiconductor materials composed of nanometer-sized grains is considered simple and straightforward. Still additional complications are bound to arise as the chemical and physical properties of the nanocrystalline semiconductors are generally very different from those of their bulk counterparts. This is not only due to the increased number of atoms on the grain surfaces taking part in the various processes but also due to the changes introduced by increasing concentration of vacancy type defects inside the grains and at their surfaces. In this paper, we review a few such investigations, each highlighting a unique aspect that could be revealed through positron annihilation studies.

2. Experimental details

The samples were synthesized through the normal solvothermal route described in detail earlier [2-9]. Appropriate amounts of the proper reagents were chosen and mixed in the right proportion, followed by rigorous stirring till the final precipitate is formed. It was filtered, washed several times in water or ethanol and then dried in vacuum in an oven at ambient temperature. The products were thereafter characterized by x-ray diffraction and the crystalline grain sizes were calculated by using the standard Scherrer equation [10]. Crystallites of larger sizes could be produced by heat-treating the synthesized sample at selected temperatures for different intervals of time. The obtained grain dimensions in all the cases were in the nanometer range, spreading over a few to a few tens of nanometers. In some cases, the sizes were also determined from high resolution electron microscopic images (figure 1, for example). The obtained nanostructures were also examined through transmission electron microscopy (TEM) and their morphology varied from simple nanoparticles to bypiramidal shaped nanocones. A few illustrative examples are shown in figures 2(a) - 2(c).

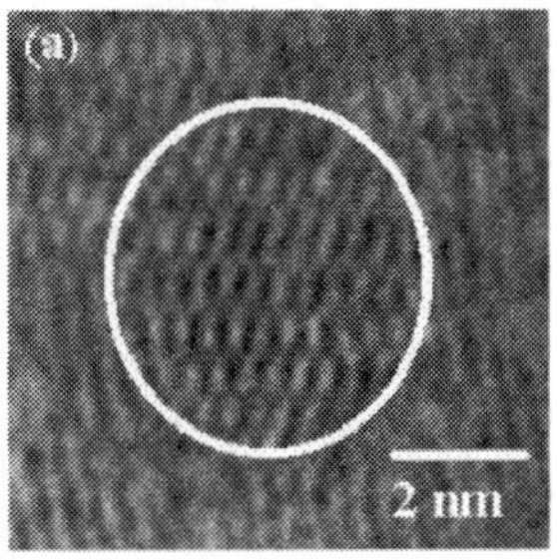

Figure 1. High-resolution electron micrograph of the nanocrystalline (grain size 4.5 nm) cadmium sulphide (CdS) sample.

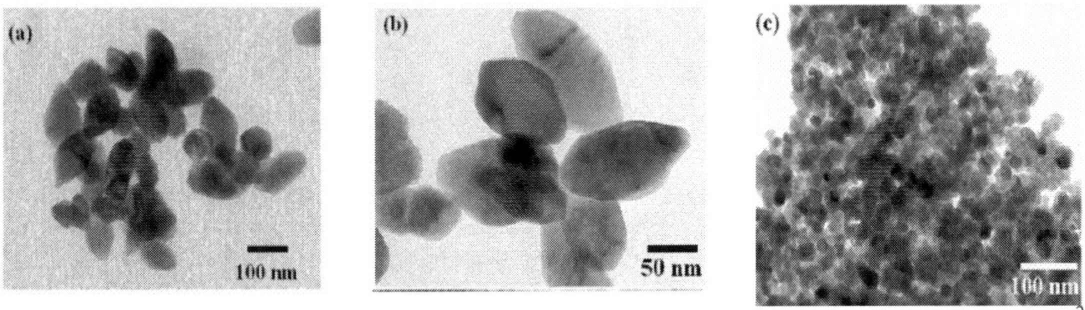

Figure 2. Transmission electron microscopic (TEM) images of the (a) undoped and (b) 1% Mn^{2+}-doped zinc oxide (ZnO) nanocone samples and that of (c) ZnO nanoparticles of diameter 15 nm.

The samples were in powder form and the one for positron annihilation experiments was taken in a glass tube. The ^{22}Na source of strength about 400 kBq deposited on a thin (~ 4 μm) Ni foil was kept immersed inside the volume of the powdered sample. The sample covered the source from all the sides in sufficient thickness to ensure positron annihilation to occur within it and the glass tube containing the source and the sample was continuously evacuated to remove traces of air and moisture otherwise likely to be trapped in it. This also helped, as demonstrated later elsewhere [8], to prevent quenching of orthopositronium atoms that provide useful information on the intergranular region within the nanomaterials. The positron lifetime data (about 1-1.5 million counts) were collected under the usual experimental settings using a gamma-gamma coincidence setup with a time resolution (full width at half maximum) of 200 ps. For Doppler broadening measurements, a high pure germanium detector with energy resolution 1.14 keV at 511 keV was used. This and another similar detector with identical energy resolution were used on either side of the source-sample assembly for recording the anticollinear annihilation gamma rays to generate the coincidence Doppler broadened spectrum (CDBS) for element-specific identification of the positron trapping sites [11,12].

3. Results and discussion

The positron lifetime data were analyzed using the programs RESOLUTION and POSITRONFIT [13] or, of late, PALSfit [14]. All the spectra gave satisfactory fit when three exponential components τ_1, τ_2 and τ_3 were resolved with intensities I_1, I_2 and I_3. The lineshape of the Doppler broadened positron annihilation gamma ray spectrum was parameterized in terms of S and W, respectively representing the area-normalized counts under the segments 0 to 0.6 keV and 1.6 to 4.0 keV on either side of the peak. The CDB spectra were analyzed following the procedure described in the original papers [11,12].

In the following sections, several aspects related to the annihilation of positrons from different sites in nanocrystalline semiconductor materials are discussed. It should be pointed out that the ability of positrons to diffuse through the material after the initial thermalization is the key factor that eventually makes them useful probes to explore the grain surfaces. The average diffusion lengths of thermalized positrons in typical materials range around 10-100 nm [15]. Hence, if the grain sizes are smaller than the thermal diffusion length in the material, large fractions of positrons will diffuse and come out of the grains before annihilation takes place. On the surfaces, positrons get trapped due to the presence of a rich concentration of vacancy type defects, which owe their origin to the highly disordered atomic arrangement. The annihilation characteristics are therefore rich sources of information on the electronic structure and properties of the grain surfaces and interfaces, which are the basic features that distinguish materials of nanocrystalline composition from the coarse-grained bulk material.

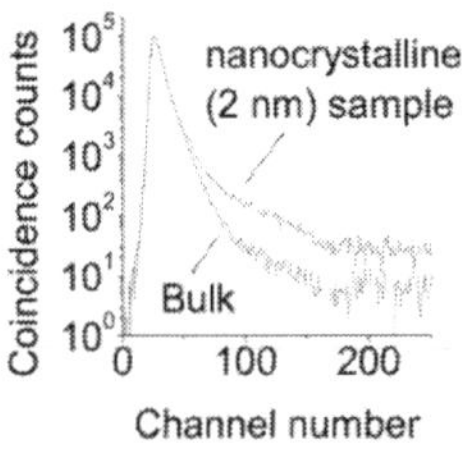

Figure 3. Peak-normalized positron lifetime spectra of CdS bulk and nanocrystalline (grain size ~ 2 nm) samples.

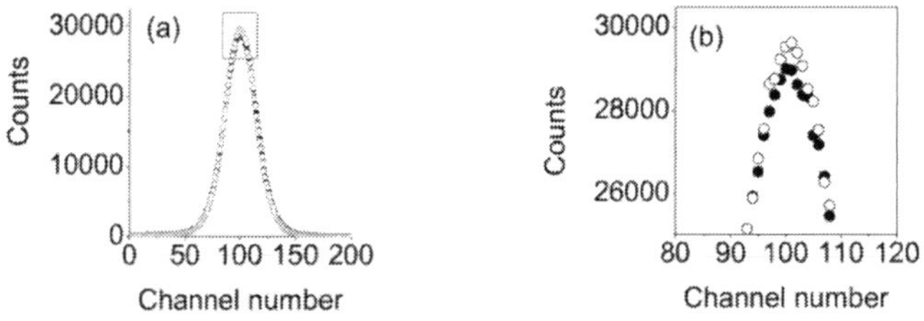

Figure 4. (a) Doppler broadened positron annihilation gamma ray spectra of CdS bulk (closed circles) and nanocrystalline (grain size ~ 2 nm) (open circles) samples. (b) The peak region inside the box in (a) is shown magnified for clarity.

3.1. Sensitivity of positron annihilation characteristics to the morphology of nanostructures

In order to demonstrate the sensitivity of positron annihilation to nanocrystalline composition of a given material, the peak-normalized positron lifetime spectra of bulk and nanocrystalline CdS are given in figure 3. The corresponding Doppler broadened gamma ray spectra are presented in figures 4(a) and 4(b). The resolved positron annihilation lifetimes and their intensities (table 1) indicated a conspicuous increase in the value of the defect-specific component τ_2 and the orthopositronium intensity I_3 in the nanocrystalline sample. It implies that the positrons entering into the grains successfully diffuse over to the grain surfaces and get trapped into the vacancies or vacancy clusters there. As the grain size decreases, the number of positrons reaching the grain surfaces will increase and the intergranular separation will become large enough to favour enhanced formation of positronium atoms.

Table 1. Positron lifetimes, their relative intensities and the S parameter in bulk and nanocrystalline (grain size ~ 2 nm) cadmium sulphide samples.

τ_1 (ps)	I_1 (%)	τ_2 (ps)	I_2 (%)	τ_3 (ps)	I_3 (%)	S
Bulk sample						
172±4	59.3±1.7	375±7	40.2±1.7	1500 (fixed)	0.5±0.05	0.431±0.001
Nanocrystalline sample						
174±4	58.9±1.8	388±4	38.2±1.8	1500 (fixed)	2.9±0.21	0.436±0.001

Yet another demonstration of the propensity of positrons to open volume regions even in nanocrystalline materials is seen in figure 5 where the positron lifetime spectrum of the sample that contained a fraction of nanotubes besides nanorods is found less rapidly decaying than that of a sample of nanorods alone [2]. Positron lifetimes can be different in samples of different morphologies of the same nanomaterial and of the same crystalline structure. This is shown schematically in figure 6.

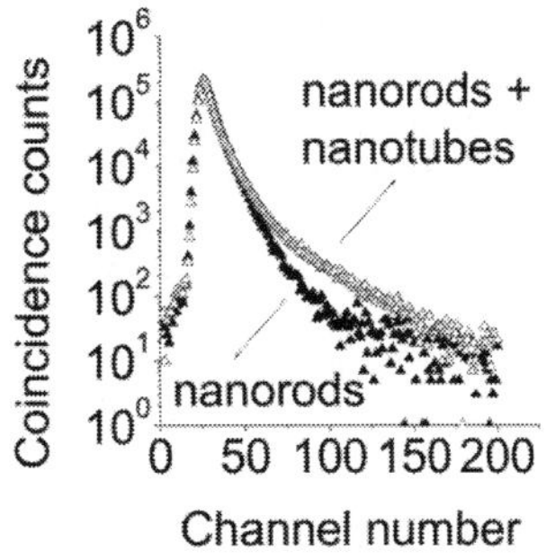

Figure 5. Peak-normalized positron lifetime spectra of two pyrite (FeS_2) nanocrystalline samples, one composed of nanorods and the other a mixture of nanorods and nanotubes.

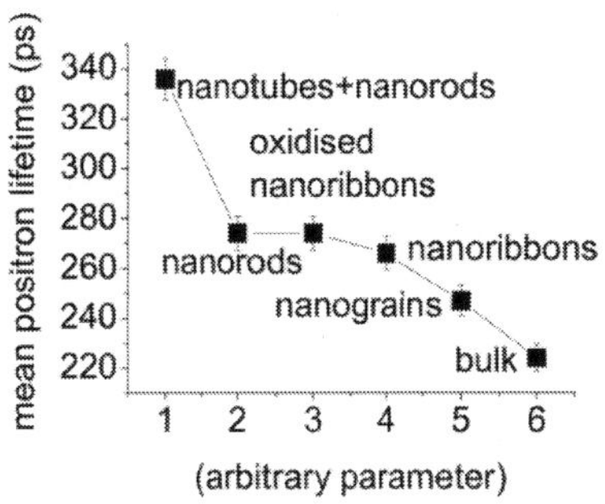

Figure 6. The mean positron lifetimes in FeS_2 nanocrystalline samples of crystallites with different morphologies.

3.2. Sensitivity of positron annihilation characteristics to nanocrystalline grain size

It is clear from the above discussion that the number of positrons diffusing to the grain surfaces will increase when the grain sizes decrease. The intensity I_2 of the lifetime τ_2 of trapped positrons will therefore increase. This has been shown in a number of experiments [5,6,9]. The magnitude of τ_2 may also increase in materials of smaller grain sizes. This is explained as due to the increase in free volume per atom (or, in other words, size of monovacancies) of grains with decrease in sizes [16]. Figure 7 illustrates the *S* parameter of zinc sulphide (ZnS) nanoparticles continuously increasing with decreasing grain size. A similar behaviour had been observed in metallic nanoparticles also [17]. The nature of variation can sometimes be camouflaged by other accompanying processes such as structural transformation and lattice expansion or contraction at very small sizes of the nanoparticles [16,18-20].

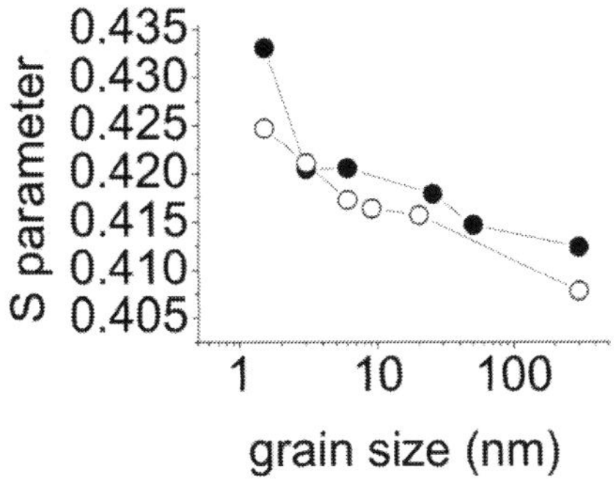

Figure 7. The *S* parameter vs grain size of the ZnS nanocrystalline samples: hexagonal (closed circles) and cubic (open circles).

3.3. Sensitivity of positron annihilation characteristics to quantum confinement effects

One of the remarkable features of significant interest in nanoparticles of extremely small sizes is the occurrence of quantum confinement effects. According to the theory put forward by Wood and Ashcroft [21], a nanoparticle consisting of a few tens or hundreds of atoms can behave as a 'super atom' with discrete energy bands instead of a continuum and the average one-electron-energy levels split at a normalized Fermi energy (ε_F) given by

$$k_B T = \frac{\varepsilon_F}{N} \tag{1}$$

where k_B is the Boltzmann constant and T the temperature. The number of atoms N in the nanoparticle is calculated assuming the same crystalline structure to hold good as that of the bulk. (The x-ray diffraction patterns have been found to depict still distinguishable peaks although with increased broadening in samples with particle sizes even as small as 2 nm.) In semiconductors, a more decisive parameter is the exciton Bohr diameter that provides a threshold value of the grain size below which the confinement regime becomes relevant. This is of special significance in optical absorption studies where a characteristic blue shift occurs implying a widening of the band gap (E_g) expressed as [22]

$$E_{g(nano)} - E_{g(bulk)} = \frac{\hbar^2 \pi^2}{2er^2}\left(\frac{1}{m_e m_0} + \frac{1}{m_h m_0}\right) - \frac{1.8e}{4\pi\varepsilon\varepsilon_0 r} \tag{2}$$

where m_e and m_h are the effective masses of the electron and the hole and m_0 is the mass of a free electron. ε_0 is the permittivity of free space and ε is the relative permittivity of the material. $\hbar$ is Planck's constant and e is the charge of the electron. The exciton Bohr radius corresponds to the grain radius signaling the onset of quantum confinement effects downward in size, i.e., $r = r_B$ at $E_{g(nano)} = E_{g(bulk)}$. It was thought worth exploring if positron annihilation parameters could exhibit any discontinuities across the said threshold size of the nanocrystalline grains. Positron lifetimes and S parameters were measured in different samples of CdS nanoparticles, each with grains of different sizes confined within a narrow distribution. As shown in figures 8(a) and 8(b), the S parameter and the mean positron lifetime (τ_m) remarkably increased in samples with grains of sizes below the confinement limit ($r_B \sim 3$ nm for CdS). The blue shift observed in the optical absorption spectra confirmed an increase of band gap from 3.7 eV to 4.3 eV in ZnS and from 2.4 eV to 3.1 eV in CdS when the grain sizes were reduced from coarse sizes to about 2.5 nm [3,4]. The widening of the band gap decreases the overlap between the positron wavefunction and that of the high momentum core electrons and consequently the positron lifetimes and S will increase, as shown in the figures.

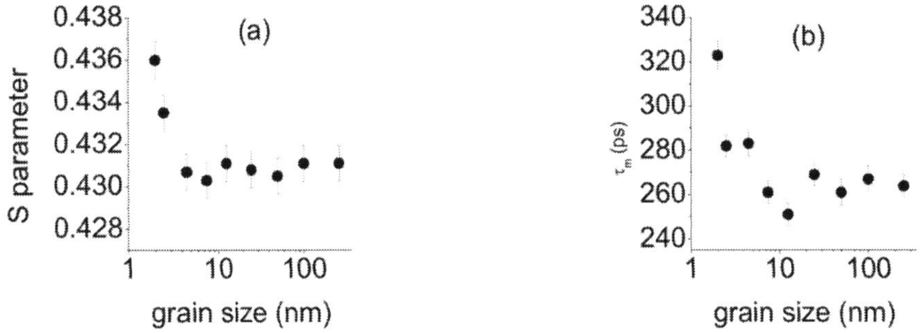

Figure 8. (a) S parameter and (b) the mean positron lifetime versus grain size of CdS nanoparticles.

3.4. Positron trapping in crystal vacancies within oxide nanocrystallites

The above discussion are mainly qualitative in nature as they were mainly centered on the perception that the thermalized positrons diffuse from the interior of the nanocrystalline grains to the surfaces and interfaces before their eventual annihilation. This relied on the possibility that no positron trapping sites existed within the grains and the positrons would freely diffuse towards the surfaces. This is true to a great extent in nanoparticles of metals and alloys but not generally in semiconductor nanosystems. The non-stoichiometric defects that are characteristic features especially of oxide and sulphide semiconductors will act as efficient trapping centres for positrons. This is verified through the application of trapping models, as discussed later. The oxygen and sulfur vacancies, by virtue of their positive charge, repel positrons whereas the cationic vacancies and/or vacancy clusters either neutral or with overall negative charge will trap them. The measured positron lifetimes will be then admixture of those due to annihilation from the crystal vacancies within the grains and in the vacancies on the grain surfaces. The charge specificity of the positrons towards the vacancies is demonstrated in figure 9 that describes the ratio curves generated from the CDB spectra of the corresponding samples and that of annealed Si of purity 99.99%. The curves of CdO, ZnO and NiO depicted a peak at the same electron momentum $p_L = 11.2\times10^{-3}\ m_0c$, that confirmed positrons encountering the same elemental surroundings in all the samples [5]. In other words, positrons are trapped in the respective cationic vacancies or vacancy clusters formed by the removal of two cations and one oxygen or sulfur ion.

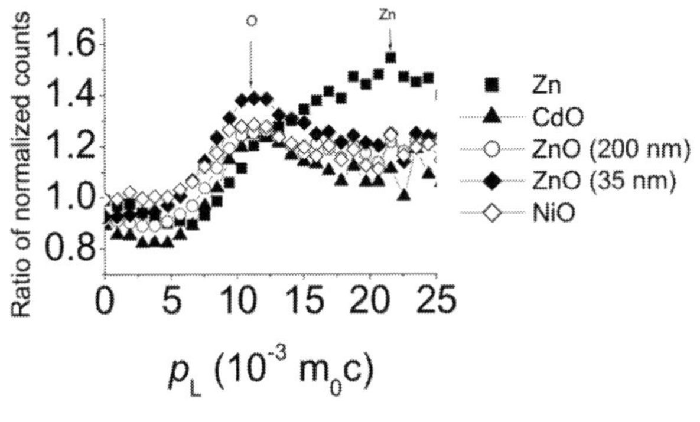

Figure 9. The ratio curves obtained from the coincidence Doppler broadened spectra of a few oxides, indicating that positrons in them encounter identical chemical environments. The corresponding curve for elemental Zn is also shown.

3.5. Positron annihilation characteristics in zinc oxide single crystalline nanobipyramids of base diameters across the thermal diffusion length

In this work, positron annihilation studies were made in ZnO nanocrystals formed with a bipyramidal configuration with the base having a regular hexagonal shape. The measured positron lifetime τ_2 and the *S* parameter, when plotted against the base diameter of the nanocrystals (figures 10(a) and 10(b)),

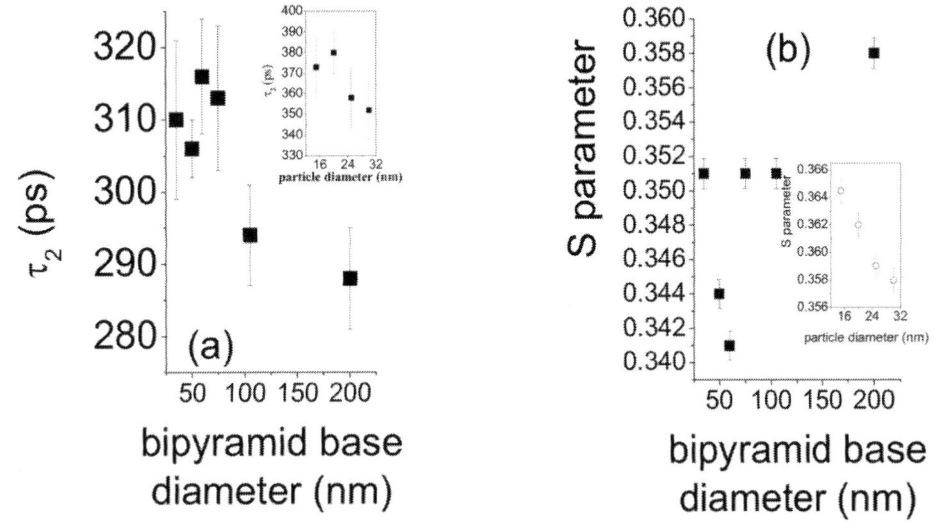

Figure 10. (a) Positron lifetime τ_2 and (b) *S* parameter versus the base diameter of the ZnO nanobipyramids. The insets show the respective variation versus diameters of ZnO nanoparticles.

clearly showed the demarcation of positrons annihilating within and partly on the surfaces of the nanocrystals. The delineation of the lifetimes admixed in a single component was necessitated in the case of nanocrystals with sizes precisely below the thermal diffusion length (~ 52 nm) of positrons in ZnO. Within this limit, the fraction of positrons reaching the crystallite surfaces reduced when their sizes increased, as shown in the insets of the figures. When the sizes of the nanobipyramids are larger than the thermal diffusion length, the positrons are entirely trapped by the crystal vacancies or their clusters within and their lifetimes changed according to the changing sizes of these trapping sites [5].

3.6. Doping effects in zinc sulphide and zinc oxide nanocrystallites

This part discusses about certain interesting aspects came to light when positron annihilation experiments were carried out on samples of nanorods and nanoparticles of ZnS and nanobypiramids of ZnO doped with different concentrations of Mn^{2+} ions. Doping was achieved during the solvothermal synthesis of the samples and the substitution of Zn^{2+} ions by Mn^{2+} ions were verified through energy dispersive analysis of x-rays (EDAX). The exact concentration of Mn^{2+} incorporated in the ZnO lattice could be estimated only in samples of increased doping although the process of intake could be asserted through complementary techniques even in the case of lightly doped samples.

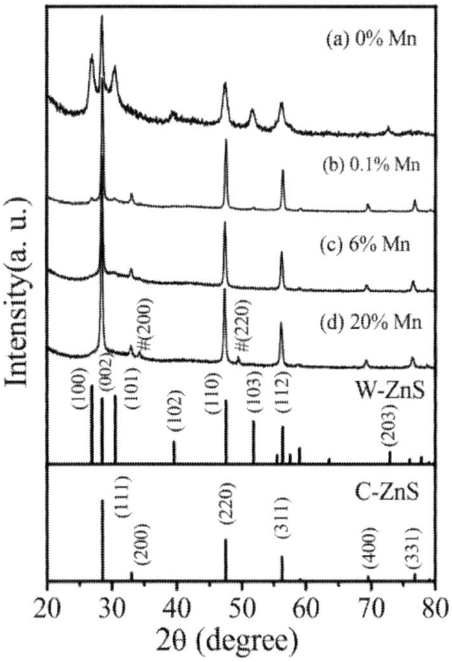

Figure 11. X-ray diffraction patterns of Mn^{2+}-doped (0 at.%, 0.1 at.%, 6 at.% and 20 at.%) ZnS nanorods. Standard patterns of hexagonal and cubic ZnS crystals are also shown.

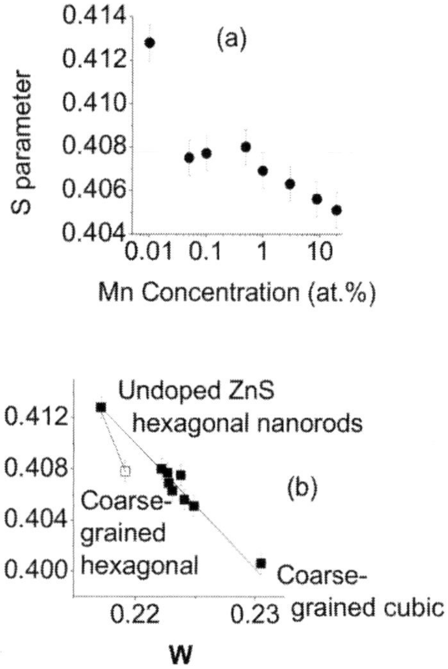

Figure 12. (a) S parameter versus concentration of Mn^{2+} ions and (b) S – W plot of ZnS nanorods.

3.6.1. Structural transformation induced in ZnS nanorods due to Mn^{2+}-doping. From the x-ray diffraction patterns shown in figure 11, a transformation of the hexagonal wurtzite structure of ZnS nanorods into the cubic zinc blend structure with increased doping by Mn^{2+} ions is evident and the same was investigated through positron annihilation studies as well. The nature of variation of the S parameter with doping and the S vs. W plot are shown in figures 12(a) and 12(b).

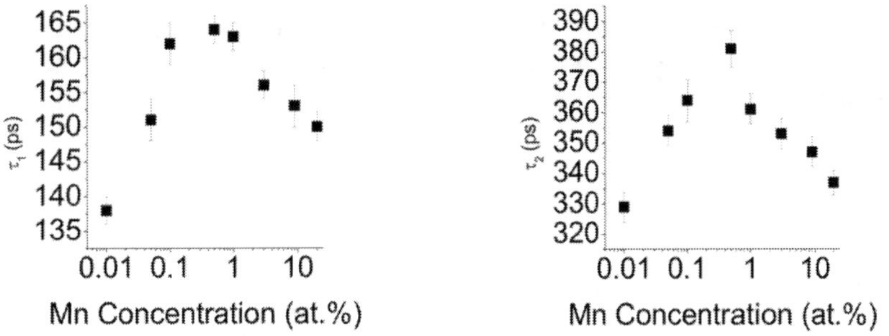

Figure 13. Positron lifetimes τ_1 and τ_2 versus Mn^{2+} concentration in ZnS nanorods.

These results convincingly proved the ability of positron annihilation techniques to monitor such structural changes occurring in nanomaterials [6,7], notwithstanding the complexities involved in the analysis of the positron lifetime data as discussed later. The said transformation was followed by a substantial increase of the diameters of the nanorods from 12-15 nm to 100-200 nm. This stage was indicated by a reduction of the positron lifetimes τ_1 and τ_2 (figure 13) and is explained as due to the decrease in the number of positrons diffusing and reaching the nanorod surfaces [6].

3.6.2. Mn^{2+}-doping in zinc sulphide nanoparticles. While the structural transformation discussed above indicated about the strain induced in the zinc sulphide lattice due to substitution of Zn^{2+} by Mn^{2+} of larger ionic radius, successive replacement of Zn^{2+} by Mn^{2+} did not cause such transformation in cubic nanoparticles. On the contrary, positron lifetimes and the *S* parameter showed changes that differed significantly from those of the nanorod samples (figures 14 and 15). The difference is attributed to the nature of positron trapping defects within the interior of the nanosystems and their evolution during the doping process.

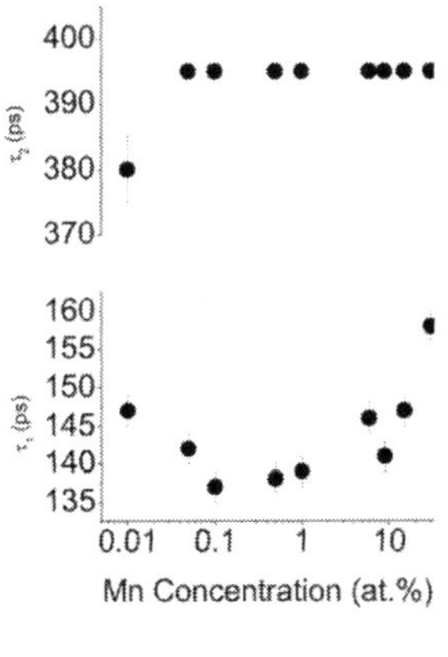

Figure 14. Positron lifetimes τ_1 and τ_2 versus Mn^{2+} concentration in ZnS nanoparticles.

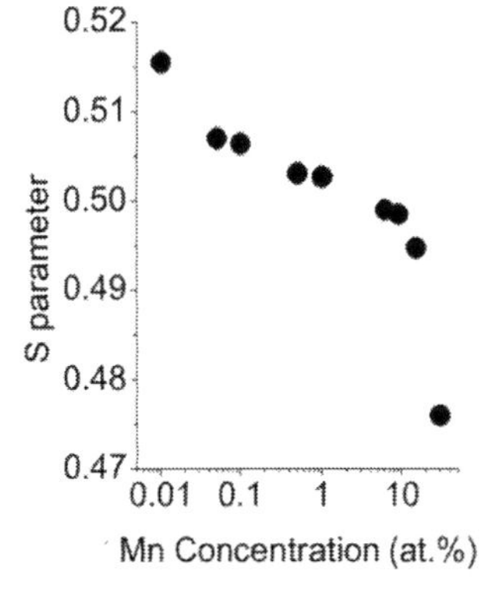

Figure 15. *S* parameter versus Mn^{2+} concentration in ZnS nanoparticles.

In the previous case of nanorods, the CDB spectra did not show any shift towards the spectrum of elemental Mn (figure 16), which implied that positrons did not reach within the proximity of the doped ions. This suggested that the vacancy-type defects that trapped positrons were zinc monovacancies (V_{Zn}) and the Mn^{2+} ions recombined with these vacancies. Positrons were then left with essentially one type of trapping centres, which are the nanorod surfaces. Whereas in the case of Mn^{2+}-doped ZnS nanoparticles, a characteristic shift of the peak of the CDB spectra towards that of Mn is visible (figure 17). The measured positron lifetime τ_2 (= 380 ps) in the nanoparticles was larger than that (= 329 ps) in nanorods. These observations suggest the presence of vacancy cluster of the type V_{Zn}-V_S-V_{Zn} in the undoped nanoparticle sample and the filling of one of the zinc vacancies by the doped Mn^{2+} ion [8].

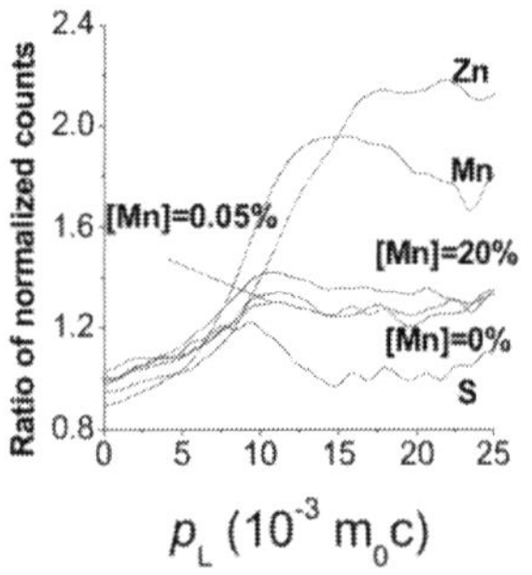

Figure 16. The ratio curves obtained from the CDB spectra of Mn^{2+}-doped ZnS nanorod samples.

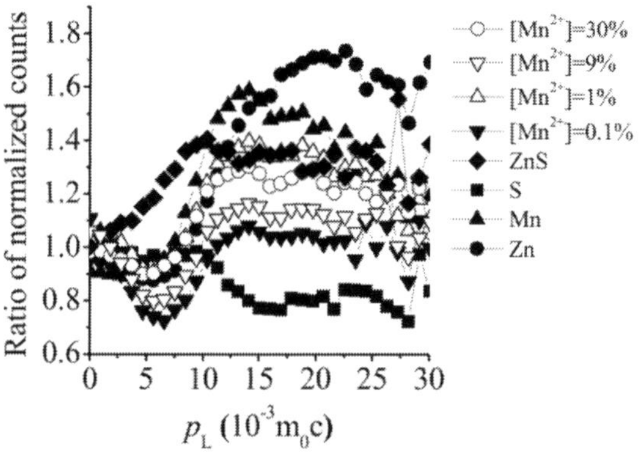

Figure 17. The ratio curves obtained from the CDB spectra of Mn^{2+}-doped ZnS nanoparticle samples.

3.6.3. Mn^{2+}-doping effects in zinc oxide bipyramidal nanocones. In the case of doping the wurtzite zinc oxide crystals with manganese, the introduction of Mn^{2+} with an ionic radius of 0.8 Å in place of Zn^{2+} which is 10% smaller than it in size will lead to the invariable impact on its structure. The associated strain in this case is found to manifest as a restricting effect on the size of the nanobypiramids (also called nanocones). For the base diameters of the nanocones decreased when the dopant concentration increased (table 2). This is explained as follows. The increase of dopant concentration increases the strain, which also leads to increase in the internal energy of the crystallite. The easiest approach to minimize the energy is to keep the Mn^{2+} ions on the surfaces of crystallites as much as possible, and this is possible when the surface to volume ratio is large. The sizes or diameters of the crystallites are therefore decreased. The *S* parameter showed a sharp increase (figure 18) and the positron lifetimes too were found sensitive to this anomalous change (figure 19). The characteristic peak of the CDB spectra shifted towards that of elemental Mn, indicating the trapping of positrons in defects decorated by Mn^{2+} ions (figure 20). As in the case of ZnS, here too a decrease of the positron lifetime in the initial stages of doping proved the conversion of a V_{Zn}-V_O-V_{Zn}-type trivacancy to a V_{Zn}-V_O-type divacancy by the filling of one of the zinc vacancies by the doped Mn^{2+} ion. The positron lifetimes thereafter increased by sensing the reduction in base diameters of the nanocones [9] as more and more number of positrons reach out to the nanocone surfaces prior to annihilation.

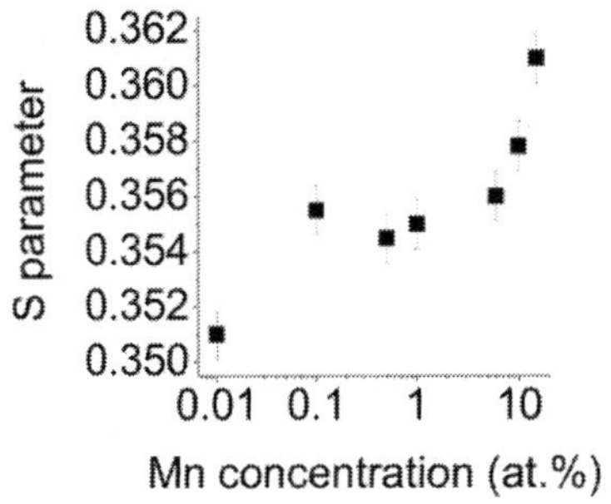

Figure 18. *S* parameter versus concentration of Mn^{2+} ions in ZnO nanocones.

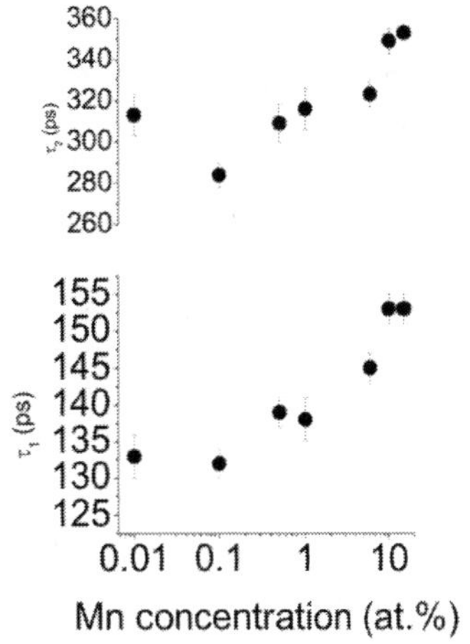

Figure 19. Positron lifetimes τ_1 (bottom) and τ_2 (top) versus concentration of Mn^{2+} ions in ZnO nanocones.

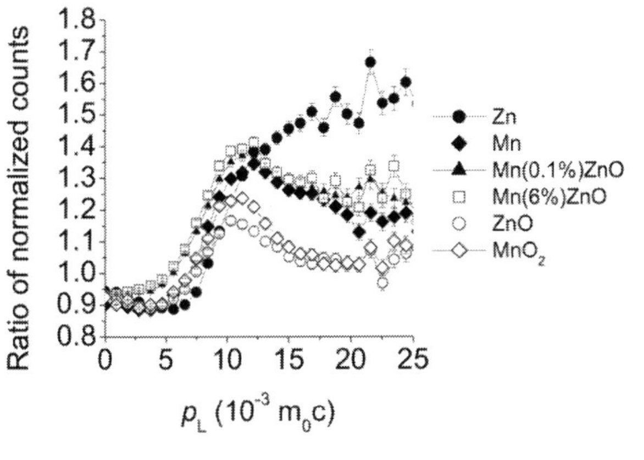

Figure 20. The CDB spectra of Mn^{2+}-doped ZnO nanocone samples, indicating the shift in the peak of the spectrum of ZnO towards that of Mn on doping.

Table 2. The average diameters of the base of the ZnO nanocones of samples doped with different concentrations of Mn^{2+} ions, as determined from transmission electron microscopic images.

Concentration of Mn^{2+} ions (at.%)	Bipyramidal base diameter (nm)
0	75
0.1	75
0.5	75
1	55
6	40
10	35
15	25

3.7. The relevance of τ_3 and I_3

Normally in bulk semiconductors, the density of electrons is large enough to forbid the formation of positronium atoms. In nanocrystalline materials, however, such a possibility cannot be ruled out as the intercrystalline (or intergranular) region is distinguishable with electron density low enough to favour the formation of positronium by the positrons present there. This feature is found remarkable as the "pick-off" lifetime of the orthopositronium atoms also exhibited systematic variation as functions of the experimental variable chosen as basis for the investigation of the properties of the nanosystem concerned. For example, in the case of the FeS_2 nanorod sample discussed in section 3.1., the intensity I_3 was 1.2% whereas the same was found as large as 6.0% in the sample that contained nanotubes. A look at the variation of τ_3 and I_3 as a function of the grain size in ZnS (both hexagonal and cubic) reveals gradual decrease (figures 21(a) and 21(b)) and is understood in terms of the decreasing number of positrons reaching the intercrystalline region after their initial entry to the grains [3]. When the variation is monitored as a function of the doping concentration, there too a definite trend is visible as, for example, shown in figures 22 for the case of ZnS nanorods. The lifetime τ_3 shows a gradual rise whereas the intensity I_3 continuously fall as more and more Mn^{2+} ions are input into the ZnS nanosystems [6-9]. A common reason for the reduction of I_3 is the presence of increasing number of Mn^{2+} ions at the intercrystalline region. This is even more evident from the positron lifetime spectra of figure 23, which on analysis showed a drop in I_3 from 2.9% to 0.5% on inclusion of 20 at.% Mn^{2+} ions in ZnS nanorods. The behaviour of τ_3 is case-specific and is discussed in the respective contexts [6-9]. Like, for example, the ZnS nanorods reducing the diameters [6] or the ZnO nanobypiramids shrinking the bases [9] during increased doping, the volume of the intercrystalline region increases to result in increased lifetime for the positronium atoms before "pick-off" annihilation takes place.

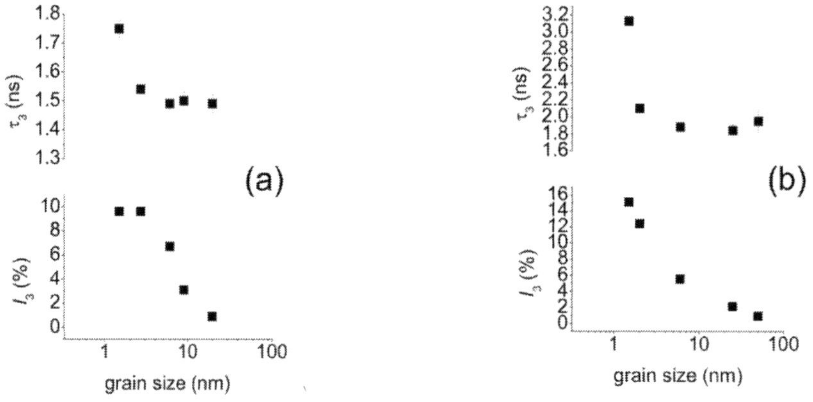

Figure 21. The orthopositronium lifetime τ_3 and intensity I_3 versus grain size in (a) hexagonal and (b) cubic ZnS nanoparticles.

Figure 22. The orthopositronium lifetime τ_3 and intensity I_3 versus Mn^{2+} concentration in the ZnS nanorods. The data corresponding to the undoped ($x = 0$) sample are shown against 0.01%.

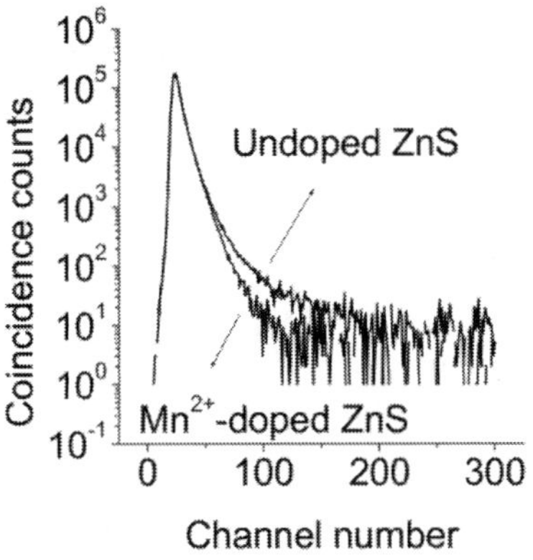

Figure 23. Peak-normalized positron lifetime spectra of undoped and 20 at.% Mn^{2+}-doped ZnS nanorod samples.

An important question that was recently raised in this particular context is the wisdom in choosing a vacuum surrounding to maintain the source-sample assembly during the experiments. But indirectly this also helped to answer yet another question regarding the location of the positron trapping defects, i.e., inside or outside the nanocrystallites. This is answered using the observations of the following experiment. Positron lifetime and Doppler broadening measurements were performed in one sample of ZnS nanoparticles with grain size about 2 nm in both vacuum and air separately and the peak-normalized positron lifetime spectra are shown in figure 24. The resolved positron lifetimes and their intensities in the two cases are summarized in table 3. Due to the extremely small size of the particles, even a second positronium lifetime (τ_4) as long as 7.29 ns could be obtained with an appreciable intensity of 10.0% besides a lower "pick-off" lifetime component τ_3 of 1.75 ns with 6.6% intensity. On the other hand, when measurements were carried out in air, the longest positronium component completely disappeared and the "pick-off" lifetime τ_3 changed only marginally with intensity 10.7%. The shorter lifetimes τ_1 and τ_2 and their intensities (when normalized to $I_1 + I_2 + I_3 = 100\%$) remained almost unchanged [8]. This shows that a significant fraction of positrons indeed diffused out to the surfaces of nanometer-sized grains and formed positronium in the intergranular region and the positronium lifetime and intensity were quenched by the presence of external gases. Further, the observation that the shorter lifetimes and their intensities were not significantly affected implied effective positron trapping by vacancy clusters within the grains. In other words, it offers proof to the presence of open volume defects inside the nanoparticles. The sharpness of the Doppler broadened lineshape of the spectrum recorded with the source-sample assembly in air is found reduced due to the quenching effect and the S parameter dropped by about 2.2% (i.e., $S_{air}/S_{vacuum} = 0.98$).

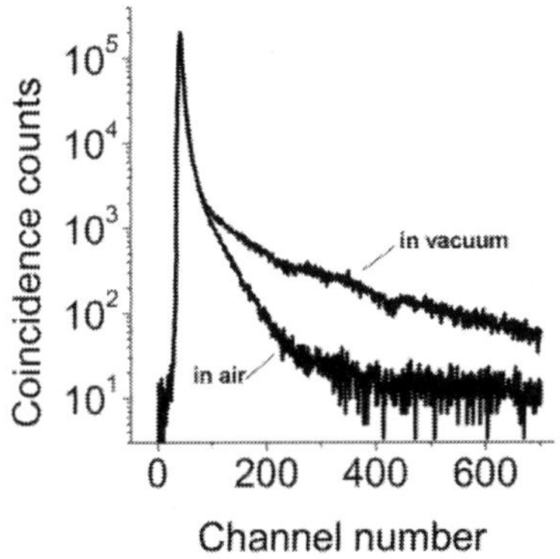

Figure 24. Peak-normalized positron lifetime spectra of ZnS sample of nanoparticles with average size 2 nm when the measurements were carried out in vacuum and in air.

Table 3. Positron lifetimes and intensities in the ZnS sample of nanoparticles with average size 2 nm when the measurements were carried out in vacuum and in air.

	τ_1 (ps)	I_1 (%)	τ_2 (ps)	I_2 (%)	τ_3 (ns)	I_3 (%)	τ_4 (ns)	I_4 (%)
In vacuum	153 ±2	55.1 ±1.3	412 ±13	28.3 ±1.1	1.75 ±.0.10	6.6 ±0.11	7.29 ±0.2	10.0 ±0.3
In air	161 ±2	54.4 ±1.1	422 ±7	34.9 ±1.0	1.91 ±.0.10	10.7 ±0.1	-	-

3.8. The application of trapping models

The issue of the exact number of types of positron trapping sites in nanocrystalline materials is crucial in rightly interpreting the positron lifetimes and intensities. In semiconductor nanosystems, apart from the defects at the surfaces and interfaces, the non-stoichiometric defects within the nanosystems also play an important role and it needs to be specially addressed. In such cases, either of the two lifetimes τ_1 and τ_2 will include the contributions from the crystal defects within the grains in addition to the defects at the surfaces and interfaces. The most common case is that τ_2 is correctly resolved as an independent component representing either one of the two types of defects and τ_1 is an admixture of the positron lifetime in the other type of defects besides the bulk and the Bloch state lifetimes. In this case, the validity of this assumption is cross-checked by calculating the value of τ_1 using the relation

$$\tau_1^{cal} = \frac{1+\frac{\kappa_1}{\lambda_1}\left(1+\frac{\kappa_2}{\lambda_b-\lambda_2+\kappa_1}\right)}{\lambda_b+\kappa_1+\kappa_2} \tag{3}$$

where the annihilation rate $\lambda = \tau^{-1}$ in the respective cases and the positron trapping rates have the form

$$\kappa_1 = \frac{\tau_1(\lambda_b - I_2\lambda_2) - I_1}{\tau_d - \tau_1} \tag{4}$$

and

$$\kappa_2 = \frac{I_2}{I_1}(\lambda_b - \lambda_2 + \kappa_1) \tag{5}$$

This model worked in the case of nanoparticles of CdS and ZnS and nanobypiramids of ZnO [3-5]. The same also held good in the cases of Mn^{2+}-doped ZnO nanocones [9].

Doping of ZnS nanorods and nanoparticles presented a slightly different picture [6,8]. In the former, the normal two-state trapping model gave the best fit to the results, according to the equation

$$\frac{1}{\tau_1^{cal}} = \frac{1}{\tau_b} + \frac{I_2}{I_1}\left(\frac{1}{\tau_b} - \frac{1}{\tau_2}\right) \tag{6}$$

where the second term on the right hand side of the equation gives the rate of trapping of positrons into the defects. This, in corroboration with the observations on the variation of the individual positron lifetimes and their intensities [6,7], gave credence to the view that the Zn^{2+}-vacancies within the interior of the nanorods are fully occupied by Mn^{2+} ions. However, identical doping in samples of ZnS nanoparticles gave results, which were entirely different. The above models failed in this case as τ_2 contained the lifetimes of positrons in both types of defects. The sensitivity of the positron lifetime technique to the individual type of defects is lost in this case and a cumulative trapping rate κ_{12} is defined such that the shorter lifetime τ_1 is calculated as

$$\frac{1}{\tau_1^{cal}} = \frac{1}{\tau_b} + \kappa_{12} \tag{7}$$

where

$$\kappa_{12} = \frac{1}{\tau_b}\left(\frac{\tau_m - \tau_b}{\tau_2 - \tau_m}\right) \tag{8}$$

The mean lifetime of positrons $\tau_m = (\tau_1 I_1 + \tau_2 I_2 + \tau_3 I_3) / (I_1 + I_2 + I_3)$. An exception to this analysis was encountered in the case of sample doped with the highest concentration (30 at.%) of Mn^{2+} ions in which the formation of Mn-Mn clusters gave rise to one more additional trapping site for positrons and the results were fitted with equations (3), (4) and (5).

The curves illustrating the variation of the trapping rates and thereby the applicability of these models are shown in figures 25(a), 25(b) and 25(c).

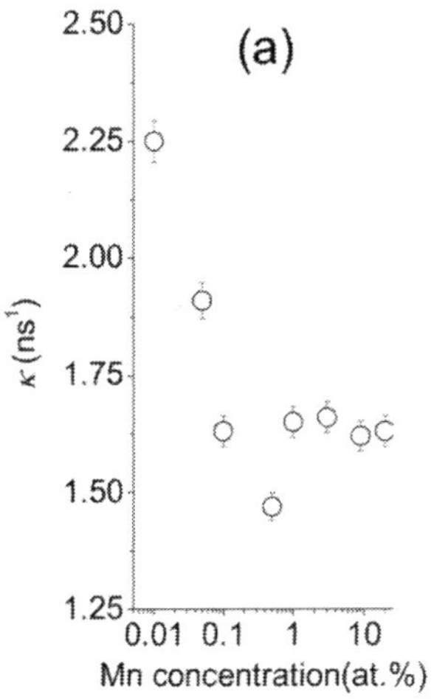

Figure 25(a). Positron trapping rate κ (equation (6)) in the ZnS nanorod samples at different Mn^{2+} doped ion concentration. The data corresponding to the undoped ($x = 0$) sample are shown against 0.01%.

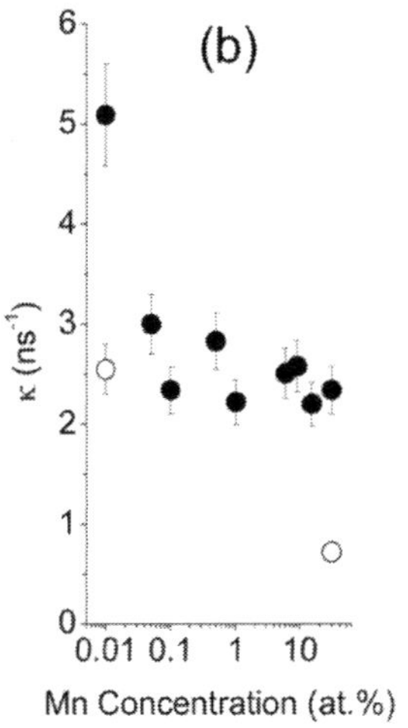

Figure 25(b). Positron trapping rate κ_1 (equation (4); open symbols) and κ_2 (equation (5); closed symbols) or κ_{12} (equation (8); closed symbols) in the ZnS nanoparticle samples at different Mn^{2+} doped ion concentration. The data corresponding to the undoped ($x = 0$) sample are shown against 0.01%.

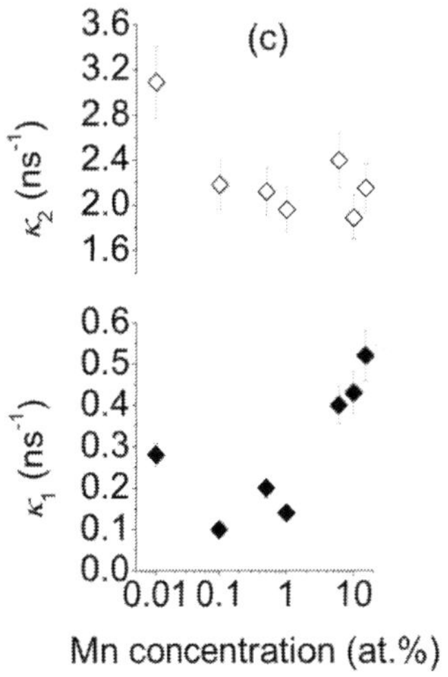

Figure 25(c). Positron trapping rate κ_1 (equation (4); closed symbols) and κ_2 (equation (5); open symbols) in the ZnO nanocone samples at different Mn^{2+} doped ion concentration. The data corresponding to the undoped ($x = 0$) sample are shown against 0.01%.

4. Summary and conclusions

The results reported in this paper and the inferences drawn from them prove unambiguously that positron annihilation techniques can be used for the investigation of innumerable features and processes involving semiconducting materials with nanometer dimensions. It is obvious from these studies that positron annihilation offers a wholly non-destructive method for characterizing the vacancy-type defects present both within and outside the nanocrystalline grains. Positron annihilation can complement as a viable spectroscopic tool to support the information available form the other conventional techniques. The success in obtaining distinguishable differences among results of experiments on nanosystems of different morphologies and sizes is a remarkable advantage.

Acknowledgements
The author wishes to thank all the collaborators involved in the experiments discussed in this paper.

References

[1] Krause-Rehberg R and Leipner H S 1999 *Positron Annihilation in Semiconductors – Defect Studies* (Berlin: Springer-Verlag) pp 1–126
[2] Kar S, Chaudhuri S and Nambissan P M G 2005 *J. Appl. Phys.* **97** 014301
[3] Biswas S, Kar S, Chaudhuri S and Nambissan P M G 2006 *J. Chem. Phys.* **125** 164719 & 2006 *Virtual Journal of Nanoscale Science & Technology* **14** (20)
[4] Kar S, Biswas S, Chaudhuri S and Nambissan P M G 2005 *Phys. Rev.* B **72** 075338.
[5] Ghoshal T, Biswas S, Kar S, Chaudhuri S and Nambissan P M G 2008 *J. Chem. Phys.* **128** 074702 & 2008 *Virtual Journal of Nanoscale Science & Technology* **17** (8)
[6] Kar S, Biswas S, Chaudhuri S and Nambissan P M G 2007 *Nanotechnology* **18** 225606
[7] Nambissan P M G, Kar S, Biswas S and Chaudhuri S 2007 *Physica Status Solidi* (c) **4** 3889
[8] Biswas S, Kar S, Chaudhuri S and Nambissan P M G 2008 *J. Phys.: Condens. Matter* **20** 235226
[9] Ghoshal T, Kar S, Biswas S, De S K and Nambissan P M G 2008 (Unpublished)
[10] Cullity B D 1978 *Elements of X-ray Diffraction* (Reading, MA: Addison-Wesley) p 102 and p 356
[11] Asoka-Kumar P, Alatalo M, Ghosh V J, Kruseman A C, Nielsen B and Lynn K G 1996 *Phys. Rev. Lett.* **77** 2097
[12] Nagai Y, Nonaka T, Hasegawa M, Kobayashi Y, Wang C L, Zheng W and Zhang C 1999 *Phys. Rev. B* **60** 11863
[13] Kirkegaard P, Eldrup M, Mogensen O E and Pedersen N J 1981 *Comput. Phys. Commun.* **23** 307
[14] Olsen J V, Kirkegaard P, Pedersen N J and Eldrup M 2007 *Physica Status Solidi* (c) **4** 4004
[15] Bergersen B, Pajanne E, Kubica P, Stott M J and Hodges C H 1974 *Solid St. Commun.* **15** 1377
[16] Chattopadhyay P P, Nambissan P M G, Pabi S K and Manna I 2001 *Phys. Rev.* B **63** 054107
[17] Mukherjee M, Chakravorty D and Nambissan P M G 1998 *Phys. Rev.* B **57** 848
[18] Nambissan P M G, Upadhyay C and Verma H C 2003 *J. Appl. Phys.* **93** 6320
[19] Chakraverty S, Mitra S, Mandal K, Nambissan P M G and Chattopadhyay S 2005 *Phys. Rev.* B **71** 024115
[20] Mitra S, Mandal K, Sinha S, Nambissan P M G and Kumar S 2006 *J. Phys. D: Appl. Phys* **39** 4228
[21] Wood D M and Ashcroft N W 1982 *Phys. Rev.* B **25** 6255
[22] Brus L E 1983 *J. Chem. Phys.* **79** 5566

Journal of Physics: Conference Series **265** (2011) 012020 doi:10.1088/1742-6596/265/1/012020

Positron annihilation study of yttria-stabilized zirconia nanopowders containing Cr_2O_3 additive

I Procházka[1,*], J Čížek[1], O Melikhova[1], J Kuriplach[1], T E Konstantinova[2] and I A Danilenko[2]

[1] Charles University in Prague, Faculty of Mathematics and Physics, V Holešovičkách 2, CZ-180 00, Praha 8, Czech Republic

[2] Galkin Donetsk Institute for Physics and Engineering, National Academy of Sciences of Ukraine, Luxemburg Street 72, 83114 Donetsk, Ukraine

E-mail: ivan.prochazka@mff.cuni.cz

Abstract. Yttria-stabilized zirconia compacted nanopowders, doped with trivalent chromium oxide, were studied by means of high-resolution positron lifetime and coincidence Doppler broadening techniques. The observed data suggest that positrons annihilate mainly in vacancy-like defects at grain boundaries or in larger open volumes most likely located at triple points. The results also show that an addition of Cr_2O_3 leads to a decrease in grain size.

1. Introduction

Zirconia (ZrO_2) has a high melting point (2700° C), a low thermal conductivity and a variety of other advantageous properties [1]. Thus, zirconia exhibits a big potential of use in a wide area of applications, especially at high temperatures. Favorable properties of ZrO_2 are still enhanced if zirconia-based materials for applications are produced starting from powders of nanoscopic size.

However, the *pure* zirconia, which is monoclinic at room temperature, undergoes a transformation to the tetragonal phase at $\approx 1100°$ C temperature. Such a transformation is connected with a considerable volume shrinkage leading even to a creation of cracks within ZrO_2 structure. Obviously, this effect inhibits any use of the pure zirconia in high-temperature applications.

Thus, a stabilization of the high-temperature ZrO_2 phase is required to make a full use of advantageous zirconia features in industry. It is well known, that the stabilization of the high-temperature phase of zirconia can be achieved by an addition of ≈ 3 mol.% (≈ 5.7 wt%) of the trivalent yttrium oxide Y_2O_3 (yttria) so as to form a solid solution of Y_2O_3 in the ZrO_2 lattice [1]. Materials, in which the stabilization of the high-temperature phase is attained by means of an yttria addition, are conventionally called yttria-stabilized zirconia (YSZ).

YSZ nanomaterials represent a complex system with a considerable amount of open-volume defects of a sub-micrometer size. First, the introduction of Y^{3+} ions into the ZrO_2 lattice leads to a violation of stoichiometry. This results in a large amount of oxygen vacancies and

* Corresponding author.

vacancy–solute complexes. Second, grain boundaries (GB's) in nanostructured YSZ materials and defects associated to GB's (vacancy-like defects, triple points at GB intersections) become of increasing importance. Third, the pores of a nanometer size are expected to occur in YSZ materials made of nanopowders. These nanosized structures undoubtedly underlay macroscopic properties of zirconia-based materials. Detailed microstructural information should thus be regarded as a factor of crucial importance for tailoring YSZ materials for practical applications.

Some other metal-oxide additives are being examined to improve properties of YSZ's, in particular to prevent grain growth during the preparation of nanopowders. Recently, Korduban et al. [2] has suggested that an addition of a small amount (≈ 1 mol.%) of Cr_2O_3 diminishes the particle size of the YSZ nanopowders prepared by the co-precipitation technique. Indeed, an effect of a Cr_2O_3 addition on defects nature and behaviour in such a 'ternary' oxide YSZ system, is to be investigated and compared to the 'binary' YSZ one, i.e. the YSZ with no Cr_2O_3 additive. It should also be mentioned that in YSZ chromium atoms predominantly exist in the form of trivalent ions though other charge states (like Cr^{5+}) cannot be fully excluded [3]. This indicates that Cr^{3+} ions will be accompanied by oxygen vacancies like in the case of Y^{3+} ions.

Positron annihilation spectroscopy (PAS) is nowadays a well-established efficient tool for defect studies [4; 5]. Several PAS investigations of zirconia-based nanomaterials were already performed before [6–9]. Despite of these studies, there is still a lack of sufficiently complete and precise experimental data on YSZ nanomaterials which would allow for an unambiguous interpretation. Recently, we have carried out extensive investigations of various pressure-compacted YSZ nanopowders by means of the positron lifetime (LT) and coincidence Doppler broadening (CDB) techniques [10; 11]. The detailed *ab-initio* theoretical calculations of positron parameters for the perfect ZrO_2 lattice and a variety of defect configurations conceivable in the YSZ have been also conducted by us [12; 13] in order to assist the interpretation of the PAS experiments. As a continuation of our preceding efforts, we report in the present Contribution on a LT and CDB study of several ternary YSZ systems with a different Cr_2O_3 content, for which there are no PAS studies available yet.

2. Experimental

2.1. Samples

The binary and ternary YSZ nanopowders were prepared by the co-precipitation technique starting from water solutions of $ZrO(NO_3)_2$, $Y(NO_3)_3$ and $CrCl_3 \cdot 6H_2O$ taken at stoichiometric compositions. Further details of the preparation procedure are given in Refs. [2; 14; 15]. The final oxide nanopowders were obtained by calcination at 600° C/1 h in air. All the nanopowders contained 3 mol.% of Y_2O_3. The resulting binary and ternary YSZ nanopowders were uniaxially pressed under 500 MPa into tablets of 10÷12 mm in diameter and ≈5 mm thickness. Specimens of the four different content x_{Cr} of Cr_2O_3 were investigated: $x_{Cr} = 0, 0.3, 2.9, 5.0$ mol.% of Cr_2O_3. These specimens are below abbreviated as Z3Y, Z3Y-$C_{0.3}$, Z3Y-$C_{2.9}$ and Z3Y-$C_{5.0}$, respectively. Their tetragonal structure was examined by X-ray diffraction and the mean grain size of the binary YSZ nanopowder (≈ 20 nm) was estimated by this technique as well.

A monocrystal of the cubic ZrO_2+9 mol.% Y_2O_3 (denoted hereafter as Z9Y, purchased from Crystec, GmbH) was also investigated to facilitate data interpretation. In addition, pure Zr, Y and Cr well-annealed metals were used as reference samples in CDB measurements. Testing LT measurements verified these specimens as virtually defect-free materials in which practically all positrons annihilate from the delocalised state.

2.2. Apparatus and data acquirement

The positron sources were fabricated using the carrier-free $^{22}Na_2CO_3$ activity supplied by iThemba Labs. A drop of about 1.3 MBq of the activity in water solution was dried and sealed between the 4 μm thick mylarC foils (DuPont). The source was sandwiched with two identical tablets of the material studied.

LT measurements were conducted using the fast-fast configuration of a BaF_2 spectrometer [16]. Using the above positron source–specimen sandwich, the spectrometer exhibited a time resolution of 165 ps (FWHM for ^{22}Na), $\approx 80 \div 100\,s^{-1}$ coincidence count rate and a peak-to-background ratio better than $\approx 10^3$. More than 10^7 coincidence events were accumulated in each LT spectrum. A maximum-likelihood procedure [16] was utilized to decompose the LT spectra into discrete components. The fractions of positrons annihilating in the source salt and covering foils were determined from separate LT spectra measurements with a well-annealed α-iron reference specimen. The annihilations in the source were then re-calculated to the YSZ case following to the method developed by Surbeck [17].

In CDB experiments, a two-detector (HPGe–HPGe) coincidence spectrometer [18] was utilized. The spectrometer showed an energy resolution of 1.1 keV (FWHM) at the 511 keV energy, a coincidence count rate of $\approx 500\,s^{-1}$ and a peak-to-background ratio of $\approx 10^5$. At least 10^8 counts were collected in each two-dimensional CDB spectrum. The results of the CDB measurements are expressed as Doppler-broadened profiles (DBP) of the samples studied related to the DBP of a well-annealed pure Zr reference specimen.

3. Results and discussion

The results of present LT measurements on the binary Z3Y and ternary Z3Y-$C_{x_{Cr}}$ nanopowders are summarised in table 1. The four-component LT spectrum, exhibited by the Z3Y sample, illustrates a typical behaviour observed [10] on a variety of binary YSZ nanopowders: (i) the shortest lifetime $\tau_1 \approx 0.185$ ns arises due to positron trapping in vacancy-like defects, (ii) the second lifetime $\tau_2 \approx 0.375$ ns comes from larger defects of open volumes comparable to a few vacancies and (iii) the two longest lifetimes $\tau_3 \approx 2$ ns and $\tau_4 \approx 30$ ns result from positronium (Ps) formation and subsequent ortho-Ps pick-off annihilation in voids or pores.

Table 1. Positron lifetimes τ_i and relative intensities I_i obtained for compacted nanopowders in the present work. Relative intensities were normalized so that $I_1 + I_2 + 4(I_3 + I_4)/3 = 100\,\%$. Errors (1 std. deviation) are given in the units of the last significant digit of the corresponding value.

Sample	τ_1 [ns]	I_1 [%]	τ_2 [ns]	I_2 [%]	τ_3 [ns]	I_3 [%]	τ_4 [ns]	I_4 [%]
Z3Y	0.182 (1)	27 (2)	0.379 (1)	63 (2)	1.5 (2)	0.7 (1)	30 (1)	7.1 (3)
Z3Y-$C_{0.3}$	0.191 (3)	31 (1)	0.381 (2)	69 (1)				
Z3Y-$C_{2.9}$	0.224 (1)	24 (1)	0.406 (2)	76 (1)				
Z3Y-$C_{5.0}$	0.259 (8)	20 (2)	0.393 (4)	80 (2)				

It can be expected that a substantial portion of positrons thermalised inside grains of ≈ 20 nm size can diffuse to GB's during the lifetime of positrons. It has been discussed by Guo [19; 20] that yttrium segregates at the GB interfaces in YSZ. For this reason, GB's are surrounded by negatively charged compensating layers characterized by a depletion of oxygen vacancies $V_O^{\bullet\bullet}$ (the Kröger notation [21] is used in the following). At the same time the GB cores in YSZ are positively charged [20]. It should also be noted that the potential 'barier' for positrons at GB's in YSZ due to the mutual repulsive interaction of positrons and GB's is about 0.2 eV, as deduced from [22], which does not prevent positrons from trapping at the GB core. The structure of grain boundaries in sintered cubic and tetragonal YSZ materials has recently been studied in detail by Matsui et al. [23]. Among others, it is shown that Y segregation in tetragonal YSZ materials increases with temperature (above 1100 °C). In this respect, it is unclear to what extent the segregation of yttrium (and thereby GB charging) occurs also in our samples that were not sintered (see also [20] and references therein). There are no experimental data regarding possible Cr segregation at GB's in YSZ. We just note that the ionic radius of the Cr^{3+} ion is very similar to that of the Zr^{4+} ion, indicating a lower tendency to segregation compared to Y. The topic of cation segregation at GB's in YSZ and related oxygen vacancy depletion is rather complex and we shall not discuss it further here.

In any case, we suppose that the positrons which did not form Ps annihilate at GB's or their close vicinity from trapped states at the two kinds of defects: (i) vacancy-like defects present at GB's, and (ii) larger open-volume defects situated probably in triple GB's junctions (triple points). Simple spatial considerations and the assumption of uniformly distributed vacancy-like defects in the GB layer lead to the proportionality of the I_2/I_1 ratio to d^{-2}, where d stands for the mean grain size. The proportionality constant for YSZ's was estimated from a fit to the data of Ref. [10] for binary YSZ nanopowders and it is used below to estimate the grain size of YSZ-$C_{x_{Cr}}$ nanopowders.

Contrary to the binary YSZ system (sample Z3Y), the LT spectra of all the ternary systems Z3Y-$C_{x_{Cr}}$ exhibited only two components, see table 1. The shortest lifetime τ_1 gradually increases with the Cr_2O_3 content up to $\approx$0.26 ns. This may be attributed to an increase of the open volume of vacancy-like defects surrounded by Cr atoms in the GB vicinity, probably due to a smaller atomic radius of Cr compared to Zr or Y.

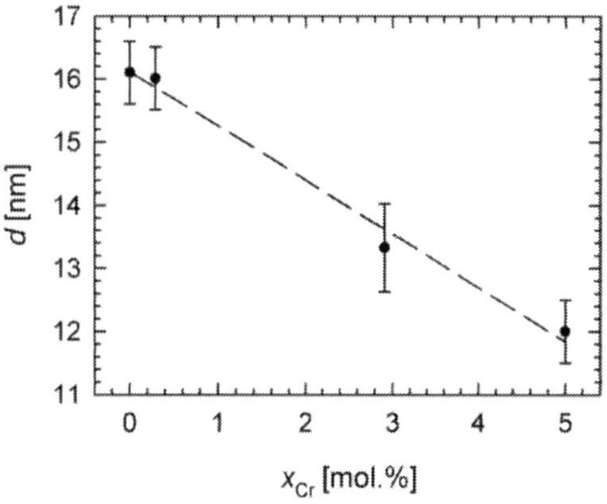

Figure 1. Estimated mean grain size d vs. Cr_2O_3 content x_{Cr}. The dashed line represents the linear regression of data points.

The longer lifetime τ_2 in the Z3Y-$C_{x_{Cr}}$ does not show any monotonic dependence on the Cr_2O_3 content suggesting thus that this component arises from positron trapping in free volumes among grains (most likely triple points), comparable to a few vacancies. The intensity ratio I_2/I_1 clearly increases with the increasing Cr_2O_3 content. We have used the above described procedure to estimate the mean grain size d of the ternary Z3Y-$C_{x_{Cr}}$ systems from the observed I_2/I_1 data. In figure 1, the estimated mean grain size d is plotted as a function of Cr_2O_3 content x_{Cr}. It is seen from the figure that an increased Cr_2O_3 content leads to a smaller grain size, which is in accordance with findings [2] that chromium dopants inhibit particle clustering during calcination and shift the onset of particle coalescence to higher temperatures.

There are no lifetimes above 1 ns and, hence, no ortho-Ps annihilations were observed in Z3Y-$C_{x_{Cr}}$. This can be understood as an effect of paramagnetic centers associated with Cr'_{Zr} dopants which suppress the Ps formation below the detection limit (less than 1 %) even for $x_{Cr} = 0.3$ mol.% .

The DBP ratio curves related to the well-annealed pure Zr for the binary Z3Y and ternary Z3Y-$C_{x_{Cr}}$ compacted nanopowders are plotted in figure 2. Ratio curves exhibit a pronounced maximum at electron momentum $p \approx 15\times10^{-3}\, m_0c$, which represents a contribution of positrons annihilated by the oxygen $2p$ electrons [8]. Contrary to the Z3Y, indeed, no enhancement at low momentum region ($p \approx 0$) is observed in Z3Y-$C_{x_{Cr}}$ specimens due to an absence of the para-Ps contribution. DBP ratio curves for YSZ binary and ternary nanopowders are located well below that for the Z9Y monocrystal indicating thus that the fraction of positrons annihilated by the oxygen electrons is reduced in the Z3Y-$C_{x_{Cr}}$ nanopowders. The CDB response of the annealed pure Cr was also included in figure 2. It exhibits a peak centered almost at the same position as that for the $2p$ oxygen electrons. Hence it is difficult to resolve an effect of the Cr addition (if any) on DBP ratio curves.

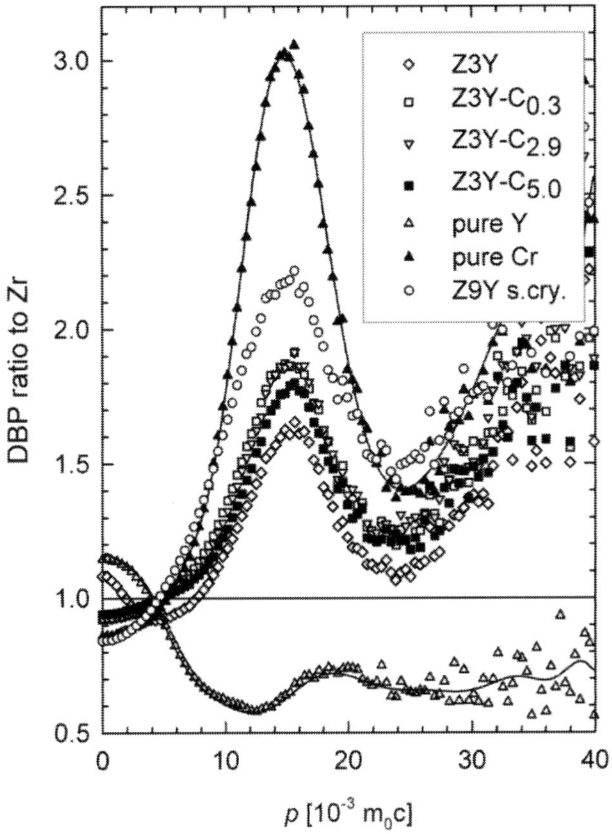

Figure 2. Measured DBP ratios as functions of the electron momentum p. The solid lines are drawn to guide eyes only.

4. Summary

The ternary yttria-stabilized zirconia pressure-compacted nanopowders with a different content of Cr_2O_3 were studied in the present work using positron annihilation spectroscopies: the positron lifetime and coincidence Doppler broadening techniques. The results of this study can be summarized in the following items:

- Positrons annihilate (i) in the vacancy-like defects situated at grain boundaries, or (ii) at GB triple points.
- The addition of the trivalent chromium oxide leads to a decrease in grain size.
- The paramagnetic centers introduced due to an addition of the Cr atoms suppress the Ps formation in voids and pores below the detection limit.

Acknowledgements

The financial support of the present work by the Ministry of Education, Youths and Sports of the Czech Republic (CR) – the scientific plan No. 0021620834, the Czech Scientific Foundation

– contract No. GA 106/06/0270, and the Grant Agency of the Academy of Sciences of the CR – projects Nos. KAN300100801 and KJB101120906 is highly acknowledged. The present work was also supported from the National Academy of Science of Ukraine (scientific plans Nos. A 116/06 H and A 106U006933).

References

[1] See, e.g., http://www.stanfordmaterials.com

[2] Korduban A M, Yashschishyn I A, Konstantinova T E, Danilenko I A, Volkova G K and Glazunova V A 2007 *Func. Mater.* **14** 454

[3] Bykov I P, Brik A B, Glinchuk M D, Bevz V V, Kalinichenko E A, Konstantinova T E and Danilenko I A 2008 *Phys. Sol. State* **50** 2311

[4] Hautojärvi P and Corbel C 1995 *Positron Spectroscopy of Solids* ed Dupasquier A and Mills A P (Amsterdam: IOS Press) p 609

[5] Krause-Rehberg R and Leipner H S 1999 *Positron Annihilation in Semiconductors: Defect Studies* (*Springer Series in Solid-State Sciences* vol 127) (Berlin, Heidelberg: Springer Verlag) p 98

[6] Ito K, Yagi Y, Hirano S, Miyayama M, Kudo T, Kishimoto A and Ujihira Y 1999 *J. Ceram. Soc. Jpn.* **107** 123

[7] Wang Z, Chen Z W, Zhu J, Wang S J and Guo X 2000 *Rad. Phys. Chem.* **58** 697

[8] Garay J E, Glade S C, Asoka-Kumar P, Anselmi-Tamburini U and Munir Z A 2006 *J. Appl. Phys.* **99** 024313

[9] Yagi Y, Hirano S, Ujihira Y and Miyayama M 1999 *J. Mater. Sci. Lett.* **18** 205

[10] Čížek J, Melikhova O, Kuriplach J, Procházka I, Konstantinova T E and Danilenko I A 2007 *Phys. Stat. Sol. (c)* **4** 3847

[11] Procházka I, Čížek J, Kuriplach J, Melikhova O, Konstantinova T E and Danilenko I A 2008 *Acta Phys. Polon. A* **113** 1495

[12] Melikhova O, Kuriplach J, Čížek J, Procházka I, Anwand W, Brauer G, Konstantinova T E and Danilenko I A 2007 *Phys. Stat. Sol. (c)* **4** 3831

[13] Melikhova O, Kuriplach J, Čížek J, Procházka I and Brauer G 2009 *Mater. Sci. Forum* **607** 125

[14] Konstantinova T, Danilenko I, Pilipenko N and Dobrikov A 2004 *9th Cimtec – World Ceramic Congress. Ceramics: Getting into the 2000's – Part A* ed Vincenzini P (Faenza, Italy: Techna Group s.r.l.) p 305

[15] Slipenyuk A M, Glinchuk M D, Bykov I P, Ragulya A V, Klimenko V P, Konstantinova T E and Danilenko I A 2004 *Ferroelectrics* **298** 289

[16] Bečvář F, Čížek J, Lešták L, Novotný I, Procházka I and Šebesta F 2000 *Nucl. Instrum. Methods Phys. Res. A* **443** 557

[17] Surbeck H 1977 *Helv. Phys. Acta* **50** 705

[18] Čížek J, Bečvář F, Procházka I and Kočík J 2004 *Mater. Sci. Forum* **445–446** 63

[19] Guo X and Wang Z 1998 *J. Eur. Ceram. Soc.* **18** 237

[20] Guo X and Waser R 2006 *Prog. Mater. Sci.* **51** 151

[21] Kröger F A 1974 *Chemistry of Imperfect Crystals, Vol. 2.* (Amsterdam: North-Holland)

[22] Guo X and Zhang Z 2003 *Acta Mater.* **51** 2539

[23] Matsui K, Yoshida H and Ikuhara Y 2008 *Acta Mater.* **56** 1315

Simulation of properties of positrons trapped at Cu nanoparticles in Fe matrix

J Kuriplach and O Melikhova

Department of Low Temperature Physics, Faculty of Mathematics and Physics, Charles University, V Holešovičkách 2, CZ-180 00 Prague 8, Czech Republic

E-mail: jan.kuriplach@mff.cuni.cz

Abstract. Positrons may get trapped in copper nanoparticles embedded in iron, which has been recently well evidenced experimentally. In this contribution we treat this effect theoretically and present an improved approach to simulations of properties of positrons trapped in copper nanoparticles embedded in the iron bcc matrix. The presented approach can be used to determine the size of Cu nanoparticles in model Fe-Cu alloys using coincidence Doppler broadening spectroscopy.

1. Introduction

Cu nanoparticles embedded in the Fe matrix represent an interesting system that can be examined by means of positrons. The reason is that the positron affinity (PA) of Cu is below that of Fe and Cu nanoparticles constitute positron potential wells where positrons can be trapped if nanoparticles are large enough. The experimental evidence of such trapping was reported in [1]. This effect has also practical consequences as the Fe-Cu system serves as a model alloy for reactor pressure vessel steels where Cu-rich precipitates are believed to play the key role in the process of embrittlement (see e.g. [2]).

In principle, nanoparticles could be identified by comparing the measured and calculated positron annihilation response. Recently, an effective computational approach was developed [3; 4], which allows us to calculate characteristics of positrons trapped at Cu nanoparticles as a function of their size. This approach is based on the atomic superposition (ATSUP) method [5; 6] and takes into account the positron affinity difference of the host and particles, which is ordinarily not included in the ATSUP method. In particular, it is possible to determine the positron localization in Cu particles, positron lifetime and binding energy to particles and also high momentum parts of the electron-positron momentum distribution, the last property being a 'fingerprint' of positron trapping in Cu nanoparticles. In this contribution the above mentioned approach is further improved by handling properly boundary conditions, which is an important point especially for very small Cu particles. The dependence of the W-parameter on the Cu nanoparticle size is presented, which could be used to 'measure' the size of Cu nanoparticles in Fe.

2. Computational methods

At the beginning a suitable supercell for calculations has to be selected. The supercell is based on the Fe bcc lattice. Cu particles are introduced into the supercell by replacing appropriate Fe

Journal of Physics: Conference Series **265** (2011) 012021 doi:10.1088/1742-6596/265/1/012021

atoms by Cu. Here we shall study the 'most compact' particles only, i.e. those corresponding to the increasing number of nearest neighbor shells of a reference atom in the bcc lattice. Hence, we consider particles having 9, 15, 27, 51, 59, 65, 89, 113 and 137 Cu atoms. It is well known that small Cu particles in the Fe matrix exhibit the bcc structure with the lattice constant very close to that of Fe and Cu particles are coherent with the matrix (see e.g. [2]). Therefore, atomic relaxations around Cu particles are neglected in our approach.

Next, we shall briefly repeat the principle of the computational/simulation method used here according to [4]. The method consists of three steps performed within the framework of the ATSUP technique:

(i) First, the energy (E_1) of positrons in the defect-free matrix material (bcc iron) is determined.

(ii) In the second step, the positron potential in the vicinity of atoms constituting the particle is shifted (by ΔV_2) in order to have the positron energy equal to E_1. This corresponds to the hypothetical case when the positron affinity difference (ΔA_+) of the host and particle is zero and the positron is delocalized simultaneously in the host and particle. In this way a reference energy level for the positron potential of atoms constituting the particle is found.

(iii) Finally, the positron potential is further shifted by the energy corresponding to ΔA_+ (i.e. the resulting shift $\Delta V_3 = \Delta V_2 - \Delta A_+$) and, thereby, the positron potential well with the correct depth is created.

Regarding step (ii) we note that in supercells used for calculations the number of atoms in the particle must be much lower than that of the matrix. In addition, the potential shift ΔV_2 for particle atoms is performed at spheres centered at atom nuclei. There is a lower limit for radius (R) of such spheres, which corresponds to the situation when the spheres fill up completely the particle volume (i.e. no interstitial space is left among spheres).

In our previous work [4] we have determined the parameters of the model, i.e. ΔA_+ and R for the Fe-Cu system on the basis of *ab initio* calculations. The resulting values are $\Delta A_+ = 0.5$ eV and $R = 1.6$ Å. Readers are referred to [4] for the details of the procedure. We shall use the above parameters in the following.

In order to calculate positron characteristics for regular Cu particles in the Fe matrix in [4], we have employed 2662 atom supercells (i.e. $11 \times 11 \times 11$ bcc cells) considering periodic boundary conditions (PBCs) for the positron wave function. In the present contribution we check in detail whether such supercell size is sufficient to obtain positron characteristics independent of the supercell size. The reason is that for small Cu particles the corresponding positron potential well is rather shallow and the size of the chosen supercell might not be sufficient.

Such an effect has been reported by Korhonen *et al.* in [7]. The solution is to use a larger supercell or to use PBCs in combination with the so called antiperiodic boundary conditions (ABCs) [7], as explained below, which also saves some computational time. Under PBCs, for a supercell characterized by e.g. d_x dimension[1] along the x-axis direction the positron wave function (ψ) is not changing its value and sign when considering two positions in space that are shifted just by d_x

$$\psi_{PBC}(x + d_x, y, z) = \psi_{PBC}(x, y, z)\,. \tag{1}$$

In the case of ABCs, we however adopt the following rule

$$\psi_{ABC}(x + d_x, y, z) = -\psi_{ABC}(x, y, z)\,. \tag{2}$$

For small supercells PBCs cause too small positron localization, whereas ABCs result in an overestimate of the positron localization in an open volume defect under study [7]. Then, if an

[1] For simplicity we consider here a rectangular supercell only.

effective wave function constructed as

$$\psi_{eff} = \frac{1}{\sqrt{2}}\sqrt{|\psi_{PBC}|^2 + |\psi_{ABC}|^2} \tag{3}$$

is considered, it gives usually apparently more precise positron wave function which matches well – in the close vicinity of the studied defect – the wave function obtained using PBCs for much larger supercell. We shall employ this approach in the present study.

We should yet note that the resulting positron energy corresponding to the wave function ψ_{eff} is calculated as the arithmetic mean from the values of energies obtained using PBCs and ABCs described above. The high momentum parts of the electron-positron momentum distribution are calculated according to the scheme presented in [8]. The positron density used to calculate the positron lifetime and momentum distribution is taken to be $n_+ = |\psi_{eff}|^2$. The range of momenta (p) used for the calculation of the W-parameter is $15 \times 10^{-3}\, m_e c < |p| < 25 \times 10^{-3}\, m_e c$. The positron localization in a Cu particle si determined as an integral of n_+ over the region where the positron potential is adjusted according to points (ii) and (iii) above. Electron-positron correlations are treated according to a gradient correction scheme developed by Barbiellini *et al.* [9]. Further computational details are specified in [4]. We further refer to literature [10] concerning the theoretical background of positron calculations.

3. Results and discussions

At the beginning we should state that there is no trapping in Cu particles with 9 atoms. This can be deduced from figure 1 where the dependencies of the positron localization and positron binding energy on the supercell size are plotted. The supercell size is determined by the multiplicative factor that is used to enlarge the (2 atom) bcc unit cell of Fe in the three crystallographic directions. The range of such factors goes from 7 to 25, which corresponds to supercells from 686 to 31250 atoms. Consequently, the supercell size is given in 'units' of 'bcc^3' in the figure. One can clearly see that the positron localization and binding energy approach zero with the increasing supercell size regardless of boundary condition used ('average' means the arithmetic mean from PBC and ABC values). This proves that 9 atom Cu particles in iron matrix do not trap positrons. We have discussed this fact already in our work [4], but here an additional proof is given.

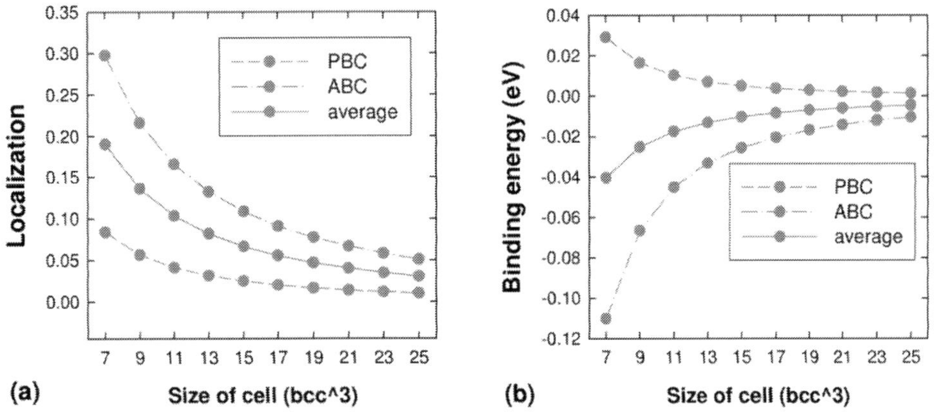

Figure 1. The dependence of (a) the positron localization and (b) positron binding energy on the supercell size for the 9 atom Cu particle in Fe matrix.

Next, we investigate the 15 atom Cu particle. The results are presented in figure 2 in the way similar to the 9 atom Cu particle. In this case the positron localization and binding energy converge to some non-zero value with the increasing supercell size. For large supercells the localization is about 38 %, whereas the binding energy is still rather small ~0.046 eV. In figure 2c we also show the positron lifetime which approaches a value being slightly above a calculated Fe bulk value of 113 ps. The just discussed behavior unambiguously show positron trapping at the 15 Cu atom particle in the Fe matrix though the corresponding positron potential well is rather shallow and the positron is not perfectly confined in the particle.

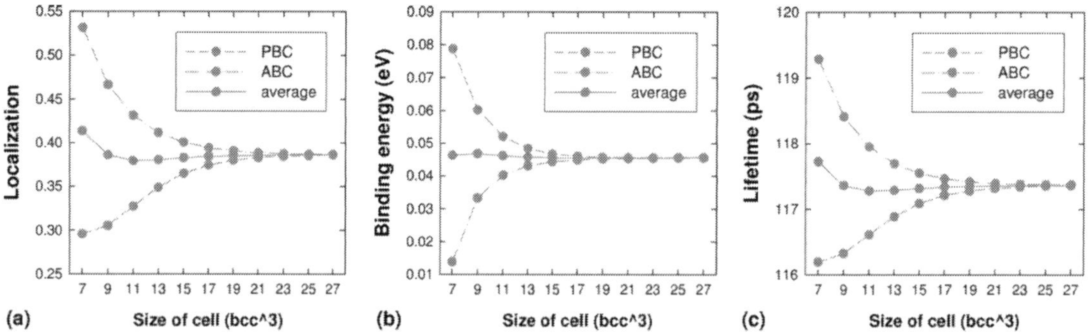

Figure 2. The dependence of (a) the positron localization, (b) positron binding energy and (c) positron lifetime on the supercell size for the 15 atom Cu particle in Fe matrix.

If the differences among examined boundary conditions are checked, it is possible to conclude that for small supercells such differences are rather large and corresponding averaged values still differ non-negligibly from converged ones. In what follows we shall consider that for 11×11×11 bcc supercell the averaged values for all quantities investigated give already the converged values almost precisely. This further indicates that using the same supercell in our previous work [4] was reasonable except for very small particles (like 15 atom one) where averaged values still deviate apparently from corresponding PBC values.

After studying in detail two particles we now proceed with dependencies for the series of selected particles. Figure 3 shows the dependence of the calculated W-parameter on the Cu particle size. One can clearly see that without considering the true positron affinity difference

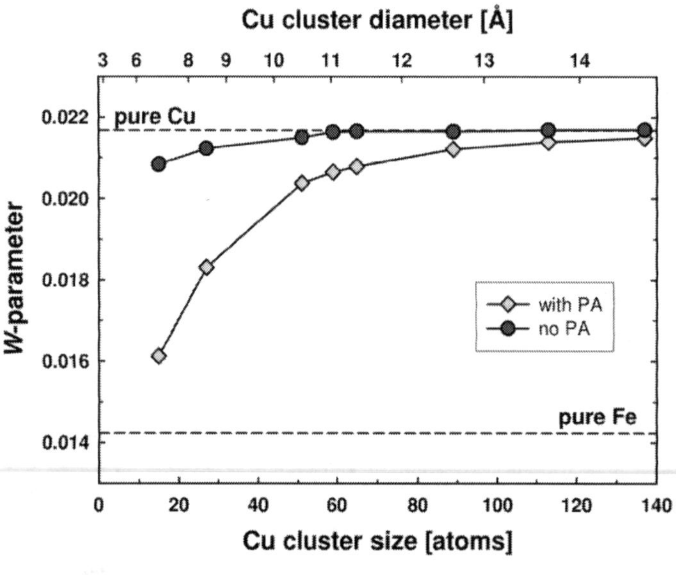

Figure 3. The dependence of the W-parameter on the Cu particle size (diameter) for two computational approaches.

('no PA' dependence in the figure) between the particle and matrix the W-parameter is very close to that for pure Cu even for very small particles, which is connected with the fact that the positron localization for such particles is overestimated. On the other hand, when the positron affinity difference is taken into account ('with PA' dependence) in the way described above, the W-parameter gradually increases, as expected, and becomes closer to the pure Cu value for apparently larger particles compared to the previous case.

This behavior could be, in principle, used to study/determine the (average) Cu particle size in real alloys using coincidence Doppler broadening spectroscopy [11]. The necessary condition would be that saturation trapping occurs in such particles, which does need to be necessarily true in reality. Otherwise the fraction of positrons annihilating at particles would have to be known, which is hard to obtain from experiment. However, simulations of the Cu precipitation process could be of help in this respect [12].

An alternative approach for determining the size of embedded Cu nanoparticles in Fe based on the momentum smearing effect has been proposed at this Workshop [13].

4. Conclusions

Cu nanoparticles embedded in the bcc Fe matrix were studied from the viewpoint of positron annihilation spectroscopy. In particular, we presented a computational/simulation method to determine positron characteristics corresponding to positrons trapped at such nanoparticles. Further work is in progress in order to employ such simulations for estimates of the Cu particle size in real Fe-Cu alloys.

Acknowledgement

We are grateful to M J Puska for his ATSUP code that served as a basis for further developments. This work was supported by the Ministry of Schools, Youths and Sports of the Czech Republic through the research plan MSM 0021620834 and project COST OC 165.

References

[1] Nagai Y, Hasegawa M, Tang Z, Hempel A, Yubuta K, Shimamura T, Kawazoe Y, Kawai A and Kano F 2000 *Phys. Rev. B* **61** 6574

[2] Phythian W J and English C A 1993 *J. Nucl. Mater.* **205** 162

[3] Kuriplach J, Bečvář F, Čížek J and Procházka I 2004 *Mater. Sci. Forum* **445–446** 132

[4] Kuriplach J 2005 *Acta Phys. Polon. A* **107** 784

[5] Puska M J and Nieminen R M 1983 *J. Phys. F: Met. Phys.* **13** 333

[6] Seitsonen A P, Puska M J and Nieminen R M 1995 *Phys. Rev. B* **51** 14057

[7] Korhonen T, Puska M J and Nieminen R M 1996 *Phys. Rev. B* **54** 15016

[8] Kuriplach J, Morales A L, Dauwe C, Segers D and Šob M 1998 *Phys. Rev. B* **58** 10475

[9] Barbiellini B, Puska M J, Korhonen T, Harju A, Torsti T and Nieminen R M 1996 *Phys. Rev. B* **53** 16201

[10] Puska M J and Nieminen R M 1994 *Rev. Mod. Phys.* **66** 841

[11] Alatalo M, Kauppinen H, Saarinen K, Puska M J, Mäkinen J, Hautojärvi P and Nieminen R M 1995 *Phys. Rev. B* **51** 4176

[12] Verheyen K, Almazouzi A, Malerba L, Domain C, Kuriplach J and Van Petegem S 2004 *Proceedings of the 2nd International Conference on Multiscale Materials Modelling* ed Ghoniem N M (Los Angeles, CA: UCLA)

[13] Tang Z *et al.* 2008 Electron momentum-density distributions of Cu precipitates in Cu-Fe alloys: a quantum size effect in momentum space ; PSD-08 workshop presentation (unpublished)

Formation of light-emitting silicon nanoclusters in SiO_2

P J Simpson[1], C R Mokry[1] and A P Knights[2]

[1] Department of Physics and Astronomy, The University of Western Ontario, London, Canada N6A 3K7

[2] Department of Engineering Physics, McMaster University, Hamilton, Canada L8S 4L7

E-mail: psimpson@uwo.ca

Abstract. Silicon nanoclusters/nanocrystals (Si-nc) in an SiO_2 matrix exhibit strong visible luminescence, and so are of interest in the pursuit of a silicon-based light emitter for optoelectronics. We have investigated the formation of Si-nc by implanting excess Si at 90 keV into SiO_2 films and then annealing to form nanoclusters by precipitation and ripening. The use of ion implantation provides control over composition and so allows us to optimize the light output. Positron annihilation provides information on vacancy-type defects produced during implantation. Our results suggest that defects may play a key role in Si-nc formation. The depth and size distributions of Si-nc are obtained by transmission electron microscopy, and are correlated with light emission measured by photoluminescence.

1. Introduction

There is considerable interest in making all-silicon integrated circuits with optoelectronic functions. The key element missing from this strategy is an efficient silicon-based light emitter. Silicon nanoclusters/nanocrystals (Si-nc) embedded in SiO_2 exhibit strong luminescence at room temperature and so have attracted considerable research interest. Si-nc can be fabricated by a variety of methods including cosputtering [1], plasma enhanced chemical vapour deposition [2], laser ablation [3], ion implantation [4], and molecular beam epitaxy [5]. We have used ion implantation of Si into SiO_2 because (i) implantation is widely used in silicon processing, (ii) it provides the ability to precisely control the distribution and concentration of silicon, and (iii) it does not introduce any undesirable impurities.

There are multiple mechanisms for luminescence from Si-nc, including quantum confinement [6], interface states [7], and oxide defects [8]. The dominant luminescent mechanism varies with the Si-nc size [9]. In implanted samples, the stoichiometry of the oxide layer varies as a function of depth, which leads to a varying Si-nc size distribution versus depth. In this study, transmission electron microscopy was used to determine the Si-nc size distribution as a function of depth from the surface to the bulk silicon interface. The size distribution of Si-nc from the surface to the oxide/substrate interface has been previously reported by Brongersma *et al.* with the sizes ranging from 2-5 nm [10]. We found that the Si-nc size distribution does not correlate strongly with the implanted ion distribution as we would expect, but is correlated more closely with the distribution of vacancies generated during implantation. This may be indicative of the role played by vacancies in the Ostwald ripening process by which the Si-nc evolve during annealing.

Positron Doppler broadening measurements were used to probe defects, particularly vacancies introduced by ion implantation and their reduction after high temperature anneals. PAS data correlate with the location of Si-nc formed in the SiO_2 film after a long anneal, presumably due to positron trapping into silicon/SiO_2 interfaces, which are efficient traps [11]. In this report we focus primarily on the positron annihilation results; electron microscopy and photoluminescence data are reported in more detail elsewhere [12].

2. Experiment

Oxide films of thickness 430nm were thermally grown in wet O_2 at 1000°C on a boron-doped (100) silicon substrate. The films were implanted with Si^- ions at 90 keV. The projected range predicted by stopping and range of ions in matter (SRIM) simulations [13] is 130 nm. Ion fluences ranged from $6 \times 10^{16} - 1.4 \times 10^{17}$ ions/cm², resulting in excess silicon (i.e. compared with stoichiometric SiO_2) of 9 – 20 at. %. Growth and ripening of Si-nc was achieved by high temperature anneals in a Jipelec Jetfirst rapid thermal annealer (RTA) or in a furnace, at 1070°C or 1170°C, both in high-purity nitrogen. Finally, to reduce the effects of non-radiative electron-hole recombination and dangling bonds, a 1 hour furnace anneal in forming gas (5% H_2 95% N_2) at 450°C was performed. The diffusion coefficient for Si in SiO_2 is small [14]. Rutherford Backscattering measurements of the silicon distribution in some of our samples before and after annealing confirmed that there is very little redistribution of the implanted silicon.

Positron annihilation Doppler-broadening spectra were obtained using the University of Western Ontario positron accelerator, using a range of positron energies from 0.5 to 30 keV. Samples were measured using PAS to examine the evolution of defects in the oxide films due to annealing, to enable comparison with PL data. Data shown in figures 1 and 3 typically represent an average of 6 measurements each containing 100 000 counts. For the investigation of difference spectra shown in figure 2, approximately 100 measurements, each of 100 000 counts, were acquired at selected positron energies. Spectra were collected using a high-purity Ge detector of volume ~210 cm^3, with a resolution at 511 keV of 1.3 keV.

3. Results

A variety of defect species can be formed in SiO_2 by ion implantation [15]: peroxy radicals, E' centres, non-bringing-oxygen hole centres (NBOHC), and vacancy-type defects. Peroxy radicals are found more commonly in oxygen-rich SiO_2, making them less likely in our silicon-rich films [16]. E' centres are found more frequently in silicon-rich SiO_2 [17], but they are positively charged making them inefficient positron traps. NBOHC are a result of the electronic stopping of the implanted ions (i.e. energy loss to electrons of the medium) [18], and have been reported to result in a reduction in the positron annihilation S parameter [19]. Vacancy-type defects are caused by the nuclear stopping of the implanted ions (i.e. direct nuclear collisions) and result in an increase in the S parameter. If the vacancy-type defect is large enough positronium can form and be observed [20].

Our implants were performed at 90 keV, at which energy the two stopping powers are nearly equal (dE/dx = 1.26 (nuclear) vs. 1.39 (electronic) MeV cm^2/mg [13]), meaning that both types of defects should be present, and should exert opposing influences on the S parameter. In figure 1, PAS measurements for three different implant doses show an S parameter value higher than the starting material, and increasing with increasing ion fluence. This suggests that the PAS results are influenced most strongly by the creation of vacancy-type defects.

Previous reports suggest this may only be the case for high implant doses: Ghislotti *et al.* [16] performed a similar experiment with similar results at high doses, but for lower implant doses the S parameter was reduced by the implant, indicating the presence of NBOHC defects. Knights *et al.* [19] and Simpson *et al.* [20] have previously reported a reduced S parameter due to Si implantation into SiO_2, but for much lower doses than those described here.

Ghislotti *et al.* [16] suggested that the increased S parameter that they observed for high dose implants was due to the increase in silicon concentration. However we note that (i) the S parameter is higher than that for bulk silicon, and therefore more consistent with open-volume defects, (ii) the peak *S* parameter (in their data and in ours) is located at a depth which is better correlated with the vacancy-type defect profile predicted by SRIM [13] than with the peak of the implanted silicon, and (iii) the observation of positronium (see below) suggests the presence of open volume defects of significant size.

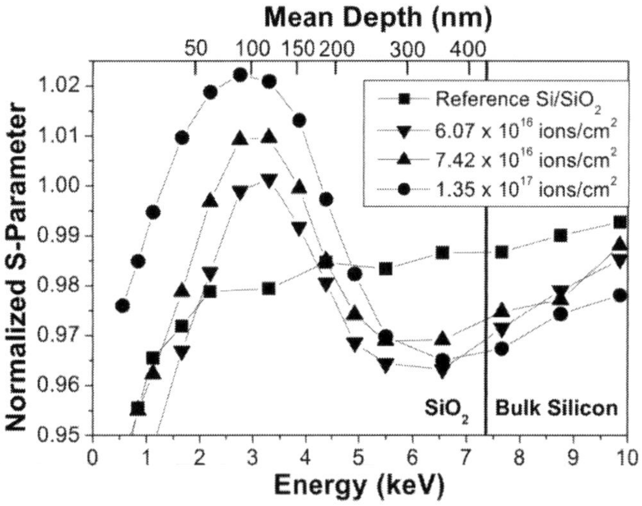

Figure 1. *S* parameter data obtained from 430 nm thick SiO_2 films implanted with Si at 90 keV to various fluences. Solid lines are a guide to the eye.

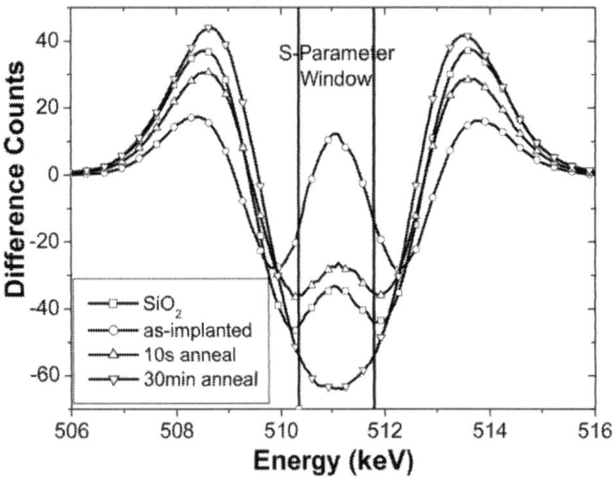

Figure 2. Annihilation spectra obtained at a positron energy of 3 keV, comparing the reference SiO_2 with the as-implanted, 10s anneal and 30min anneal at 1170°C, for a sample implanted to a fluence of 1.35×10^{17} ions/cm². Data have been smoothed for clarity.

Examining the Doppler-broadened annihilation spectra in detail is revealing. Figure 2 shows difference spectra obtained by subtracting the spectrum of 'bulk' defect-free silicon from various spectra of interest. Considering first the reference SiO_2 spectrum we propose that the narrow peak in the central region is due to 2-gamma annihilation of positronium. This suggests that the as-grown SiO_2 film contains open volumes of sufficient size for Ps to form within them. We do not see in our spectra any evidence for 3-gamma annihilation of orthopositronium, which suggests that any orthopositronium that is formed annihilates by pick-off. Considering next the spectrum of the as-implanted sample, we note an increase in the narrow Ps peak, which suggests an increase in the size and/or number of open volumes large enough to accommodate the formation of Ps. This is consistent with the fact that the very high implantation dose (1.35 × 10^{17} ions/cm²) produces a very large number of atomic displacements. Annealing for 10 seconds at 1170°C restores the annihilation spectrum to very nearly that of the as-grown film prior to implantation. Further annealing for 30 minutes, yields a spectrum that shows no evidence at all of 2-gamma positronium annihilation.

It is of interest to examine how these results correlate with the evolution of photoluminescence during a similar annealing sequence [12]: the peak wavelength of photoluminescence evolves to its final value within ~2 minutes of annealing or less, and thus correlates strongly with the fact that a 10 second anneal is sufficient to restore the positron annihilation spectrum almost to its pre-implant appearance. If we assume that the mechanism responsible for the photoluminescence is quantum confinement, then the peak wavelength is an indication of the size of the nanocrystals formed, and this then suggests that the nanocrystals evolve to their final size very rapidly during annealing. The intensity of the photoluminescence evolves more slowly, requiring ~10 minutes of annealing to reach its final value, suggesting that the longer annealing removes defects which compete with the luminescence mechanism by providing sites for non-radiant recombination of electron-hole pairs.

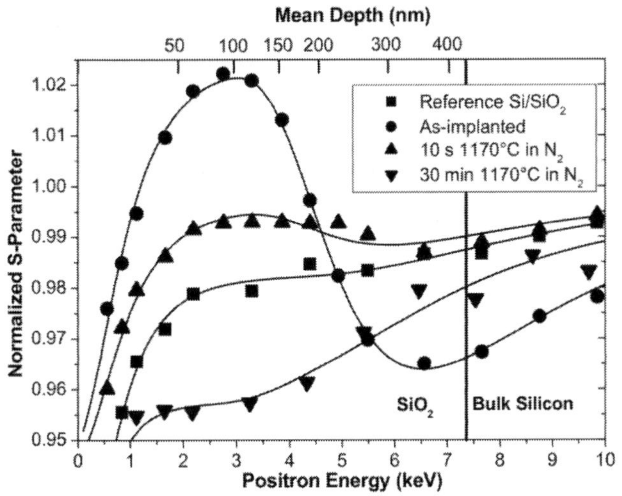

Figure 3. *S*-parameter data showing the evolution of the *S* parameter during annealing for a 430 nm thick SiO_2 film implanted at 90 keV to a fluence of 1.35 × 10^{17} ions/cm². Solid lines show results of modeling using POSTRAP5.

The positron annihilation spectrum for a sample annealed for 30 minutes shows no 2-gamma Ps annihilation, suggesting the elimination of sites with sufficient open volume for Ps formation. Transmission electron microscopy shows that the mean distance between Si-nc is approximately 7 nm, which is comparable with reported values [20] for the positron diffusion length in SiO_2. So it may be the case that in annealed films the positrons trap primarily in the interfaces between the Si-nc and the surrounding SiO_2 matrix, and that longer annealing creates greater ordering of these regions, with reduced open volume.

The positron data shown in figure 3 have been modelled using POSTRAP5 [22]. The data for the as-implanted sample cannot be fitted well without including *both* a layer of vacancy-type defects in the near surface region (0 to 285 nm) and a deeper layer (285 nm to 430 nm) with an *S* parameter lower than that of the unimplanted film. The depth of the vacancy-type layer is in good agreement with the profile of vacancy-type defects predicted by SRIM [13] simulations, shown in figure 4. Results of fitting are summarized in table 1.

We have obtained transmission electron microscopy images as a function of depth into the sample [12], and from these we extract the nanocrystal size and concentration vs. depth summarized in table 2. Of particular interest is the fact that the nanocrystal size distribution is not well correlated with the profile of implanted silicon (predicted by SRIM). This is the case despite the low diffusivity of Si in SiO_2 (which is supported by our Rutherford Backscattering measurements of the silicon distribution before and after annealing). There is however, strong correlation between the profile of vacancies, predicted by SRIM and supported by positron annihilation, and the depth distribution of nanocrystal size. This may suggest that vacancies play a key role in the diffusion required to form the nanocrystals. Further experiments are underway to better understand this phenomenon.

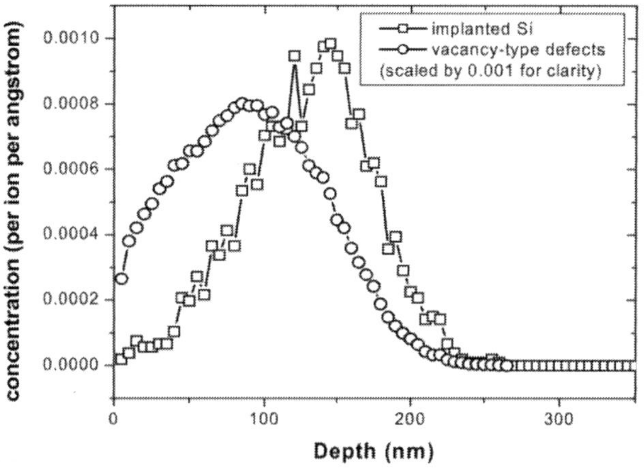

Figure 4. Prediction from SRIM simulation [13] of the depth distribution of implanted ions (90 keV Si into SiO_2) and of vacancy-type defects. The vacancy-type defect concentration has been divided by 1000 for clarity.

Table 1. Normalized *S*-parameters obtained from POSTRAP5 by fitting PAS data collected from the 430 nm thick SiO_2 film implanted at 90 keV to a fluence of 1.35×10^{17} Si ions/cm². Data were normalized so that the *S*-parameter for the Si substrate was equal to 1.00 in all cases. Parameters are reported to 3 digits of precision, although the coupling between related parameters in the modeling limits the real accuracy to roughly 0.01; values reported provide the best fit but slightly different parameters yield acceptable fitting also.

Sample	Surface	Implanted Region	SiO_2 film	Si/SiO_2 Interface
Reference SiO_2 film	0.917	-	0.983	0.972
As-implanted film	0.950	1.02	0.834	0.979
10s anneal at 1170°C	0.947	0.995	0.979	0.958
30min anneal at 1170°C	0.907	0.958	0.984	0.954

Table 2. The median and average diameter of Si-nc as a function of depth compared with the normalized vacancy-type defect and implant concentrations simulated by SRIM for an implant at 90 keV and a fluence of 1.35×10^{17} ions/cm². TEM images analyzed are those from ref [12]. The median and average diameters obtained may be skewed by the inability to observe nanoclusters below ~1 nanometre diameter.

TEM Image Depth Range (nm)	Median Diameter (nm)	Average Diameter (nm)	Normalized Vacancy-type Defect Concentration	Normalized Implant Concentration
0 – 45	1.71	1.77 ± 0.05	0.66	0.08
45 – 90	1.79	1.82 ± 0.05	0.99	0.44
90 – 135	1.82	1.86 ± 0.05	1.00	1.00
135 – 180	1.61	1.67 ± 0.04	0.54	0.98
180 – 225	Too small to distinguish		0.11	0.33

4. Conclusions

We have previously found that silicon nanocrystals grown by silicon implantation at 90 keV and annealing exhibit a broad photoluminescence peak centred at ~800 nm [12]. Positron annihilation was used to probe the defects created by implantation, and their recovery during annealing. We found that for the very high implantation doses used here, vacancy-type defects dominate the positron annihilation behaviour, resulting in an increased *S* parameter, contrary to results for low-dose implantation into SiO_2. The depth profile of the vacancy-type defect layer needed to model the data is in good agreement with that predicted by SRIM simulations. Examination of annihilation spectra shows that the increased *S* parameter caused by implantation is due to the formation of positronium, and its 2-gamma annihilation. We find that the nanocrystal size distribution obtained from electron microscopy [12] correlates with the profile of implant-induced vacancies, and not with the profile of implanted ions, suggesting that vacancies may play a role in the silicon diffusion required to form nanocrystals.

References

[1] Furukawa S and Miyasato T 1988 *Phys. Rev. B* **38** 5726
[2] Takagi H, Ogawa Y, Yamazaki Y, Ishizaki A and Nakagirl T 1990 *Appl. Phys. Lett.* **56** 2379
[3] Hayashi S, Tanimoto S, Fujii M and K. Yamamoto K 1990 *Superlattices Microstruct.* **8** 13
[4] Iwayama T S, Nakao S and Saitoh K 1994 *Appl. Phys. Lett.* **65** 1814
[5] Lu Z H 1995 *Nature (London)* **378** 258
[6] Lehman V and Gösele U 1991 *Appl. Phys. Lett.* **58** 865
[7] Koch F, Petrova-Koch V and Muschik T 1993 *J. Lumin.* **57** 271
[8] Prokes S M 1993 *Appl. Phys. Lett.* **62** 3244
[9] Wolkin M V, Jorne J, Fauchet P M, Allan G and Delerue C 1999 *Phys. Rev. Lett.* **82** 198
[10] Brongersma M L, Polman A, Min K S, Boer E, Tambo T and Atwater H A 1998 *Appl. Phys. Lett.* **72** 2577
[11] Nielsen B, Lynn K G, Chen Y-C and Welch D O 1987 *Appl. Phys. Lett.* **51** 1022
[12] Mokry C R, Simpson P J and Knights A P 2006 *Proceedings SPIE* vol 6343 pp 63432P-63438P (Photonics North Conference, Québec City, June 2006)
Mokry C R, Simpson P J and Knights A P 2008 submitted to *J. Appl. Phys.*
[13] www.srim.org
[14] Van Petegem S, Dauwe C, Van Hoecke T, De Baerdemaeker J and Segers D 2004 *Phys. Rev. B* **70** 115410
[15] Griscom D L 1985 *J. Non-Cryst. Solids* **73** 51
[16] Ghislotti G, Nielsen B, Asoka-Kumar P, Lynn K G, Mauro L F D, Corni F and Tonini R 1997 *Appl. Phys. Lett.* **70** 496
[17] Kalnitsky A, Ellul J P, Poindexter E H, Caplan P J, Lux R A and Boothroyd A R 1990 *J. Appl. Phys.* **67** 7359
[18] Fujinami M and Chilton N B 1993 *Appl. Phys. Lett.* **62** 1131
[19] Knights A P, Simpson P J, Allard L B, Brebner J L and Albert J 1996 *J. Appl. Phys.* **79** 9022
[20] Asoka-Kumar P, Lynn K G and Welch D O 1994 *J. Appl. Phys.* **76** 4935
[21] Simpson P J, Spooner M, Xia H and Knights A P 1999 *J. Appl. Phys.* **85** 1765
[22] Aers G C 1990 *Slow Positron Beams for Solids and Surfaces, AIP Conf. Proc.* **218** 162

Journal of Physics: Conference Series **265** (2011) 012023 doi:10.1088/1742-6596/265/1/012023

Intermolecular free volumes and intramolecular defects in *n*-alkanes

T Goworek*, M Pietrow, J Wawryszczuk, R Zaleski and B Zgardzińska

Department of Nuclear Methods, Institute of Physics, Marie Curie-Skłodowska University, 20 031 Lublin, Poland

E-mail: tomasz.goworek@poczta.umcs.lublin.pl

Abstract. Ortho-positronium lifetime and intensity were measured as a function of temperature for a homologous series of normal alkanes. Free volume shape and concentration of defects are different for various phases. The lifetimes observed were compared with model predictions for an empty gap in the crystalline structure (at low temperatures), finite cylinder or cuboid (rotator phase) and sphere (liquid). The role of intramolecular defects (nonplanar conformers) was discussed. A distinct difference in temperature dependence of lifetime spectra when melting is preceded by the rotator phase or C phase was found. Attention was paid to equality of lifetimes in the rotator phase and in liquid in a broad range of carbon chain lengths.

1. Introduction

Positron annihilation lifetime spectroscopy (PALS) is a method of investigation of "free" volumes in condensed matter, basing on the effects of trapping positrons or positronium in such low electron density spaces. In metals, e^+ lifetime value serves to identify the structural imperfections, like vacancies, dislocations, vacancy clusters, etc. In insulators, trapped *ortho*-Ps lifetime allows to estimate the trap sizes. In 1972, Tao [1] proposed the relation between *o*-Ps lifetime τ and the radius *R* of "bubble" forced by Ps in the liquid. Naturally, it was assumed that the geometry of free volume was spherical. The final form of that "spherical model" was given by Eldrup [2] and applied to the case of vacancies in plastic crystals. Such crystals are composed of molecules not too far from sphericity, thus accepting the Tao's idea seems a good approximation. However, already in the series of papers by Eldrup et al. [3,4] the radius has no the meaning of a real sphere, but rather that of Wigner-Seitz, i.e. of an equivalent sphere with the volume equal to that of the elementary crystal cell divided by the number of molecules per cell. The Tao-Eldrup model found soon a wide application to free volumes in polymers, which certainly are neither spheres, nor prisms. In the Tao-Eldrup equation

$$\tau = 0.5\text{ns}\left[1-\frac{R}{R+\Delta}+\frac{1}{2\pi}\sin\frac{2\pi R}{R+\Delta}\right]^{-1} \quad (1)$$

R is an equivalent radius of a sphere giving the same *o*-Ps lifetime as observed for real free volume (Δ means the parameter describing penetration of Ps wave function into the bulk matter; Δ = 0.166 nm). The *ortho*-Ps decay rate in the bulk in this equation and the next one is assumed 2 ns^{-1}.

* Corresponding author.

In 1996, Jasińska et al. [5] published an equation similar to (1) assuming the geometry of a rectangular box, which seems to be useful in cases of long channels or gaps in the layered structures (one or two dimensions extending to infinity):

$$\tau = 0.5\text{ns}\left[1 - \prod_{i=1}^{3}\left(\frac{a_i}{a_i + 2\Delta} + \frac{1}{\pi}\sin\frac{\pi a_i}{a_i + 2\Delta}\right)\right]^{-1}, \quad (2)$$

where a_i are the sides of a cuboid. Finally, we introduced the cylindrical geometry [6] and modified the equations to describe correctly the voids exceeding 1 nm in size, by taking into account the population by Ps some higher lying states in a potential well (**E**xtended **T**ao-**E**ldrup model, ETE [6,7]):

$$\tau = \left[N\sum \lambda_i \exp(-E_i / kT) + \lambda_T\right]^{-1}, \quad (3)$$

where N is the normalizing factor, λ_i are the *pick-off* decay constants of *o*-Ps from various levels of energy E_i in the well, λ_T is the decay constant of intrinsic three quantum annihilation.

The present status of PALS enables to discuss not only the size (equivalent radius) but also the shape of free volumes. In the paper below, we would like to give some examples of positronium in free volumes of various shapes, which can be found in a homologous series of *n*-alkanes.

2. Structure of *n*-alkanes

Normal-alkanes (*n*-alkanes, paraffins) are saturated hydrocarbons C_nH_{2n+2} (further "C_n" for shortness) in the form of carbon chains. The low temperature form of alkane molecules is a straight zigzag chain (*all-trans*). Alkane crystals have the lamellar structure; within the lamella the molecules are packed in parallel, perpendicularly (odd numbered chains), or obliquely (even numbered) to the lamella surface. In the range of carbon atoms $9 \leq n \leq 39$ in odd numbered alkanes and $22 \leq n \leq 40$ in even numbered ones, a "rotator phase" exists in the interval from 1 to 12 K before melting. In the rotator phase the translational symmetry of the crystal is preserved, yet the molecules can rotate around their long axes (orientational disorder). The reduction of ordering is accompanied by appearance of nonplanar conformers; their concentration rises with the chain length. For odd numbered chains with $n \geq 25$ and even ones with $n \geq 34$ next solid phases appear. They retain the lamellar structure although differ in lateral packing, tilt angle of carbon chains, stacking of lamellae etc.

3. Experimental

The PALS spectra were measured using a standard fast-slow delayed coincidence setup; the resolution time was about 250 ps. The spectra were assumed to consist of three exponential components (*p*-Ps decay, free annihilation, *o*-Ps decay) and were analysed using the LT programme [8]. Our object of interest was the lifetime τ_3 and relative intensity I_3 of *ortho*-positronium formed in the sample. The polycrystalline samples were from Sigma-Aldrich, 99% purity, and were used without further refinement. Liquid samples were degassed by the freeze-thaw technique in order to eliminate the *ortho-para* conversion in dissolved paramagnetic oxygen. The measurements were performed above the room temperature (if not stated otherwise), where the effect of excess electron trapping, distorting the Ps intensity, does not appear.

4. Low temperature phase

In the low-temperature phase ("rigid solid"), the ortho-positronium lifetime τ_3 is 1.1 – 1.5 ns, depending mainly on temperature. Figure 1 shows the temperature dependence of *o*-Ps lifetime in some solid *n*-alkanes. In rigid solid, the lifetime rises slowly with temperature. It is visible that the experimental data lie on two parallel curves; these for odd-numbered alkanes are shifted, compared to even-numbered, by about 200 ps upwards. This odd-even difference is likely to indicate that positronium in this phase locates in the interlamellar gap. The width of gap in even numbered alkanes is d_0 = 0.125 nm, while in odd numbered ones 0.195 nm [9]. Inside the lamella the molecules are tightly packed leaving no sufficient free space to accommodate the Ps atom (application of high pressure up to 1.4 GPa does not induce transitions to any other phase; above 500 MPa, the *o*-Ps lifetime is almost constant [10]). Another argument in favour of Ps location in the gap is an increase in *o*-Ps lifetime in mixed alkanes, e.g. *n*-nonadecane with 20 % of *n*-heneicosane [11]. Longer *n*-heneicosane molecules protrude from the lamellae forcing increased separation of lamellae. Another example supporting the concept of Ps location in the interlamellar gaps is the behaviour of very long chain alkanes at relatively low temperatures (near the room temperature). The X-ray scattering measurements with *n*-hexatriacontane C_{36} prove the coexistence of two or three phases [12]. We have analyzed the PALS spectra at 303 K using the MELT routine [13]. The result is shown in figure 2. Two *o*-Ps peaks can be distinguished (thus belonging to two phases). The peak at about 1.2 ns is typical of the low-temperature rigid phase appearing in all solid alkanes. The other one is located at 1.8 ns; it is a very long lifetime as for the rigid phase (and low temperature), however X-ray measurements demonstrate that the interlamellar gap in one of these phases is $d_0 = 0.36$ nm [12]. It is more than twice of that in shorter even chain alkanes, and one can expect the respective lifetime elongation.

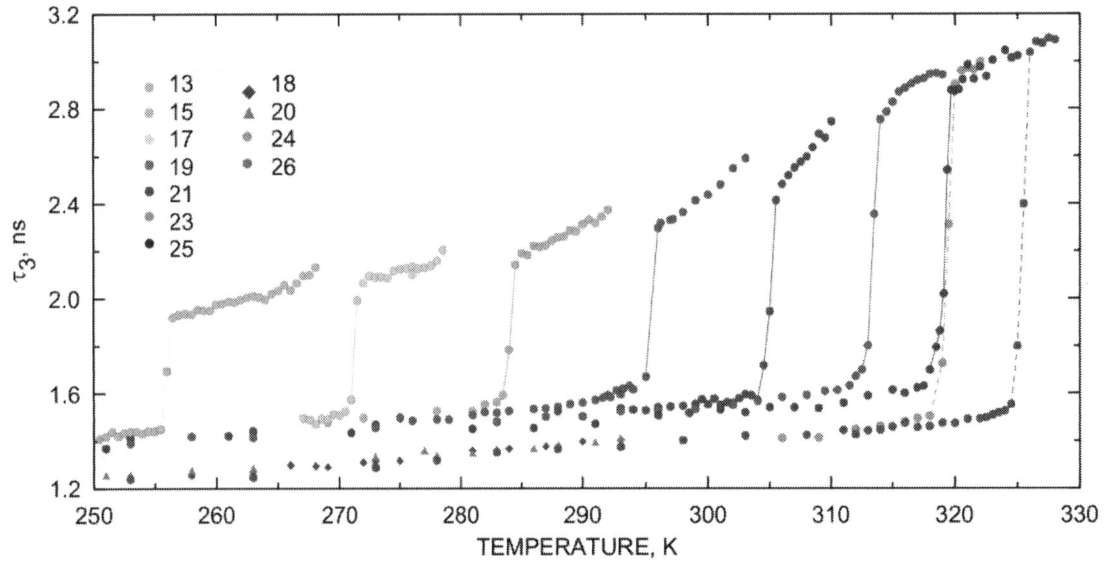

Figure 1. Temperature dependence of *o*-Ps lifetime in homologous series of *n*-alkanes (transition to liquid phase is not shown). Reddish colours are for even carbon chains, green-blue for odd ones. Solid and dashed lines are drawn at the transition rigid – rotator (no such transitions in C_{18} and C_{20}).

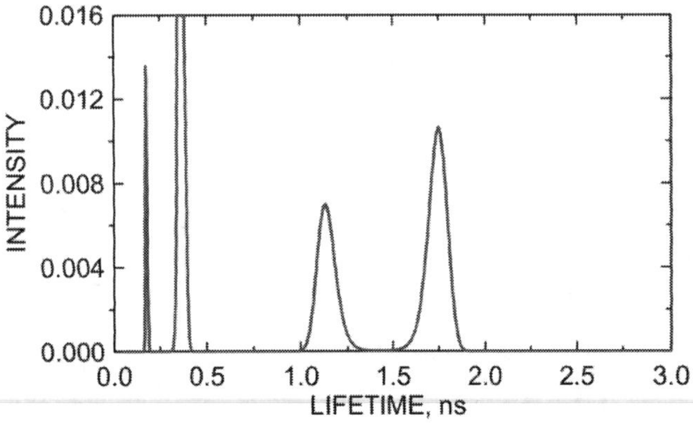

Figure 2. Distribution of positron lifetimes found by MELT routine in C_{36} at 303 K. The sample is obtained from the melt. The peaks, from left to right, belong to *para*-Ps, free e^+ annihilation, and the two last components originate from *ortho*-Ps.

All these examples show a correlation between the interlamellar gap width and *o*-Ps lifetime. In such cases, the free volume seems to be a flat empty space with two dimensions tending to infinity, the third dimension is d_0. If so, one can apply equation (2) to get the ETE estimate of lifetime. The calculated $\tau_3(d_0)$ dependence is shown in figure 3. The experimental values, at least for moderately long alkane chains, lie rather close to the model curve, the point for C_{36} is evidently too low. The source of disagreement could be the definition of gap width. The gap width, given in the papers about X-ray diffraction [14], is

$$d_0 = D - 0.127(n-1)\sin\beta - w, \quad (4)$$

where D is the spacing of lamellae (in nm), β is the staggering angle of molecular chain, w – small addition ($\approx$ 0.12 nm) for methyl groups at the end of a molecule. It is not evident whether this definition can be used in positron lifetime experiments. It can rather serve as a rough estimate; by accounting the finite size of carbon atoms and neglecting the hydrogen atoms extending at the end groups CH_3 we obtain the value of d_0 similar to that given above. However, possible misestimate of that gap should be a constant, independent of chain length. If so, the model curve in figure 3 can be shifted to the left or right only, yet the experimental data indicate that we need the modification leading to lowering of the curve inclination. In our calculations, we have assumed that the inter-lamellar gap limits have the form of ideal planes extending to infinity whereas in real crystal the lamella surface is rough, due to the presence of longitudinal shifts of molecules. Thus, in fact, positronium is not delocalized in an unlimited interlamellar space, but trapped between local distortions of the structure. If this was the source of divergence between the ETE model and experiment seen in figure 3, the reduction of τ_3 from 3 ns to 2 ns would mean Ps localization in the area of $\approx$0.8 nm in size.

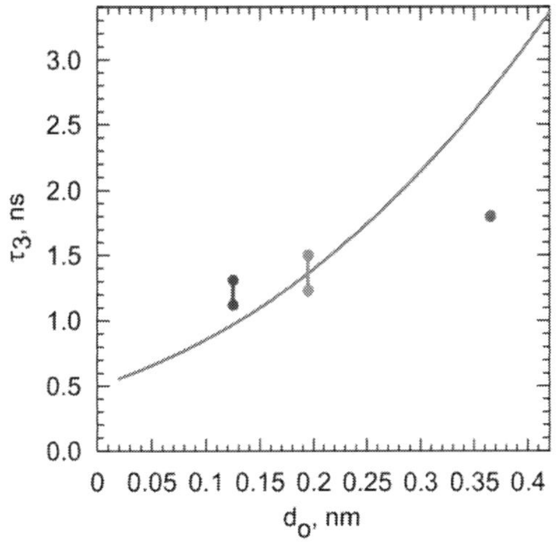

Figure 3. *o*-Ps lifetime as a function of empty gap thickness, according to the equation (2) and experimental points for rigid phase. Upper ends of bars correspond to 273 K, lower ends – 173 K.

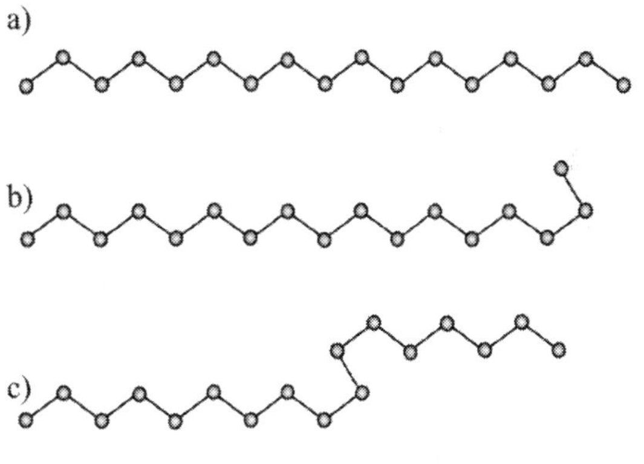

Figure 4. The most abundant conformers of chain hydrocarbons: a – *all trans*, b – *end gauche*, c – *kink*. Hydrogen atoms omitted for simplicity.

5. Rotator phase

In the rotator phase, the *o*-Ps lifetime is evidently longer than in the rigid one. In rigid crystal, we have the molecules in *all-trans* configuration (*t*), in the rotator one, nonplanar conformers begin to play an important role [15]. The main type of such conformers is that containing the *end-gauche* modification, or *gauche-trans-gauche* (*gtg*) sequence (figure 4). The latter gives a kinked shape to the molecule. The fraction of "kinked" molecules at n = 17 is about 8%, and at n = 29 approaches 70% [14]. Mismatching the nonplanar conformers and *all-trans* ones leads to creation of the free volume inside the lamellae. The shape of such a volume can be expected to be elongated; the cuboid or cylinder with the cross section independent on chain length seems to be most suitable to estimate the lifetime of *o*-Ps trapped in the vicinity of a kinked molecule. With an increase in chain length the lifetime rises, tending to the limiting value of about (3.1 – 3.2) ns. If we assume as rough estimate the value of 3.2 ns for an infinitely long channel, we obtain the square cross section of rectangular channel 0.56×0.56 nm, or the radius of cylindrical one 0.32 nm. The position of kink can move along the molecule, thus, the lifetime observed in the experiment is averaged over certain distribution. One can suppose that the average length of channel (in nm) is $a = [(n-1)\ 0.127 + w]/2 + d_0$. Taking the experimental τ_3 value and assuming the section of channel as given above, the channel length a can be calculated [16]. The result for the whole set of alkanes is shown in figure 5a. If average location of *gtg* kink is in the middle of chain, we can expect $a - d_0 = L/2$, where L is the molecule length. It is seen that $(a-d_0)/L$ approaches 1/2 for the longest chains only. For the chains shorter than n = 20 the value $(a-d_0)$ stabilizes at 0.2 L. This value of $(a-d_0)$ is hardly acceptable. Thus, instead of fixing the channel cross section, we fix the channel length as $L/2 + d_0$ and calculate the cross section (assumed square). The result is shown in figure 5b. The cross section S rises from ≈0.18 nm^2 for the shortest chains up to about 0.3 nm^2 for chains with $n > 25$. This S dependence on chain length is easier to explain. In short alkanes the concentration of molecules containing (*gtg*) defect is very low, thus the kinked molecule is surrounded by *all-trans* ones. A part (half) of defected molecule is displaced sideways and the free volume is the one left after displacement. As a rough approximation one can assume the cross section of free volume the same as the cross section of a crystal cell per one molecule (cross section of a column occupied by *all-trans* molecule). This section is almost constant in a broad range of n [14] and amounts 0.196 to 0.198 nm^2, which is close to S calculated from our experimental data for $C_{11} - C_{17}$. With the chain length extending, the concentration of kinked molecules rises and for $n > 30$ almost all molecules are distorted. The empty volume near them is now a cluster of channels running along the molecules and the effective section increases up to a certain maximal value, found here as S = 0.30 – 0.31 nm^2. Note that experimental points in figure 5 are given for both even and odd numbered chains, having different d_0.

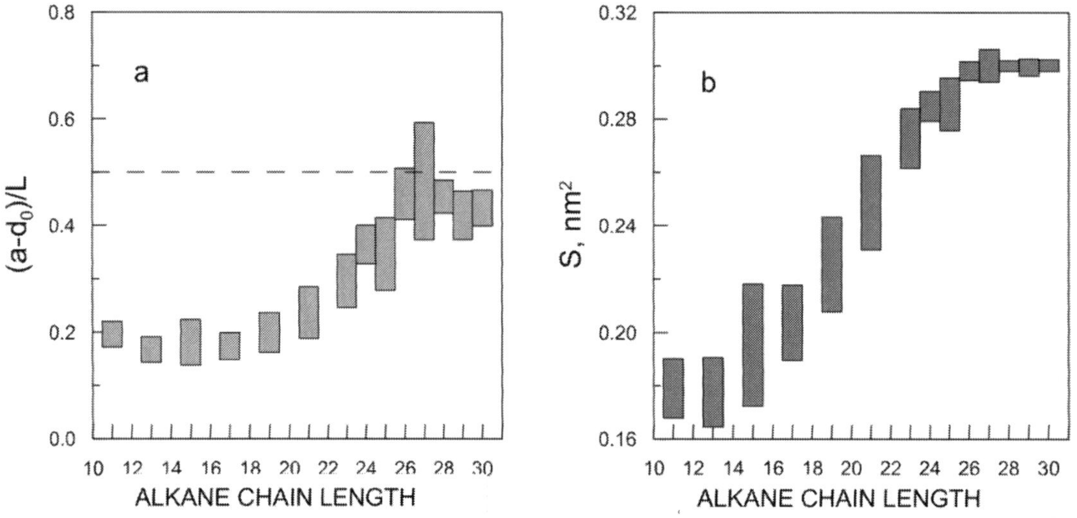

Figure 5. a) The ratio of empty channel length found using equation (2) to the molecule length for a set of alkanes. The channel is assumed rectangular with sides 0.56 × 0.56 nm. b) Channel cross section found using equation (2) assuming the channel length $L/2 + d_0$. Lower and upper end of bar on both drawings relate to beginning and end of the rotator phase range.

Naturally, there is a certain distribution of free volumes depending on kink location and on mutual orientation of kinked molecules (if in large concentration). This should be reflected in the distribution of lifetimes, which is confirmed by LT analysis [8] of high statistics spectra.

In fact, there are five versions of the rotator phase. In some alkanes, depending on temperature, one observes phases RI and RII for $22 \leq n \leq 26$, and phases RIII, RIV for $27 \leq n \leq 30$. The transitions from one rotator phase to the other are not visible in PALS measurements. The change of τ_3 lifetime (or I_3 intensity) is related to the structure modification; more radical rebuilding of the structure should result in more distinct changes in τ_3. A certain measure of structural modification can be the enthalpy of transition. The enthalpy of alkane melting is of the order from 60 kJ/mol for C_{25} to 144 kJ/mol for C_{40}, of the transition rotator-rigid crystal from 28 kJ/mol to 47 kJ/mol, while e.g. for transition RI-RII in C_{25} it is only 0.2 kJ/mol.

An interesting feature of the rotator phase is weak dependence of positronium formation intensity on the molecule length. The values of I_3 for the whole range of existence of the rotator phase are drawn in figure 6, in which the dashed line divides the regions of rotator RI and rotator RIV phases. It seems that in phase RIV the I_3 value is constant, while in RI it slightly decreases with shortening of the carbon atom chain. However, it may have nothing in common with the crystalline structure, being the result of very high concentrations of non-planar conformers. In one of our early papers [17], we have shown that the addition of spatially mismatched molecules (e.g. anthracene to *p*-terphenyl host) in the amount of 1% is enough to obtain maximal (saturation) I_3 intensity. The fraction of "kinked" molecules in cases of large n is much higher than 1% [14], thus at such concentrations I_3 should be maximal possible.

The range of rotator phase narrows with increasing of the chain length. In C_{38} it is still visible in the $I_3(T)$ curve as a knee less than 1 K wide, in C_{40} it cannot be found (at least by the PALS method [18]).

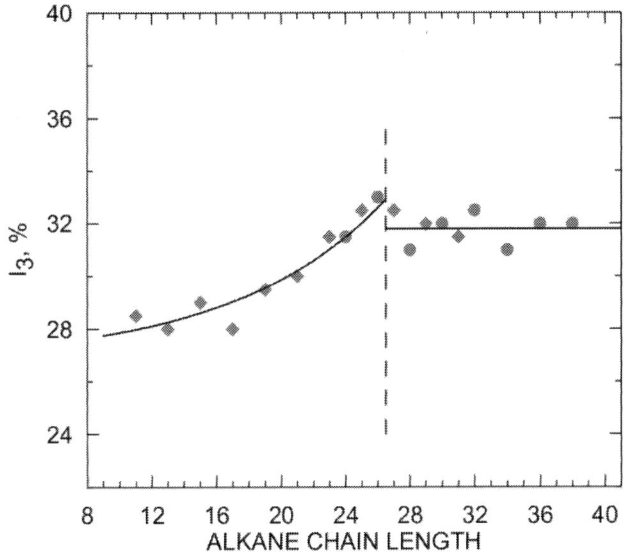

Figure 6. Intensity of *o*-Ps component in the rotator phase as a function of carbon chain length. Vertical dashed line separates the regions of R_{II} and R_{IV} phases.

6. High temperature solid phases

When the chain contains over 25 carbon atoms, new solid phases, preceding the rotator phase, appear. In odd numbered alkanes with $n \geq 25$, there are two phases, usually denoted as IV and V. Phase IV is clearly visible as the range where τ_3 is increased by about 0.2 ns. An example of $\tau_3(T)$ dependence in C_{27} is shown in figure 7. The region of phase IV extends from 320.5 to 325.5 K. An increase in τ_3 in that phase is due to increased concentration of *end-gauche* defects, shortening slightly the molecules. Phase V is not seen by the PALS method. Similarly to different rotator phases, the enthalpy of transition from low-temperature phase I to phase V is only 0.3 kJ/mol.

In even numbered alkanes, new phases appear from $n = 36$ upwards. In such alkanes, the structure is much more complex than in shorter ones. The crystalline phase depends on the way of sample preparation. Different phases are obtained when the sample is crystallized from the solution, different when obtained from the melt. The phase transition points given in literature are inconsistent [19]. It concerns not only the high temperature phases, but also the structure at low temperatures. In the melt-grown samples of C_{36} up to three phases coexist at room temperature [20]; an example of the room temperature structure consisting of two phases was already mentioned in section 4.

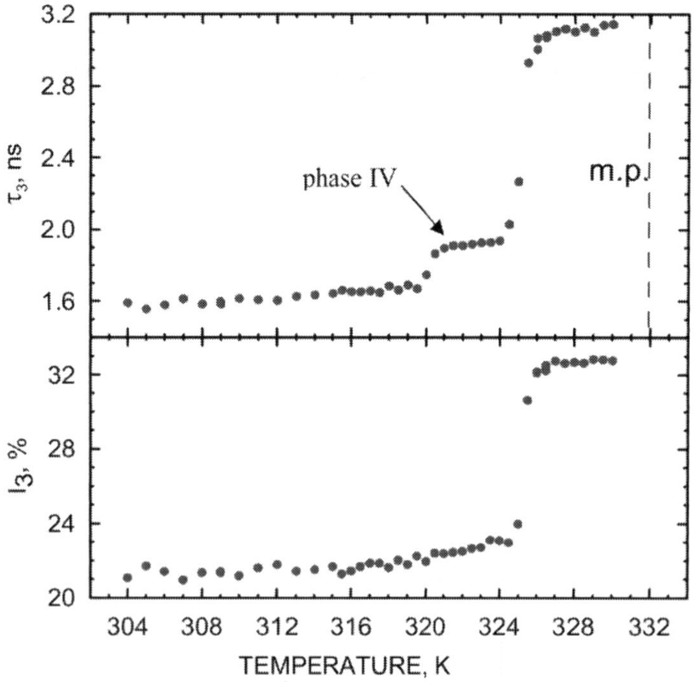

Figure 7. Temperature dependence of *o*-Ps lifetime and intensity in C_{27}.

If the sample is obtained by crystallization from the solution, the PALS method allows to observe the transition from the low-temperature monoclinic structure to the (also monoclinic) phase called C [21] or M_{101} for n longer than 36. In such positronic measurements, the spectra in the region near that transition and in the range of phase C look differently than those in shorter chain paraffins. Up to C_{34}, the transitions rigid → rotator and rotator → liquid are sharp, step-wise, while between the transition points the spectrum parameters are constant, as shown for C_{32} in figure 8. In C_{38} and next homologues, where the rotator phase is substituted by phase C, the lifetime begins to rise already below the transition point, the step in I_3 values extends over several degrees. In phase C, the $\tau_3(T)$ curve has a high slope; the transition to C phase is visible in τ_3 as a change of slope only. Such a smooth change of sample properties was seen by Kim et al. [22] in the temperature dependence of *gauche* bond density, i.e. the number of such bonds per one molecule, *m(g)*, for alkanes longer than C_{36}. For these longest chains, the step in *m(g)* at the phase transition point is entirely lost and described by the authors as "premelting". This term seems to be exaggerated; in PALS spectra, the transition point is still well observable (moreover, melting looks more sharp than rigid→C transition).

Contrary to the rotator phase, in which the *o*-Ps lifetime is the same as in liquid, this lifetime in phase C is shorter, changing with temperature, and the next phase transition, i.e. melting, is well marked by a stepwise rise of τ_3 by about 0.2 ns, i.e. to the value observed in all shorter chain structures. Figure 8 shows the $\tau_3(T)$ dependence for C_{40}; similar dependences can be seen in other even alkanes in C phase, C_{38}, C_{44}. In PALS measurements, an increase in *o*-Ps lifetime is usually a signal of growing disorder. In long chain alkanes we expect that the number of defects rises with chain length. Relatively shorter lifetime in phase C suggests rather a reverse effect, which is confirmed by the results of Kim et al.; their estimate of *m(g)* for C_{36} is 0.7, whereas for C_{50} it reduces surprisingly to about 0.4.

Hexatriacontane C_{36} is an intermediate case, still containing the rotator phase, with phase C already appearing (both in the range about 2.5 K wide).

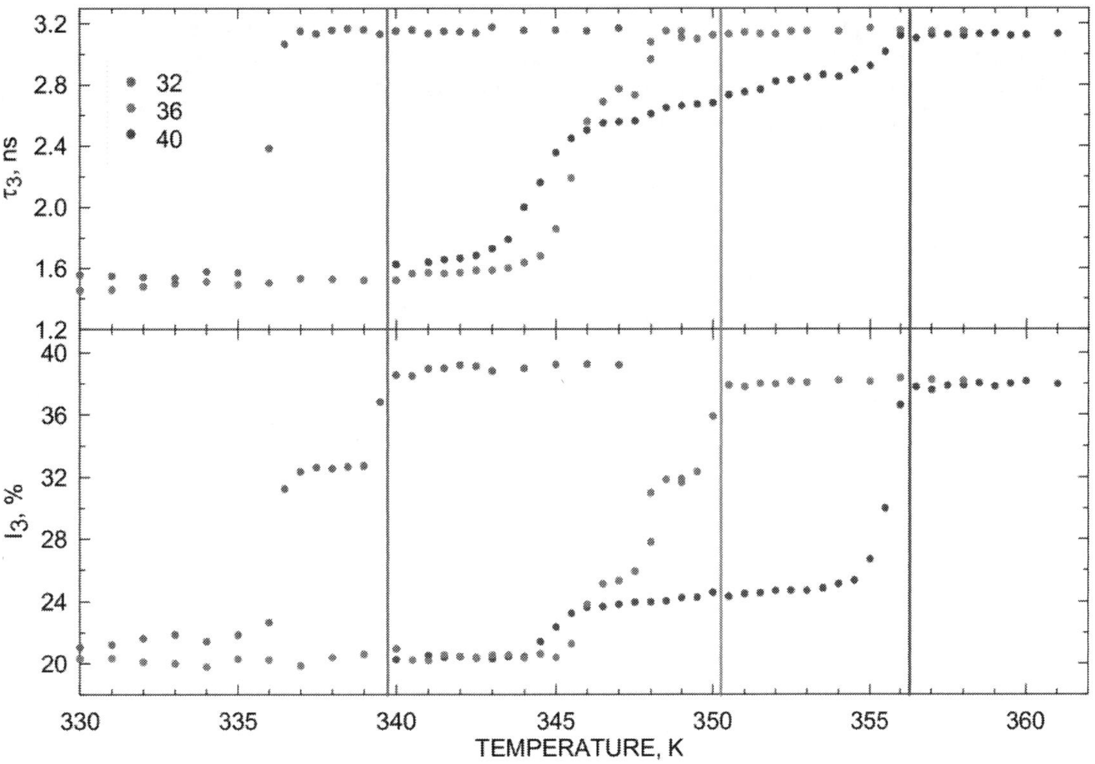

Figure 8. Temperature dependence of *o*-Ps lifetime and intensity for three characteristic long chain *n*-alkanes: C_{32}, C_{36}, C_{40}. Vertical lines indicate the melting points of respective alkanes.

The intensity I_3 in phase C is rather low, resembling that in the low-temperature rigid phase. The step at the rigid→C transition diminishes with chain length; that intensity increase ΔI_3 in C_{36} amounts 4.4 %, in C_{44} 1.5 %, and in C_{50} it disappears. All the data about lifetime and intensity indicate that phase C, appearing at higher temperatures and in the case of chains longer than in the range of existence of the rotator phase, is better ordered than the last one. This supposed (partial) restoration of order needs confirmation by independent methods.

The transition temperature rigid→C is weakly depending on the chain length and locates at about 75°C, thus the range of phase C stretches with increasing chain length; in C_{44}, it amounts about 11 K.

The existence of a next phase at cooling the melt was reported recently [23,24,25]. This is the effect of surface freezing, extending up to 3 K above the bulk solid-liquid transition. Such a monolayer structure cannot be visible in the measurements using positrons from β^+ decay, due to deep penetration of energetic positrons, which interact with the bulk and not with the surface. Moreover, the enthalpy of transition liquid – frozen surface is extremely small (≈0.01 kJ/mol).

6. Liquid

Ortho-Ps lifetime in liquid *n*-alkanes, near the melting point is almost independent on chain length (except for the shortest ones) and is found to be (3.1 – 3.2) ns. The positronium bubble is naturally spherical, and according to (1) its radius is 0.38 nm.

To date, it has been commonly assumed that the mechanism of Ps formation is entirely different for solid and liquid phases. In the former, Ps is a "seeker", which forms in free volumes existing *a priori*, their dimensions are determined by the imperfect structure of solid. In the liquid, Ps is a "digger" producing the free volume by itself, owing to the pressure exerted by zero-point motion of the confined Ps particle. However, according to our experimental data, the lifetime for $25 \leq n \leq 38$ on both sides of the melting point is the same (figure 8)! If in the rotator phase the free volume is given in advance, and in the liquid the size of bubble is determined mainly by surface tension, such an identity of lifetimes cannot be a random effect in such a broad range of chain lengths.

The rotator phase contains certain elements of disorder, however, incomparable to the structure of liquids. The disorder in solid is introduced by *gauche* bonds. Kim et al. [22] estimate that in the rotator phase of C_{36} $m(g)$ is about 0.7 only; in the closest vicinity of melting point that number rises to about 2, thus in the middle of rotator phase range, an average linear *trans* segment contains more than 18 C-C bonds. Raman spectroscopic investigations of liquid alkanes by Brambilla and Zerbi [26] demonstrate the length of *trans*-planar sequences, which can appear (almost independently on chain length) as about 5 C-C bonds (0.625 nm), i.e. comparable to the "bubble" size. Thus, in the rotator phase the molecules retain a roughly linear form and long range order; in liquid alkanes, the ordering can be neglected (even if close to the melting point certain bundling of *trans* fragments can exist [26]).

In the same sample, where *o*-Ps lifetime τ_3 does not feel the transition from rotator to liquid, the intensity I_3 changes substantially at transition, in a stepwise form. In the melting point the intensity I_3 rises by about 10%. An increase in intensity at melting the phase C is even larger. The *o*-Ps lifetime is determined by the molecular structure around the bubble surface, while the intensity is related to the radiochemical processes in the bulk. A question arises whether the molecular structure surrounding the bubble is the same as that in the unperturbed bulk medium.

Acknowledgements
This work was supported by the grant of Polish Ministry of Academic Education and Research N202 181 31/3928.

References

[1] Tao S J 1972 *J. Chem. Phys.* **56** 5499
[2] Eldrup M, Lightbody D and Sherwood J N 1981 *Chem. Phys.* **63** 51
[3] Eldrup M, Pedersen N J and Sherwood J N 1979 *Phys. Rev. Lett.* **43** 1407
[4] Lightbody D, Sherwood J N and Eldrup M 1983 *Mol. Cryst. Liq. Cryst.* **96** 197
[5] Jasińska B, Kozioł AE and Goworek T 1996 *J. Radioanal. Nucl. Chem.* **210** 6178
[6] Goworek T, Ciesielski K, Jasińska B and Wawryszczuk J 1997 *Chem. Phys. Lett.* **272** 91
[7] Goworek T, Ciesielski K, Jasińska B and Wawryszczuk J 1998 *Chem. Phys.* **230** 305
[8] Kansy J 1996 *Nucl. Instrum. Methods Phys. Res. A* **374** 235
[9] Craievich A, Doucet J and Denicoló I 1985 *Phys. Rev. B* **32** 4164
[10] Zgardzińska B, Wawryszczuk J and Goworek T 2006 *Chem. Phys.* **320** 207
[11] Zgardzińska B 2008 *Ph.D. Thesis,* Marie Curie-Sklodowska University, Lublin, Poland
[12] Asano T 1985 *J. Phys. Soc. Jpn.* **54** 1403
[13] Shukla A, Peter M and Hofmann L 1993 *Nucl. Instrum. Methods Phys. Res. A* **335** 310
[14] Craievich A, Denicoló I and Doucet J 1984 *Phys. Rev. B* **30** 4782
[15] Maroncelli M, Song Ping Qi, Strauss H L and Snyder R G 1982 *J. Am. Chem. Soc.* **104** 6237
[16] Goworek T, Zaleski R and Wawryszczuk J 2003 *Chem. Phys.* **295** 243
[17] Goworek T, Rybka C and Wawryszczuk J 1978 *phys. stat. sol. (b)* **89** 253
[18] Goworek T, Pietrow M, Zaleski R and Zgardzinska B 2009 *Chem. Phys.* **355** 123
[19] Mina F, Asano T, Mondieig D, Würflinger A and Josefiak C 2004 *J. Phys. IV France* **113** 35
[20] Reynhardt E C, Fenrych J and Basson I 1994 *J. Phys.: Condens. Matter* **6** 7605
[21] Wang S, Tozaki K-I, Hayashi H, Inaba H and Yamamoto H 2006 *Thermochim. Acta* **448** 73
[22] Kim Y, Strauss H L and Snyder R G 1989 *J. Phys. Chem.* **93** 7520
[23] Ocko B M, Wu X Z, Sirota E B, Sinha S K, Gang O and Deutsch M 1997 *Phys. Rev. E* **55** 3164
[24] Li H Z and Yamamoto T 2001 *J. Chem. Phys.* **114** 5774
[25] Kawamata M and Yamamoto T 1997 *J. Phys. Soc. Jpn.* **66** 2350
[26] Brambilla L and Zerbi G 2005 *Macromolecules* **38** 3327

Journal of Physics: Conference Series **265** (2011) 012024 doi:10.1088/1742-6596/265/1/012024

Comparison of open volumes in silica based thin films produced by different precursors

S Mariazzi[1], L Toniutti[1], R Checchetto[1], S Dirè[2] and R S Brusa[1]

[1] Dipartimento di Fisica, Università di Trento, Via Sommarive 14, I-38050 Povo, Trento, Italy
[2] Dipartimento di Ingegneria dei Materiali e Tecnologie Industriali, Università di Trento, Via Mesiano 77, I-38050 Povo (TN), Italy

E-mail: mariazzi@science.unitn.it

Abstract. Silica thin films were deposited by spin coating on p-type (100) Si substrate and depth profiled with Positron Annihilation Spectroscopy (DP-PAS). A first series of samples were produced from precursor gel with molar ratio $HSi(C_2H_5O)_3$ [=TES]: $(CH_3)HSi(C_2H_5O)_2$ [=MDES] : H_2O : EtOH = 1 : 1 : 5 : 5. Samples were thermally treated in air at temperatures up to 600 °C. The structure and the chemical environment of the open volumes of these samples were compared with the ones of a previous studied second series of films obtained by hydrolysis and condensation reaction of Tetraethylorthosilicate (TEOS). The effects of sol precursors and thermal treatment temperature on open volumes were studied by Doppler broadening spectroscopy and detection of 3γ-annihilation of ortho-positronium. The introduction of CH_3 groups by MDES produces an open disordered structure whose decoration of H atoms gives rise to a very high S parameter (1.076) but with very low positronium 3γ-annihilations.

1. Introduction

In the last years, silica-based porous materials have been widely studied for their huge range of applications such as low refractive index materials for optical coatings, low-dielectric constant (low-k) films for microelectronic applications and membranes for separation, catalysis and chemical sensing [1].

Materials with low refractive index are interesting for possible applications ranging from optical materials and display devices, to solar cell coatings in order to increase transmittance and decrease the reflection. One promising way of reducing the refractive index is the reduction of the material density by introduction of porosity [2]. The introduction of porosity leads also to a reduction of the dielectric constant of a given material. Actually, insulators with a dielectric constant lower than the conventional SiO_2 (k~4) are used to improve the performance of integrated circuits [3, 4].

Nano-porous amorphous silica films deposited on porous support (pore sizes larger than some microns) exhibit good gas filtering characteristics. In these films the presence of a network of sub-nanometer pores separate the components of a gas mixture by a molecular sieving effect. Applications range from the field of production of pure molecular hydrogen [5-7] to the separation of CH_4 in biogases [8]. Silica based materials are also used in the purification of deoxyribonucleic acid (DNA) from biological samples. DNA molecules are selectively adsorbed by Coulombic force while contaminants are washed away by organic solvents [9].

In these applications the porosity plays a fundamental role, therefore the characterization of the distribution, the size, the shape and the interconnectivity of the pores is extremely important. Few analytical techniques are available to extract this information [3]. DP-PAS has been shown to be one of the most sensitive tools to probe pores identifying their distribution, size and interconnectivity [4, 10, 11]. Positrons injected in a solid, after thermalization with the medium and a diffusion path, annihilate in two 511 keV gamma rays with an electron of the host. Due to its positive charge positron is mainly localized in the open volumes of the solid where the annihilation with the low momentum electrons produces a narrowing of the 511 keV annihilation line [12].

In insulators, due to their open structure, positron can also form a positronium (Ps) atom. In vacuum Ps is formed for 1/4 as para-positronium (p-Ps) that annihilates in two gamma rays with a lifetime of 125 ps and 3/4 as ortho-positronium (o-Ps) that annihilates in three gamma rays after 142 ns. In a solid o-Ps can also annihilate with an electron of the medium (pick-off process) in two gamma rays and its lifetime is shortened to a few nanoseconds. In presence of interconnected porosity, o-Ps can out-diffuse towards the surface and eventually be emitted into the vacuum [11].

The probability of pick-off annihilation and the shape of the 511 keV gamma annihilation peak is influenced by the pore sizes and by the chemical termination of the pore walls [13, 14].

Because o-Ps and positron are very sensitive to the local environment of the annihilation site, it is fundamental to know the annihilation properties to define the open volumes dimension and concentration. The lack of this information can hinder a direct correlation both of the 3γ - 2γ ratio and the width of 511 keV annihilation peak with the open volume size.

In this work we present DP-PAS measurements on silica based porous thin films deposited by spin coating on Si substrate. In particular the morphology and the chemical environment of the open volumes are analyzed and compared for two series of samples obtained using as precursors Triethoxysilane (TES)-Methyldiethoxysilane (MDES) and Tetraethylorthosilicate (TEOS), respectively. While TEOS precursor has been shown to give pure silica films after annealing above 500 °C [7], films deposited using MDES are expected to contain CH_3 groups. As CH_3 groups in SiO_2 structure are organized in long chains with different cross-linking [3], they are expected to generate more open free volumes compared to silica [4]. The walls of the free volumes would be H terminated [4, 14].

2. Experimental

Silica thin films were deposited by spin coating on p-type (100) Si substrates and thermally treated in air at temperatures up to 600 °C. The silica sols, for the deposition of the first series of film, was prepared dissolving Triethoxysilane (TES) and Methyldiethoxysilane (MDES) in absolute ethanol (EtOH). The molar ratio was TES : MDES : H_2O : EtOH = 1 : 1 : 5 : 5. Films were deposited after 24 h from the precursor preparation and, after deposition, the films were baked at 125 °C. Two samples were successively annealed in static air at 300 and 600 °C for 60 min using a heating rate of 1 °C min^{-1}. The samples treated at 125 °C, 300 °C and 600 °C will be referred as #1, #2 and #3, respectively. The thicknesses of these films ranged from 350 to 500 nm.

The second series of silica films was prepared by hydrolysis and condensation reaction of Tetraethylorthosilicate (TEOS) in presence of water and ethanol. The molar ratios were TEOS : H_2O: EtOH : HNO_3 = 1 : 6 : 6 : 0.01 (sample #4) and 1 : 6 : 6 : 0.1 (sample #5). A third sample was produced adding Polyvinylpyrrolidone (PVP) as porogen with molar ratio TEOS : H_2O: EtOH : HNO_3 : PVP = 1 : 6 : 6 : 0.1 : 1 (sample #6). After deposition, samples #4, #5 and #6 were baked at 125 °C and then thermally treated at 600 °C for 60 min using a heating rate of 1 °C min^{-1}. Data of samples #4 and #5 are the same as in Ref. [7] while sample #6 was analyzed in Ref. [15-17].

The Doppler broadened 511 keV positron annihilation line was measured as a function of the positron implantation energy E in the energy range 50 eV - 25 keV. E was related to the mean positron implantation depth z through the relation $z = (40/\rho)\, E^{1.6}$. In this relation z is in nm when the density ρ is expressed in g/cm^3 and E in keV. For silica films a density of 2.2 g cm^{-3} has been used. The measurements were carried out with a slow positron beam [18] equipped with two high purity germanium detectors with 1.4 keV resolution at 511 keV. The annihilation peak was characterized by the shape parameter *S* calculated as the ratio of the counts in the $| 511- E_\gamma | \leq 0.85$ keV central area to the total area of the peak ($| 511- E_\gamma | \leq 4.25$ keV) and the wing parameter *W* calculated as the fraction of the counts in the region ($1.6 \leq |E_\gamma -511| \leq 4$ keV). E_γ is the energy Doppler shift of the annihilation γ-ray. The *S* and *W* parameters were normalized as $S_n = S_{measured}/S_{bulk\text{-}Si}$ and $W_n = W_{measured}/W_{bulk\text{-}Si}$.

The VEPFIT program, [19] based on the solution of the diffusion equation, was used to extract from the $S_n(E)$ and $W_n(E)$ curves the characteristic (S_n, W_n) values of the films. The $S_n(E)$ and $W_n(E)$ curves of samples #1, #2 and #3, were fitted with three layers (film, interface of a few nanometers and Si substrate).

The 3γ - 2γ ratio [$R(E)$ parameter] was calculated as the ratio between the valley area (E_γ between 410-500 keV) and the 511 peak area. The $R(E)$ parameter was calibrated by measuring the Ps formation in a Ge crystal as a function of the temperature. The $R_n(\%)$ calibrated parameter is defined as $(R(E)-R_0)/(R_{100}-R_0)$. Where R_{100} (100% Ps formation) is assumed as the *R* value obtained by extrapolating to zero implantation energy the $R(E)$ curve measured in Ge at 1000 K [20] where all positrons reaching the Ge surface form Ps [21]. Because in the Ge bulk positrons do not form Ps, as R_0 (0% Ps formation) the *R* value at the highest positron implantation energy was assumed.

3. Results and discussion

In figure 1 the $S_n(E)$ curves of the measured samples are reported. The samples produced using TES and MDES as precursors (#1, #2 and #3) show very high S_n values in the film region (below *E*~6-7 keV) if compared with samples obtained by TEOS (#4, #5 and #6). The fitted S_n values characteristic of the films #1, #2 and #3 have been found to be 1.056, 1.054 and 1.076, respectively. These values are much higher also than that of the sample #6 ($S_n = 1.012$) that is rich of pores after removal of PVP porogen. In this sample the total porosity was estimated to be around 58% [17] and the pore size was evaluated to be $d \geq 2.0$ nm [15].

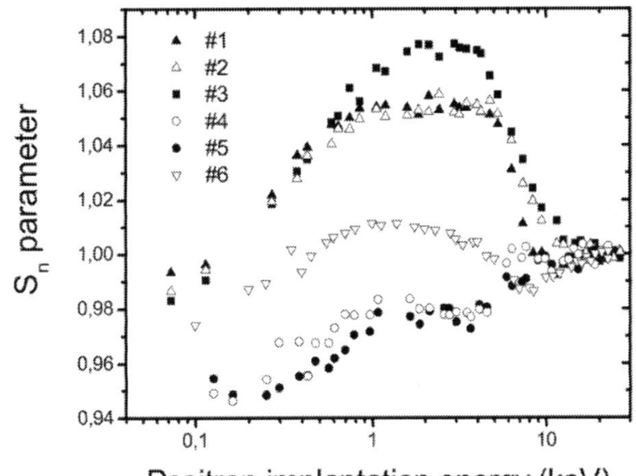

Figure 1. S_n parameter vs. positron implantation energy *E*.

$R_n(\%)$ vs. *E* curves of all the measured samples (figure 2) points out a low 3γ - 2γ ratio in the films deposited starting from TES and MDES. $R_n(\%)$ in the #1 and #2 films is lower than 1%, in sample #3, $R_n(\%)$ value is slightly higher (~2%). These $R_n(\%)$ values point out a very low o-Ps self annihilation and they are comparable with those found in films #4 and #5. In these last samples the total porosity was estimated to be less than 30% by dielectric constant measurements [7], i.e. around the half of the porosity of sample #6 corresponding to a $R_n(\%)$ ~14% [15].

In figure 3 the (S_n , W_n) values characteristic of each film are shown. On the same figure, data of Ref. [7] indicating (S_n , W_n) values of films prepared with the same composition of the sol precursors of #4 and #5 and annealed at different temperature up to 900 °C are reported. The dashed and continuous lines are linear fit through the points groups of the films realized with TEOS and TES+MDES, respectively. As discussed in Ref. [7] the position of each (S_n , W_n) value on its line is given by the number and the size of open volumes present in the film that depend on the annealing temperature, while the chemical environment probed by positrons is characterized by the line parameters. The large difference in the behavior of dashed lines and continuous line indicates a considerately different chemical environment of the annihilation site.

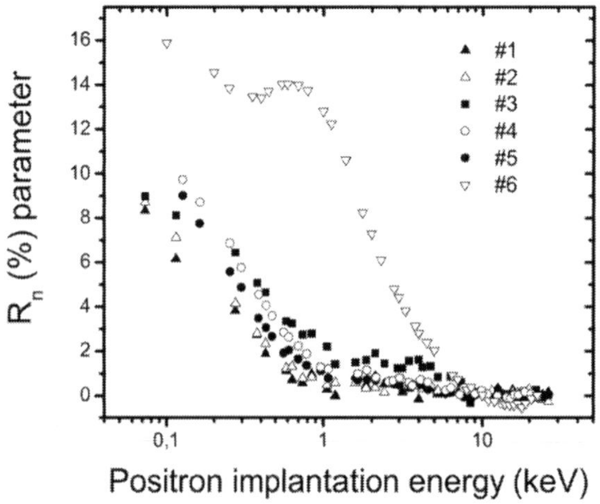

Figure 2. R_n(%) parameter vs. positron implantation energy E.

The very high S_n and very low W_n values of films #1 [(S_n , W_n) = (1.056, 0.991)], #2 [(1.054, 0.989)] and #3 [(1.076, 0.918)] are a strong indication of the presence of H atoms around the positron annihilation sites in these films [4, 22, 23]. The low S_n and high W_n values observed in the films produced with TEOS are characteristic of the presence of O around the annihilation site.

The slight increase of S_n and R_n(%) values of sample #3 with respect to samples #1 and #2 (figure 1 and 2) could be attributed to the loss, induced by the thermal treatment at 600 °C [4], of weakly bound H on the walls of the open volumes. This could call for an increase of the size of the open volumes in #3.

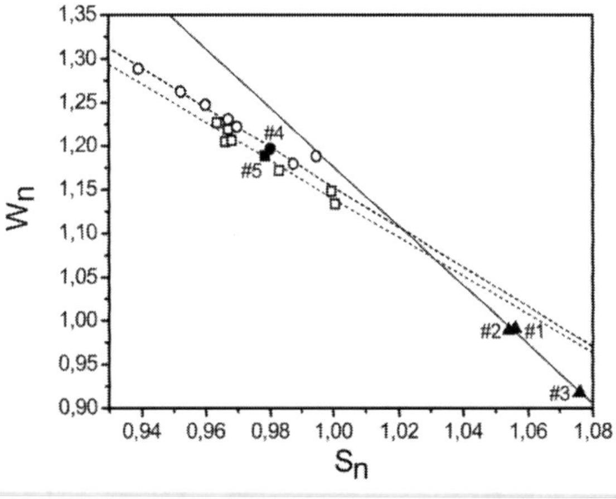

Figure 3. Open circles and open squares are the characteristic (S_n , W_n) values of films prepared with the same composition of #4 (full circle) and #5 (full square), respectively, and annealed at different temperature up to 900 °C. Dashed lines are their best fits. The continuous line is a linear fit through the experimental points of the samples #1, #2 and #3 (full triangles).

The experimental high S_n and low R_n(%) values in the #1, #2 and #3 films are compatible with the presence of small open volumes decorated by H atoms. However DP-PAS measurements are not able to exclude the presence of larger open structures where Ps is formed but o-Ps is completely quenched by pick-off. Remembering that, in silica films, from 30 % up to 80 % Ps is yield [24] and that CH_3 groups enhance the open free volumes with respect to pure silica films [4], a high Ps formation could be expected also in these films. In this case, a part of the observed high S_n value could be attributed to p-Ps self annihilation and another part to the pick-off of o-Ps with H electrons.

In order to understand if positrons annihilate directly with electrons of hydrogen atoms or after Ps formation, positron annihilation lifetime spectroscopy (PALS) and total porosity evaluation by dielectric constant measurements are planned.

References

[1] Ying J Y, Mehnert C P and Wong M S 1999 *Chem. Int. Ed.* **38** 56

[2] Maruo T, Tanaka S, Nishiyama N, Motoda K, Funayama K, Egashira Y and Ueyama K 2008 *Colloids and Surfaces A: Physicochem. Eng. Aspects* **318** 84

[3] Maex K, Baklanov M R, Shamiryan D, Iacopi F, Brongersma S H and Yanovitskaya Z S 2003 *J. Appl. Phys.* **93** 8793

[4] Brusa R S, Spagolla M, Karwasz G P, Zecca A, Ottaviani G, Corni F, Bacchetta M and Carollo E 2004 *J. Appl. Phys.* **95** 2348

[5] Verweij H, Lin Y S and Dong J 2006 *MRS Bull.* **31** 756

[6] Nenoff T M, Spontak R J and Aberg C M 2006 *MRS Bull* **31** 735

[7] Patel N, Mariazzi S, Toniutti L, Checchetto R, Miotello A, Dirè S and Brusa R S 2007 *J. Phys. D: Appl. Phys.* **40** 5266

[8] Asaeda M and Yamasaki S 2001 *Sep. Purif. Technol.* **25** 151

[9] Yu S, Geng J, Zhou P, Wang J, Feng A, Chen X, Tong H and Hu J 2008 *Anal. Chim. Acta* **611** 173

[10] Gidley D W, Frieze W E, Dull T L, Sun J, Yee A F, Nguyen C V and Yoon D Y 2000 *Appl. Phys. Lett.* **76** 1282

[11] Petkov M P, Weber M H, Lynn K G and Rodbell K P 2001 *Appl. Phys. Lett.* **79** 3884

[12] Schultz P J and Lynn K G 1988 *Rev. Mod. Phys.* **60** 701

[13] Petkov M P, Wang C L, Weber M H, Lynn K G and Rodbell K P 2003 *J. Phys. Chem.* B **107** 2725

[14] He C, Ohdaira T, Oshima N, Muramatsu M, Kinomura A, Suzuki R, Oka T and Kobayashi Y 2007 *Phys. Rev. B* **75** 195404

[15] Mariazzi S, Patel N, Toniutti L, Checchetto R, Miotello A and Brusa R S 2007 *Phys. Stat. Sol. (c)* **4** 3823

[16] Mariazzi S, Toniutti S, Patel N and Brusa R S 2008 *Appl. Surf. Sci.* **255** 191

[17] Toniutti L, Mariazzi S, Patel N, Checchetto R, Miotello A and Brusa R S 2008 *Appl. Surf. Sci.* **255** 170

[18] Zecca A, Bettonte M, Paridaens J, Karwasz G P and Brusa R S 1998 *Meas. Sci. Technol.* **9** 1

[19] van Veen A, Schut H, de Vries J, Hakvoort R A and Ijpma M R 1990 *AIP Conf. Proc.* vol 218, *Positron Beams for Solids and Surfaces*, ed P J Schultz, G R Massoumi and P J Simpson (New York, NY: AIP Press) p 171

[20] Brusa R S, Macchi C, Mariazzi S and Karwasz G P 2005 *Acta Phys. Polon.* **107** 702

[21] Mills A P, Jr. (p 432) and Lynn G K (p 609) 1983 *Positron Solid State Physics*, ed W Brandt and A Dupasquier (North Holland: Amsterdam)

[22] Brusa R S, Macchi C, Mariazzi S, Spagolla M, Karwasz G P and Zecca A 2004 *Mater. Sci. Forum* **445** 268

[23] Petkov M P, Weber M H, Lynn K G, K P Rodbell and Cohen S A 1999 *J. Appl. Phys.* **86** 3104

[24] Sasaki Y, Nagay Y, Ohkubo H, Inoue K, Tang Z and Hasegawa M 2003 *Rad. Phys. Chem.* **68** 569

Journal of Physics: Conference Series **265** (2011) 012025 doi:10.1088/1742-6596/265/1/012025

3-D defect density plots of large scale structural materials with positron annihilation spectroscopy: feasibility and optimization

V Makarashvili[1], D P Wells[1] and A K Roy[2]

[1]Department of Physics, Idaho State University, Campus Box 8106, Pocatello, ID 83209 USA
[2]Department of Mechanical Engineering, University of Nevada Las Vegas, 4505 Maryland Parkway, Las Vegas, NV 89154 USA

E-mail: makavakh@isu.edu

Abstract. We investigated the feasibility of the pair-production based positron annihilation spectroscopy (PAS) technique to generate 2-D and possibly 3-D defect density images (defect density plots) of structural materials widely used in industry. After assessing the possible PAS techniques and performing MCNPX simulations on different positron production mechanisms, we chose to employ Doppler broadening analysis of the annihilation peak coupled with the pair-production method of positron creation using bremsstrahlung photon beams from a linear accelerator. First, to optimize the important parameters of the proposed experimental setup, MCNPX simulations were performed. These parameters are: thickness of the bremsstrahlung beam hardener, detector orientation, thickness of the bremsstrahlung converter, beam energy and current, etc. These parameters were studied to maximize the number of 511 keV counts in the detector and minimize the low energy background photons that Compton scatter into the detector. A Figure of Merit (FOM) was used to quantitatively track effectiveness (or ineffectiveness) of each value for all the parameters. We conducted experiments with the simulated setup and these results were compared to MCNPX calculations.

1. Introduction

PAS techniques have long been employed to study various defect states and obtain information about electron momentum-density distributions in different materials. PAS measurements based on conventional radioactive sources or positron beams are restricted to probe only up to a few hundred μm in solids due to the low penetrability of positrons in the keV range. The feasibility of utilizing MeV positrons have been also demonstrated [1, 2], but it only extends PAS measurements to a few mm depth.

We have been developing accelerator-based in-situ pair-production techniques to produce positrons in thick samples ($\sim$ 4-40 g/cm^2, or $\sim$ 0.5-5 cm in steel). These techniques are called "Accelerator-based Gamma-induced Positron Annihilation Spectroscop" (AG-PAS). AG-PAS employs an accelerator to produce bremsstrahlung, which creates positrons inside the material via pair production. No positron emitters or photon-induced activation are involved in this process. The annihilation radiation is counted by a high resolution HPGe, which allows us to analyze the Doppler broadened 511 keV peak in terms of S and W parameters. This technique combines high penetration of MeV gamma rays with the high sensitivity of positrons to microscopic defects (mono-vacancies, di-vacancies, dislocations, etc). Hence, it creates a possibility to add defect density imaging capabilities of large scale structural materials to the conventional PAS methods. Studying these microscopic defects is very important for a number of industries in order to predict structural integrity of materials. Based on MCNPX (Monte Carlo N-Particle eXtended) simulations [3] on different positron creation mechanisms and knowledge from the previous experiments [4] we think that the AG-PAS is the most practical approach for defect studies of thick structural materials in engineering applications.

The biggest limitation of the AG-PAS experiments performed in the past [5–7] was the low repetition rate of the LINAC (200-300 Hz). However, the work presented here is performed by utilizing a 1 kHz High Repetition Rate LINAC (HRRL), which makes defect density imaging with AG-PAS more feasible. In order to demonstrate this feasibility and optimize the experimental setup, FOM analyses have been performed based on the experimental and MCNPX simulation results.

2. Methods and materials

2.1. Experimental setup

A 15 MeV electron beam from the HRRL was bent 270° before striking a 1 mm thick tungsten converter. The accelerator is capable of producing 30-60 ns pulses, up to 100 mA peak current at 1 kHz repetition rate. The generated bremsstrahlung cone was filtered (hardened) by a graphite hardener to reduce the contribution of the low energy photons in the spectrum. The hardened beam went through a 1.2 m thick concrete wall that had two collimators. The primary collimator was a 25 cm long steel cylinder with a 6 mm diameter hole in the center. The secondary (clean up) collimator was a 2.54 cm diameter hole in a 10 cm thick lead block. The highly collimated beam then struck the sample, which in this case was a 0.95 cm (3/8 inch) thick steel punch. The sample was located about 1.5 m away from the wall (figure 1).

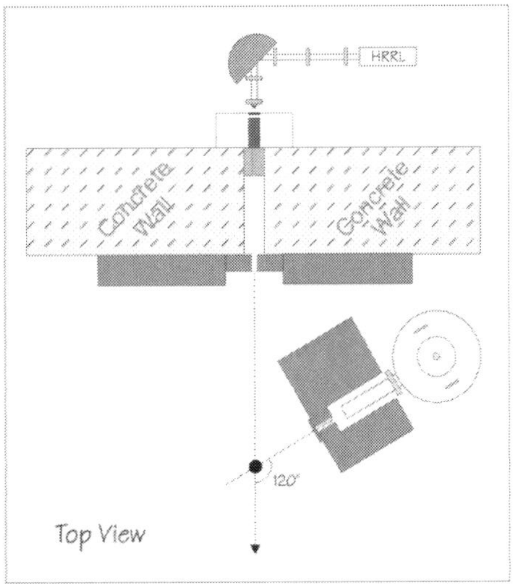

Figure 1. Experimental setup

The 15 MeV endpoint energy bremsstrahlung beam produced photo-electric effect, Compton scattering and pair production processes in the sample. Compton scattering predominantly contributed to the detected low energy background, while pair production created positrons throughout the sample volume, resulting in emission of 511 keV photons. These 511 keV photons were Doppler broadened due to conservation of momentum and contained valuable information about defects in the material. The spectrum emitted from the sample was recorded by a heavily shielded, 20% relative efficiency HPGe detector. The detector orientation was such that there was a 120° angle between the beam line and the detector field of view. The detector shielding was composed of lead bricks and had the following dimensions: 30 cm thick on the wall side,

20 cm thick on the opposite side and 10 cm thick in front, top and bottom. The shielding had 2.54 cm diameter collimator which was covered with a secondary lead hardener to reduce the low energy background even further. The detector side of the concrete wall was also shielded with 20 cm thick and 25 cm tall lead to reduce the scattered photons coming from the beam. The intensity of the bremsstrahlung beam was monitored by a NaI(Tl) detector. The signal from the NaI(Tl) detector was also used to gate the germanium detector so we can record the spectrum coming only from the accelerator pulses. 1 μCi ^{133}Ba and ^{137}Cs sources were used for calibrating the HPGe detector.

The parameters that we optimized in this setup were: thickness of the tungsten converter, thickness of the graphite hardener, orientation of the detector and the thickness of the secondary lead hardener. Experiments were conducted to optimize the graphite and lead hardeners and also to measure background rates in this experimental environment with different shielding configurations.

We introduced a FOM definition to quantitatively track effectiveness of different values of different parameters. The idea was to maximize the 511 keV count rate per unit total rate in the detector. So, the obvious definition was

$$FOM = \frac{R_{511}}{R_T} \tag{1}$$

where R_{511} and R_T are the 511 keV and total count rates respectively.

In a pulsed environment like ours, one has to take careful measures to address detector dead time and pulse pile-up issues. We know that under the conditions $T < \tau < (1/f - T)$, where T, τ and f are pulse width, detector dead time and repetition rate respectively, which is certainly satisfied in our case ($T = 30$ ns, $\tau = 10$ μs, $f = 1$ kHz), the true event rate n and the recorded count rate m are related by the following expression: [8]

$$n = f \ln\left(\frac{f}{f-m}\right) \tag{2}$$

When the dead time losses are small (desired condition for our experiment) under the condition $m << f$, equation (2) can be simplified by expanding the logarithmic term

$$n = \frac{m}{1 - m/2f} \tag{3}$$

We decided to stay at 10% pile-up level which sets the upper limits for the accelerator peak current and consequently recorded total rate. In the first approximation, when only two pulse pile-up is significant, the pile-up can be expressed in terms of m and f the following way:

$$PU = \frac{2(n-m)}{n} = \frac{m}{f} \tag{4}$$

where we used equation (3) for n. This tells us that 10% pile-up means observed 100 counts/s at 1 kHz repetition rate. Hence, we kept the total rate at about 100 counts/s and looked at R_{511} as a function of thickness for both graphite and lead hardeners. The reason we used graphite as a primary hardener is that low Z materials are effective bremsstrahlung filters, which has been demonstrated experimentally in the past [9]. This is simply due to the fact that the Compton scattering cross-section is proportional to Z, while pair production goes approximately as Z^2. Thus, the ratio of Compton interactions to pair production events in the hardener goes as $1/Z$. On the other hand, a high Z material is desired as a secondary hardener since there are no pair production events caused by the photons coming from the sample.

The beam background coming from the accelerator side of the wall was also measured for two shielding configurations of the lead next to the wall. One was the shielding described above and the other with added shielding: 30 cm thick and 50 cm tall.

2.2. MCNPX simulations
MCNPX version 2.5f was employed to perform simulation for optimizing the tungsten converter, both, graphite and lead hardeners and the detector angle. The experimental setup described above was replicated for the MCNPX geometry. The bremsstrahlung converter was optimized by simply looking at the pair production events per electron in the sample as a function of the tungsten thickness. The biggest obstacle encountered during the other simulations was the "thick penetration problem". Since one can only run $2{\cdot}10^9$ source particles in version 2.5f, it was extremely difficult to simulate the whole process in a single run and generate reliable statistics even with the weight window technique. So, we simulated the process in two steps: step one was finding the bremsstrahlung energy distribution in the sample while step two was assigning this distribution to the source photons and sending them mono-directionally (ignoring the angular divergence justified by the collimation) towards the sample in order to determine the resulting spectrum in the detector. All the other parameters were optimized by using the same FOM definition as in the experimental case.

All the MCNPX simulations were done on a cluster called BREMS (Beowulf REsource for Monte-carlo Simulations) at the physics department of Idaho State University. BREMS is made of 21 nodes: 12 nodes with two 2.0 GHz Opteron CPUs and 9 nodes with two dual-core 2.0 GHz Opteron CPUs - a total of 60 processing cores.

3. Results and discussions
3.1. Background
Let's clarify the term background first. In our optimization problem the only useful information is the 511 keV counts. We call everything else - "background". The total background can be broken down the following way

$$R_T = B_1 + B_2 + B_3 + R_{511} \tag{5}$$

where:

B_1 - is the natural background plus counts from the calibration sources

B_2 - is the beam background coming from the accelerator side of the wall

B_3 - is the Compton scattered photons from the sample

So, another way to state our problem is to minimize B_1, B_2 and B_3 for a fixed R_T. Based on the not-gated spectrum with the beam-off and the width of the gate, we estimated that in the gated spectrum $B_1 \approx$ 8-9 counts/s when R_T = 100 counts/s and there isn't much we can do about optimizing B_1.

B_2 was measured by removing the sample with the beam-on for two different shielding configurations. The results are presented in figure 2. With the added shielding, B_2 ranges about 8-13% of the R_T. We could have reduced this background if we had more room for shielding. B_2 becomes the major limitation with thick hardeners, because we need to increase the current to keep the R_T fixed.

The biggest potential for optimization is in B_3, because it constitutes about 70% of the total count rate. The both, graphite and lead hardeners and the detector angle optimizations were targeted to minimize B_3.

3.2. Tungsten converter
The simulation results for the tungsten converter are demonstrated in figure 3. The pair production events in the sample per electron as a function of tungsten thickness is shown for a 15 cm thick graphite hardener. If the tungsten is too thin there is simply not enough high Z material to produce energetic bremsstrahlung efficiently. The pair production events in the sample slowly drops with thickness, because we start to loose high energy photons via pair

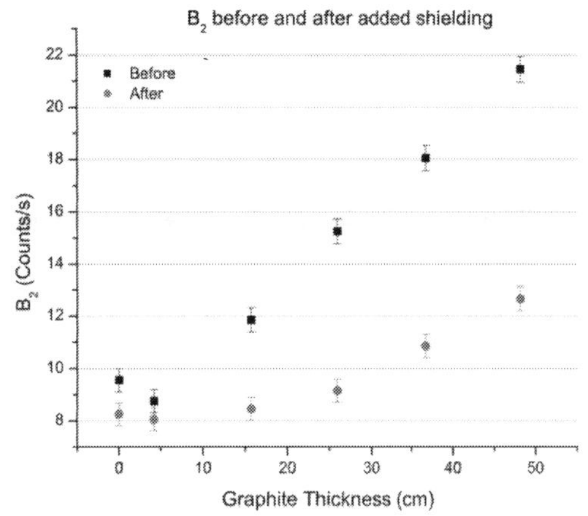

Figure 2. B_2 background for two shielding configurations. Before - lead shielding, 20 cm thick and 25 cm tall. After - lead shielding 30 cm thick and 50 cm tall.

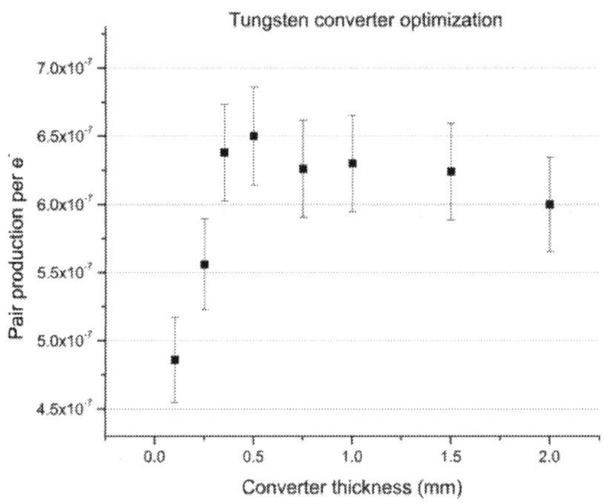

Figure 3. MCNPX simulation results for the tungsten converter. Pair production events in the sample as a function of the converter thickness.

production in the converter. This is only a part of the full picture. Increased tungsten thickness means increased bremsstrahlung and therefore increased B_2. For example, even though the graph shows that the pair production events at 0.5 mm is slightly better than at 2 mm, 2 mm tungsten produces about 3 times more bremsstrahlung than 0.5 mm. So, we conclude that around 0.5 mm tungsten is the most effective thickness for our purposes.

3.3. Graphite hardener

The effect of the graphite hardener to B_3 is demonstrated with the experimental spectra in figure 4 and figure 5 for 4.2 cm and 48.1 cm graphite respectively. In figure 5 the low background allows us to see the Compton edge of the 511 keV peak. 356 keV and 662 keV are ^{133}Ba and ^{137}Cs lines. The observed 511 keV count rate (R_{511}) was about 10 counts/s.

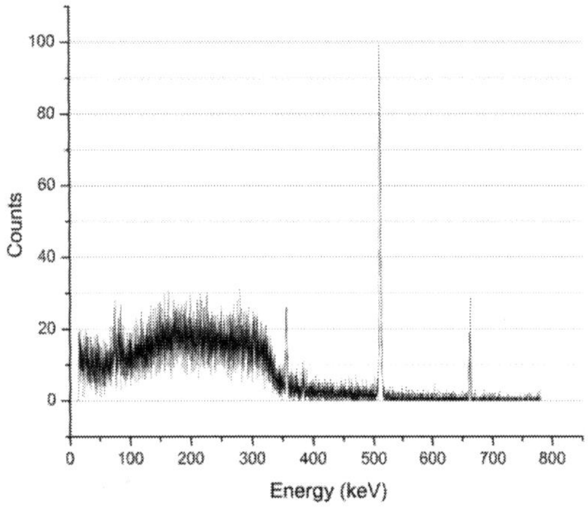

Figure 4. Experimental spectrum with a 4.2 cm graphite. Note the ^{137}Cs and ^{133}Ba lines that were measured synchronously.

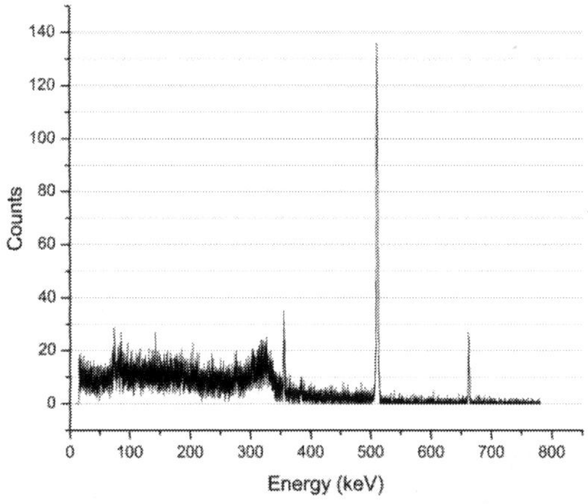

Figure 5. Experimental spectrum with a 48.1 cm graphite. Note the ^{137}Cs and ^{133}Ba lines that were measured synchronously.

Figure 6 shows the simulated bremsstrahlung spectra in the sample for different graphite thicknesses. We can clearly see the effect of the hardening. Figure 7 compares three data sets against each other. Two experimental data sets are FOM as a function of graphite thickness: upper one including only B_3, lower one including all the background. The third one is the simulation results of FOM including only B_3. The simulation and the experimental results are close to each other for thin hardeners. The reason they deviate from each other as the thickness increases is that the simulation ignores angular divergence of the bremsstrahlung beam, which becomes more and more significant for thicker hardeners. As the lower experimental data set suggests, B_2 is the major limitation for thick hardeners.

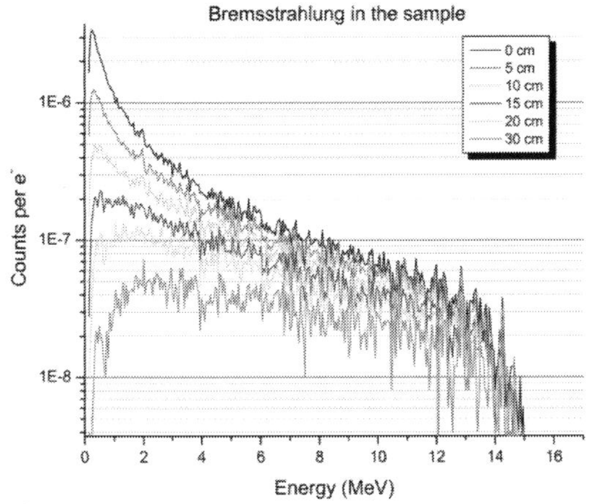

Figure 6. Simulated bremsstrahlung with different thickness of graphite hardener.

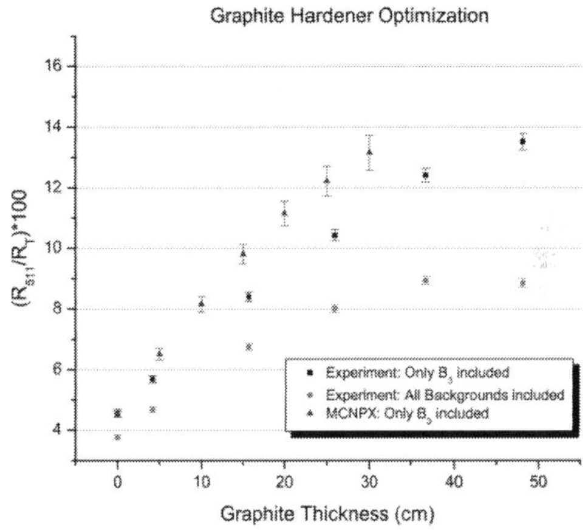

Figure 7. Experimental and simulation results for graphite hardener

3.4. Detector angle

Simulated spectra for different detector angles with a 15 cm graphite hardener are presented in figure 8. We can identify three features of each spectrum - first, they all have the Compton edge of the 511 keV peak itself at about 340 keV; second, the Compton background, B_3, decreases as the angle increases which is what one should expect based on the angular dependence of the Compton scattering cross-section; and also, the Compton edge of the 15 MeV end-point bremsstrahlung shifts left with increasing angle which is consistent with the Compton scattering kinematics.

Figure 9 is the 3D plot of FOM as a function of the detector angle and the graphite thickness. According to the results, the optimal angle appears to be at around 140°.

3.5. Lead hardener

Finally, experimental and simulation FOM analysis for the lead hardener are shown in figure 10. The reason that they strongly disagree as the thickness increases, is again the beam background B_2, which is ignored in simulations but becomes a major problem at high accelerator currents due to a concrete wall that is not sufficiently thick ($\sim$ 1.2 m) to eliminate this background (figure 2).

4. Future work

The knowledge obtained from this work will be exploited to prepare the setup for defect density imaging experiments using AG-PAS. A well-collimated bremsstrahlung beam will strike a bundle

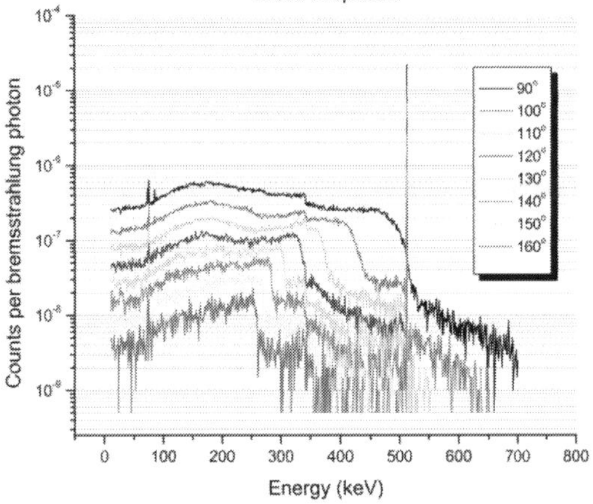

Figure 8. Simulated detector spectra for different angles

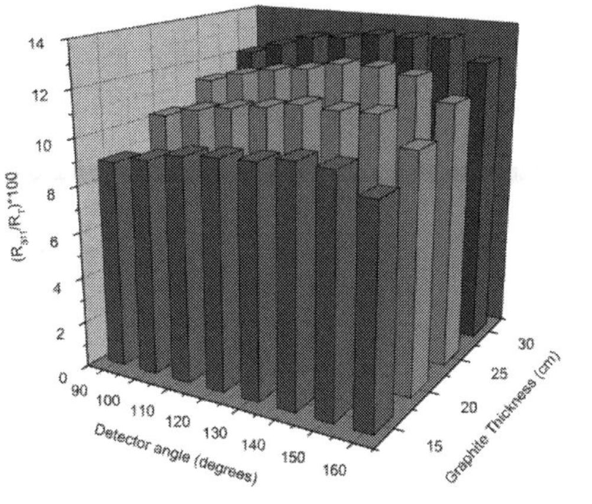

Figure 9. *FOM* as a function of hardener and detector orientation

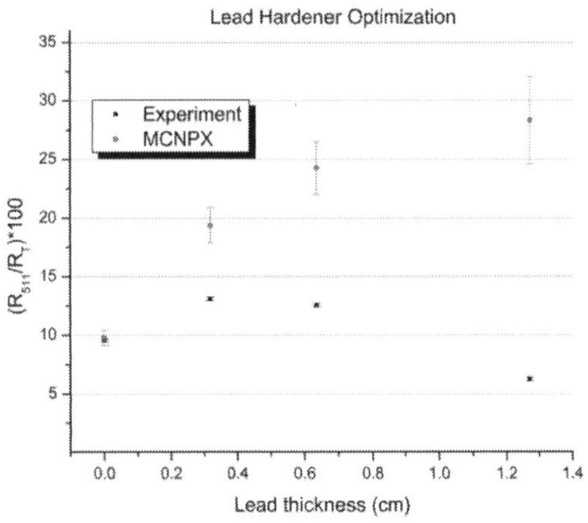

Figure 10. Experimental and MCNPX *FOM* results for lead hardener

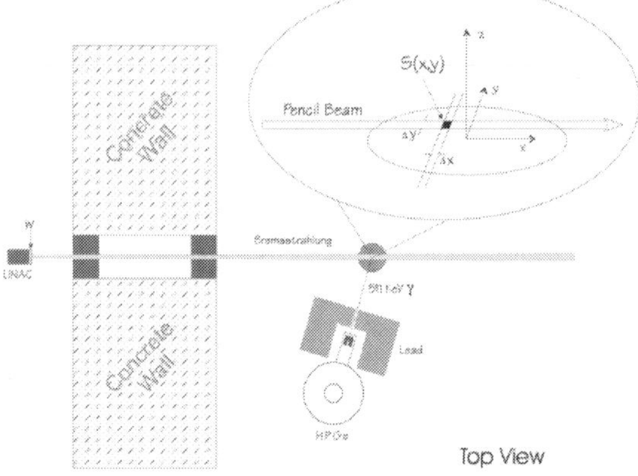

Figure 11. Simplified scheme of future experiments

of steel samples with different degrees of mechanical damage. Also a well-collimated HPGe detector will look at one irradiated voxel at a time. This way we can map S (or W) parameter as a function of the position (x, y). The simplified scheme is shown in figure 11.

5. Conclusions

We conducted experiments and MCNPX simulations to determine optimal values of important experimental parameters for defect density imaging with AG-PAS. We also uncovered some of the major limitations and difficulties we will face to produce such an image. Regardless of these limitations, based on the experimental 511 keV rates (R_{511}) obtained, we conclude that defect density mapping of large scale engineering samples with AG-PAS seems feasible.

References

[1] Hartley J H, Howell R H and Stern P A 1999 *Applications of Accelerators in Research and Industry, AIP. Conf. Proc. No. 475* (Woodbury, NY: American Institute of Physics) p 366

[2] Bauer W, Maier K, Major J, Schaefer H E, Seeger A, Carstanjen H D, Decker W, Diehl J and Stoll H 1987 *Appl. Phys. A* **43** 261

[3] Makarashvili V and Wells D P 2007 *Accelerator Applications, AccApp'07 Pocatello, ID* (American Nuclear Society) p 973

[4] Hunt A W, Spaulding R, Urban-Klaehn J, Harmon J F and Wells D P 2005 *Nuclear Instruments and Methods in Physics Research Section B: Beam Interactions with Materials and Atoms* **241** 262

[5] Selim F A, Wells D P, Harmon J F and Williams J 2005 *Journal of Applied Physics* **97** 113539

[6] Selim F A, Wells D P, Harmon J F, Kwofie J, Erikson G and Roney T 2003 *Radiation Physics and Chemistry* **68** 427

[7] Selim F A, Wells D P, Harmon J F, Scates W, Kwofie J, Spaulding R, Duttagupta S P, Jones J L, White T and Roney T 2002 *Nuclear Instruments and Methods in Physics Research, Section B* **192** 197

[8] Knoll G F 1999 *Radiation Detection and Measurements* (New York: Wiley)

[9] Toms M E 1972 *Nuclear Instruments and Methods* **99** 45

International Workshop on Positron Studies of Defects (PSD 08) IOP Publishing
Journal of Physics: Conference Series **265** (2011) 012026 doi:10.1088/1742-6596/265/1/012026

Digital positron lifetime: the influence of noise

Arnold Krille[1], Wolfgang Anwand[2] and Reinhard Krause-Rehberg[1]

[1] Department of Physics, Martin-Luther-University Halle-Wittenberg, 06099 Halle, Germany
[2] Institute of Ion Beam Physics, Research Center Dresden-Rossendorf, 01314 Dresden, Germany

E-mail: arnold.krille@physik.uni-halle.de

Abstract. In contrast to the world around where everything seems to go digital as soon as possible, positron lifetime spectrometers are kind of a "last sanctuary" for analog measurements. Only a few of the newer spectrometers use the analog-digital-converters directly after the photomultipliers and extract the timing information via computer. Judging from their results it seems as if the current available converters and the timing mathematics are only as good as the conventional analog setup in the timing resolution.

As it is decided that EPOS [1] will use digital positron lifetime, we try to find some reasons for limited timing resolution by simulating anode pulses from the photomultipliers and measuring the FWHM. We create pulses similar to current state-of-the-art 4GS/s digitizers but can control the level of noise and the bit-depth independently.

We found that especially the noise (that would come from the analog electronics in/before the converters) has a great influence on the timing resolution. Also we try to use lowpass filtering to reduce that influence with great success.

1. Introduction

While the use of computers and digital equipment more and more rules the daily live especially of scientists, the usage of computers for PALS is more or less limited to the evaluation of the final spectra. The advent of (ultra-)fast digitizers and fast enough computers opens up new possibilities of connecting the photomultipliers directly to the computer. Now all the evaluation of the anode-pulse can be done in software, thus replacing the almost 50 years old analog electronic chain.

As this change of the setup introduces new sources for errors, it is vital to know which part of the processing introduces which errors. Several Monte-Carlo-Simulations were done to simulate the different parts of the combination of photomultiplier, pre-amplifier and ADC.

Investigations on their influence on the final result have been made both independent of each other and in combination. The results show that the vertical resolution has a very big influence on the timing resolution partly because of the limited number of bits (usually 8-10 bits) available in this range of needed sampling rates of 2-8 GS/s and partly because of the noise introduced by pre-amp and ADC reducing the bit-depth to an effective value as low as 6 bits. Apart from the uncertainties from scintillator and photomultiplier this introduces a limit on the timing resolution which might come as a surprise.

Additionally noise reduction via lowpass-filtering is applied to the raw pulses to test whether this improves the timing resolution.

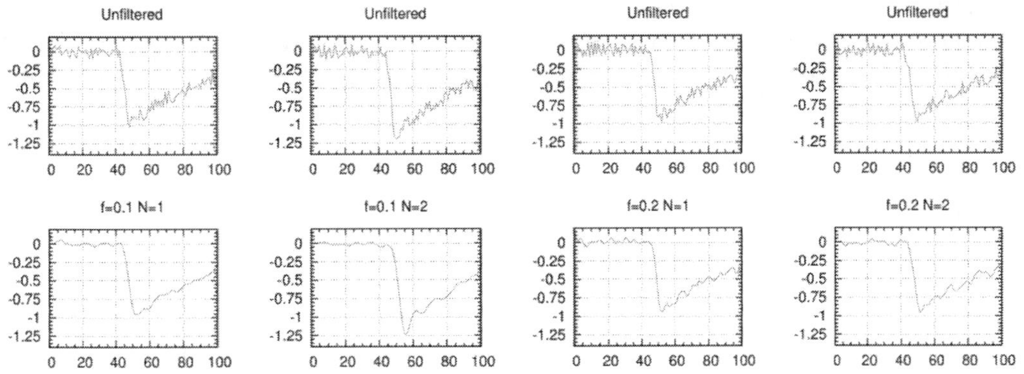

Figure 1. Raw and lowpass-filtered pulses. The upper row shows four pulses as generated for the simulations, the lower row shows the effect of various lowpass filter settings (to be explained in chapter 4). The noise added simulates 6.5 effective bits while the shape and risetime resemble pulses detected with LSO-scintillators and Hamamatsu H3378-50 photomultipliers. The horizontal time unit is samples.

2. Simulated pulses

The theory of signal processing says that the discrete event of the gamma-quantum is folded with the resolution functions of scintillator, photo-multiplier and used analog electronics. These are widely accepted to following a Gaussian-function:

$$y \sim \exp\left(-\frac{(x-x_0)^2}{2\sigma^2}\right)$$

As the δ-peak of a discrete event folded with the gaussian function results in the gaussian function itself, at least the rising slope of the photomultipliers signal can be estimated very well by this function. Note that the works of [2, 3] show that the gaussian function can also be used for interpolation on the real data. However this is slow compared to other methods because of the addtional step of fitting the Gaussian to the data.

The upper row of figure 1 shows four pulses as generated for these simulations. The shape and risetime are similar to what we measured with LSO-scintillators on Hamamatsu H3378-50 photomultipliers digitized with a 4 GS/s digitizer. To investigate whether the noise and the quantization from the vertical direction have an influence on the horizontal time resolution is the aim of this paper.

2.1. Parameters for the simulations

For the pulses several parameters can be set in the EPOS-Software [4]:

(i) *The amplitude (and distribution) of each channel individually*, used to simulate a ^{22}Na energy spectrum. The values were choosen so that the stop signal as a bit less than half the amplitude of the start signal. With the energy distribution the absolute levels were choosen to fit in the vertical full-scale without distortions. These were not changed through the simulations of this work to obtain comparable data.

(ii) *The distribution of the time-spread between the pulses.* While one might think that not shifting the pulses in time gives the best values, this would in fact give unrealistic values. In real life (even when measuring the time resolution with ^{60}Co) there is a slight deviation (a) in the time-derivation from the center of the pulse to the digitizer clock (this is implemented in the simulations as a box-distributed random number of $x \in [0, 1)$ samples) and (b) in the distance between the two signals because of geometrical, scintillator and photomultiplier uncertainties. The last one is simulated by using a Gaussion distribution with $\sigma = 0.001$ ($\equiv$ FWHM of 0.00235482 samples).

(iii) *The bit-depth of the vertical resolution.* This allows to simulate the analog-digital-conversion and its effects on the signal. Possible values range from 1 to 32 and more bits while native double resolution is also possible.

(iv) *The level of white noise added to simulate the analog electronics.* The noise added is white noise from a random number generator evenly distributed in the $[-max, max]$-range, the difference between the fullscale and the noise level is called the signal-to-noise-ratio (SNR) and calculated as follows:

$$SNR\ [\mathrm{dB}] = 20 \log \frac{A_{noise}}{A_{fullscale}}$$

In the following simulations the bit-depth and the noise level are changed according to the setup, the "energy spectrum" and the time distribution of the simulated pulses are left the same to create comparable results. This is an advantage of the simulation over reality where ^{60}Co is used for reference timing measurements with an energy spectrum different from ^{22}Na.

2.2. The true-constant-fraction algorithm

To extract the timing information the following algorithm is applied to the data (our pulses are negative, so the rising slope is actually falling):

(i) Find and interpolate the minimum of the pulses by simple 3rd order polynom interpolation (using four points around the minimum sample).

(ii) Find and average the baseline before the pulses.

(iii) Find the (true) constant fraction point-in-time on the falling slope between the baseline and the minimum on the pulses. This is again done by 3rd order polynomal interpolation. The fraction used was 20%, which proved to be stable in various simulations and real digital measurements and is also used in the analog measurements at least here in the Halle positron lab.

(iv) Subtract one cf-time from the other to get the "lifetime" between the two pulses.

As this algorithm has different details but the same results as the analog constant-fraction-modules, the name *true constant fraction* was proposed [5].

3. Simulated 4 GS/s digitizer

The time unit used for the following simulations is natively a [sample], which makes the results independent of the real sampling rate used in the digitizer. However, as the shape of the pulses is targeted at rebuilding real signals digitized with 4 GS/s, the results simulated here do not apply to sampling rates much higher or lower than this. Multiplying the resulting numbers with 250 ps gives the timing values to expect from real measurements.

Still the results will be shown with the unit [sample] as the transition to real [ps] depends on the parameters of the simulation, specifially the shape of the rising slope of the pulses. Here the values were choosen to closely resemble pulses from Hamamatsu photomultipliers with BaF_2-scintillators digitized with 4GS digitizers simulations for other sampling rates, other photomultipliers and other scintillators.

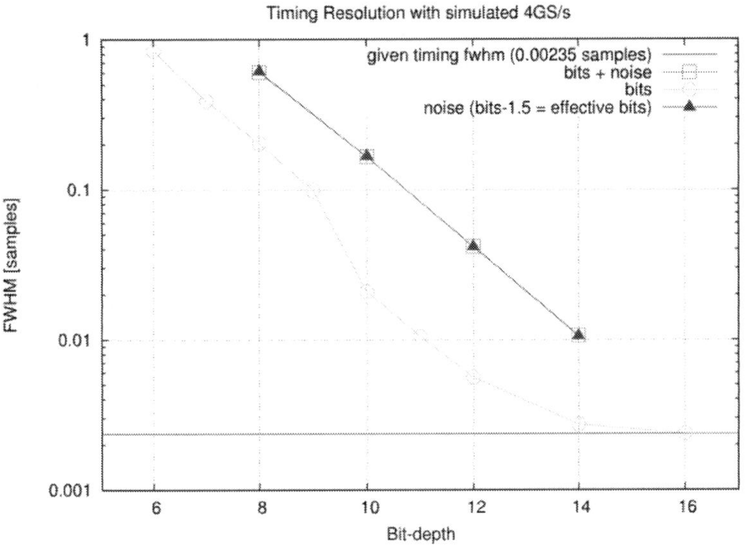

Figure 2. Timing Resolution with simulated 4 GS/s digitizer. The timing resolution (FWHM) is determined for various bit-depths with (green circles) just the bit-depth, (blue triangles) just the corresponding effective bits and (red squares) the bit-depth and the effective bits.
The solid black line shows the timing distribution used to shift the simulated pulses and is the theoretical optimum to achieve in this case.

3.1. Reducing the bit-depth
The first error-source to investigate is the vertical quantization, the bit-depth. The green circles in figure 3 show the determined timing resolutions for various bit-depths. The vertical range of signals fitted well into the vertical full scale. Given was a timing distribution with 0.00235 samples FWHM. It is clearly shown that higher bit-depths give better timing resolutions (with the same sampling rate). Calculating this to a 4 GS/s digitizer gives ~50 ps FWHM with 8-bit and ~5.2 ps FWHM with 10-bit resolution.

3.2. Effective bits-depth: noise reducing the accuracy
As the datasheets for (at least) Acqiris digitizers state only an effective bit-depth of 1.5 bits less than the real bit-depth due to noise, the next simulations were done a) with vertical double resolution and noise of the right level and b) with reduced bit-depths and the according noise levels combined. The blue triangles in figure 3 show the results of just the added noise, while the red squares are noise and bit-depth. These two show no real difference which makes the noise added from the digitizers analog electronic (pre-amps and converters) the main source for the limited timing resolution. Computed to real 4 GS sampling rate this gives 152 ps FWHM with 8-bit (6.5 effective bits) and 41 ps for 10-bit (8.5 effective bits).

4. Less noise with lowpass filter
To reduce the noise, one of the best options is to apply a lowpass filter on the raw data before the normal time-extracting algorithm is applied. A Butterworth filter from literature [6] was used for these tests. Given an input of white noise, figure 4 shows the frequency response of the filter for various cutoff frequencies and filter orders. The frequency is used relative to the sampling rate which makes the implementation of the filter module independent of the digitizer used. Figure 1 shows in the lower row the effect of the filtering on the pulses compared to the original raw pulses above.

Figure 4 shows the timing resolutions determined at different cutoff frequencies (f) and filter orders (N). All the simulations were done with double resolution and added noise of 6.5 effective bits. Clearly, the timing resolution is better than without the filtering. The best resolution with $N = 1$ and $f = 0.05$ evaluates to 75 ps for an 8bit-4GS/s-digitizer. Compared to without the filter, this is an improvement of ~ 2 in the timing resolution.

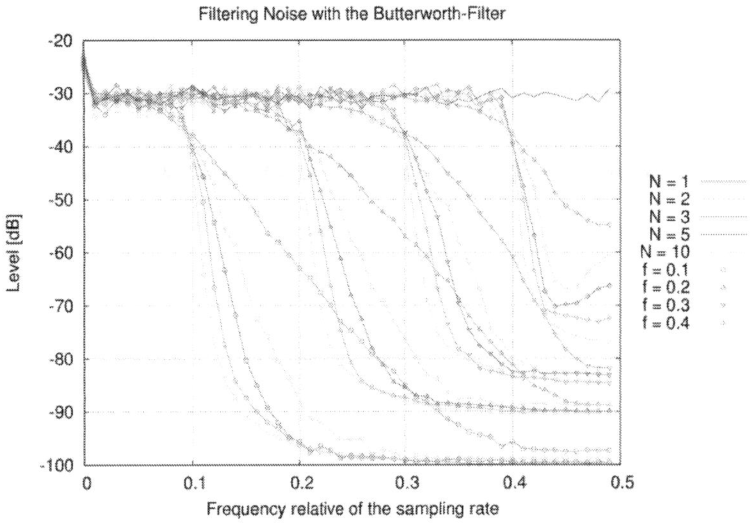

Figure 3. Frequency response of the Butterworth lowpass filter. Different filtering orders and different cutoff frequencies were used.

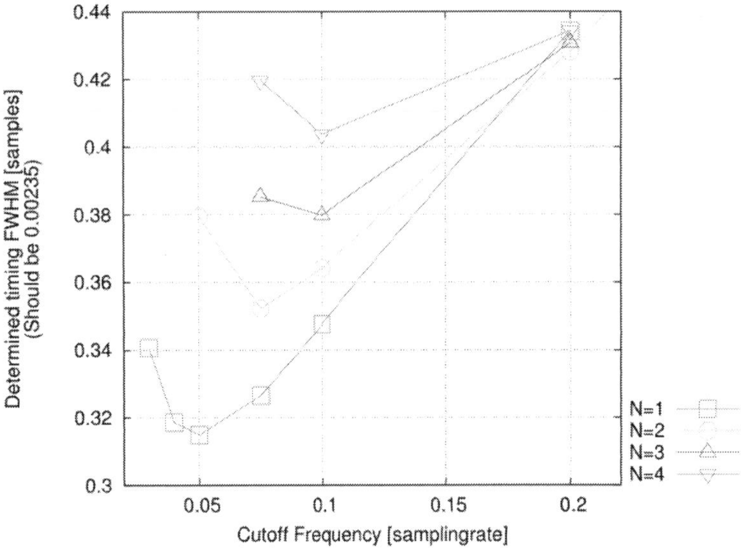

Figure 4. Timing resolution with lowpass filtered signals. For the 6.5 effective bits several combinations of filter order and cutoff frequency are tested to find the best timing resolution. It seems as if $N = 1$ and $f = 0.05$ is the best result in this case.

On first thought it seems strange that the lowest filtering order with the weakest suppression of higher frequencies (see figure 4) yields the best results. The explanation is that the Gaussian shaped slope of the pulses contains a lot of higher frequencies (an ideal δ-peak contains all frequencies with same phase). So while the higher orders are good at suppressing the high frequencies of the noise, they also suppress the high frequencies present in the rising slope which results in a more shallow slope and longer risetime. Which in turn gives a worse timing resolution.

5. Comparison of the results

Table 1 shows the three best resulting timing resolutions compared to each other for clarity. It shows that applying a lowpass filter almost removes the effect of the noise. Calculated to 4 GS/s digitizers the timing resolution of 153 ps is comparable to traditional analog positron lifetime setups [5] and to the first digital positron lifetime setups with basic constant fraction methods. However, somehow there seems to be a "magic border" at ~100 ps FWHM timing resolution.

Table 1. Comparing the results. The resulting (best) timing resolutions from the simulations are listed here. It is clear to see that lowpass filtering can almost remove the effect of the noise added from the analog electronics.

Method	**Relative Timing FWHM** [samples]	**4GS/s "real" FWHM** [ps]
Vertical quantization only (8-bit)	0.202 samples	50 ps
Noise of effective 6.5 bit	0.612 samples	153 ps
Butterworth-Lowpass $f = 0.05$ $N = 1$	0.314 samples	78.5 ps

Following our results this border can be crossed by applying optimal noise filtering.

6. Conclusion

One of the weak points of this simulation is definitely the self-choosen restriction to using [sample] as time unit. This means that the results given in [samples] are directly comparable, the transition to real life [ps] on the other hand remains to be checked.

We have shown that the main source for the limited timing resolution seems to be the noise reducing the bit-depth, especially in the 8-bit digitizers. It is pretty clear, removing the noise from the sampled data improves the timing resolution. Different strategies can be applied for this, we found that applying lowpass-filters from the dsp-world should give good results.

Acknowledgement

The work is funded by the German Federal Ministry of Education and Research under the project 05KK7NH1.

References

[1] Krause-Rehberg R, Sachert S, Brauer G, Rogov A and Noack K 2006 *Applied Surface Science* **252** 3106–3110
[2] Nissilä J, Rytsölä K, Aavikko R, Laakso A, Saarinen K and Hautojärvi P 2005 *Nuclear Instruments and Methods in Physics Research A* **538** 778–789
[3] Aavikko R, Rytsölä K, Nissilä J and Saarinen K 2005 *Acta Physica Polonica A* **107** 592–597
[4] positron.physik.uni-halle.de/epos/software URL `http://positron.physik.uni-halle.de/EPOS/Software/`
[5] Bečvář F 2007 *Nuclear Instruments and Methods in Physics Research B* **261** 871–874
[6] Stearns S D 1975 *Digital Signal Analysis* (Hayden Book Company Inc.)

Monte-Carlo simulations for timing-system of EPOS at ELBE in Research Centre Dresden-Rossendorf

M Butterling[1], M Jungmann[1], V Bondarenko[1], S Sachert[1], G Brauer[2], W Anwand[2] and R Krause-Rehberg[1]

[1] Department of Physics, Martin-Luther University, 06099 Halle, Germany
[2] Research Center Dresden-Rossendorf, PO Box 510119, 01314 Dresden, Germany

E-mail: maikbutterling@web.de

Abstract. Most of the monoenergetic slow-positron beam systems capable for positron lifetime spectroscopy are bunched from a continuous beam (e.g., the PLEPS system in Munich). The EPOS system at the ELBE radiation source will use the original time structure of the 40-MeV electron beam which is utilized for pair production. This is an infinite repetition of very short electron pulses (< 5 ps) with an adjustable repetition rate (typical 77 ns). In both cases, the time structure of the positron beam needs sharpening so that the time focus is located at the sample position. This is realized by the bunching system. Usually, double-slit bunchers are used. They are operated by a sinusoidal RF-voltage. This bunching requires an adjustment according to the final positron energy (0.5...30 keV), since the acceleration of the positrons shifts their time focus. Monte-Carlo simulations were done to investigate the effect of different methods of time focusing. One method is the variation of amplitude of the buncher RF-voltage. Another improvement is the combination of such a buncher voltage variation with a variable drift path acceleration short before the final acceleration which is operated by a DC voltage. Our simulations show that indeed a combination of a buncher variation with a drift path gives the sharpest positron pulses when both devices are supplied with individual voltages for each positron implantation energy.

1. Introduction

At Research Center Dresden-Rossendorf the 40 MeV electron beam ELBE (Electron Linac for beams with high Brilliance and low Emittance) is used to generate positrons by pair production. An advantage of this source is the time structure (bunch width about 5 ps; repetition time 77 ns) which makes ELBE to an ideal host for a bunched positron beam. The following illustration shows the EPOS system (ELBE Positron Source) in a simplified way: only those elements are included which are important for the timing system. More detailed information is given in [1].

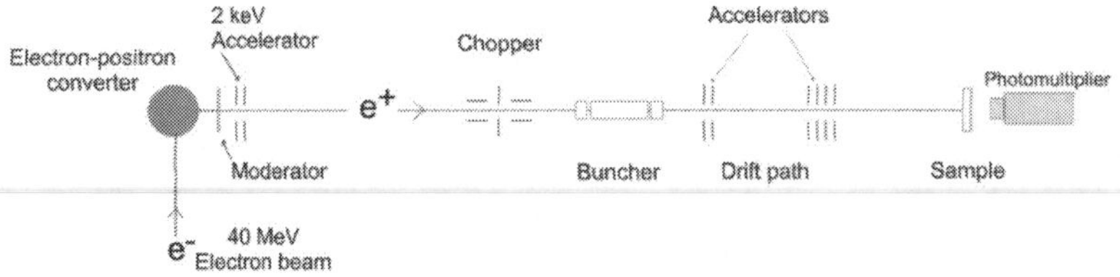

Figure 1. Simplified scheme of the EPOS beam with focus on the timing system.

The ELBE primary beam is partly converted by pair production into fast positrons. These positrons are cooled down in a moderator foil (tungsten or platinum) and are extracted to the positron lab as a monoenergetic positron beam with an energy of 2 keV. The extremely sharp time structure of the primary electron beam allows the realization of the positron lifetime spectroscopy in a unique way. For this purpose, the positron beam is treated by a system of choppers, double-slit bunchers and an additional drift path short before the final acceleration to improve the positron beam structure thus making the system very simple and effective. Without these devices the time structure of the positron beam will be broadened and positron lifetime measurements would be impossible.

2. Monte-Carlo simulation of the lifetime

Monte-Carlo simulations were done to calculate the dependency of the time structure of the positron beam for different settings (i.e. change of beam energy, variation of the buncher RF amplitude and a variable drift path). The converter has a fixed energy of 2 keV with an expected standard derivation of < 10 eV. To realize energies between 0.2 keV and 30 keV the positrons will additionally be accelerated close to the sample. The simulations calculate the timing behavior of the positron bunches. The result is the full width at half maximum (FWHM) of the positron bunch and the location of the time focus of the bunch. These parameters are used to evaluate the benefit of the two devices. Each simulation was done for a number of 2000 positrons. The following picture illustrates the correlation between positron energy, position of the time focus and the FWHM of the positron bunch at sample position for a fixed buncher RF-voltage (the simplest case):

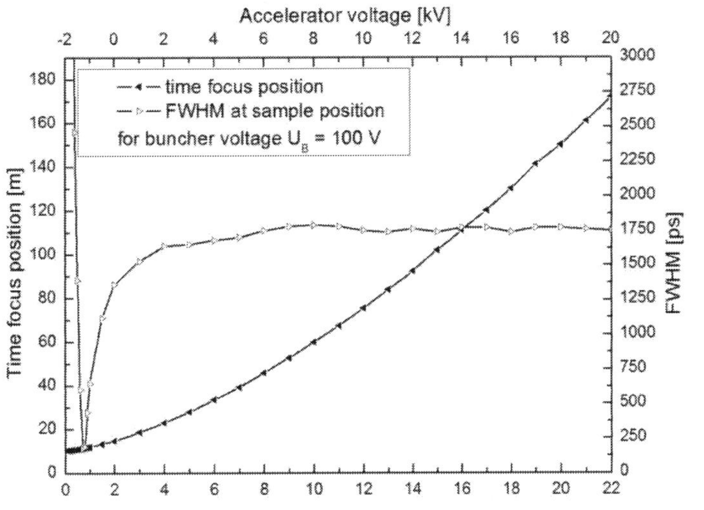

Figure 2. Time focus position and FWHM at a sample position of 11m (distance from the electron-positron converter) for different positron acceleration energies.

The result is a useless FWHM of about 1750 ps (a good value would be lower than 100 ps) due to the shift of the time focus. Figure 2 illustrates the importance of a timing-system that improves the time structure of the beam. A first method is to vary the buncher RF-voltage for each implantation energy.

3. Variation of the buncher RF-voltage

For this part of simulation the buncher RF amplitude was varied for every energy with the aim of moving the time focus to the sample position. The plot of these data (figure 3) shows that for lower energies (up to about 5 keV) a shift of the buncher RF amplitude has a great influence on the FWHM at sample position. Otherwise in this region the buncher RF-voltage has to be changed more than for higher energies (from 10 keV to 22 keV). The best value for the FWHM given by this method is nearly 60 ps.

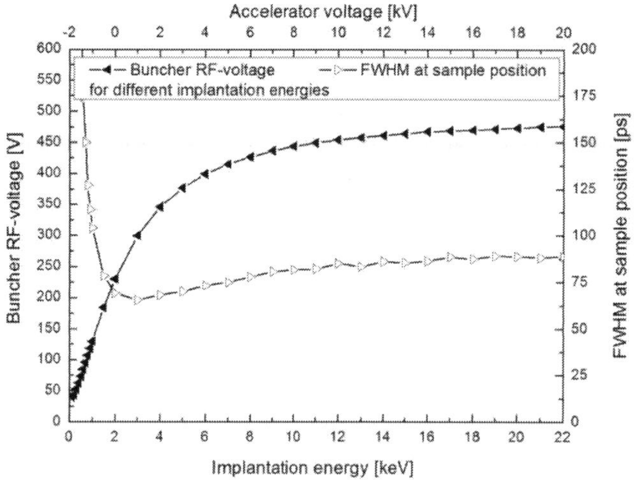

Figure 3. Optimum buncher RF amplitude and FWHM at sample position for different positron energies.

To visualize the importance of a stable buncher RF-voltage the relation between a small shift in the buncher RF-voltage and the FWHM of the positron pulse was simulated. The result (figure 4) shows that for lower positron energies (up to 2 keV) a small shift of the buncher RF-voltage has no great influence on the FWHM. A stability of the voltage of 1 or 2 V seems to be enough.

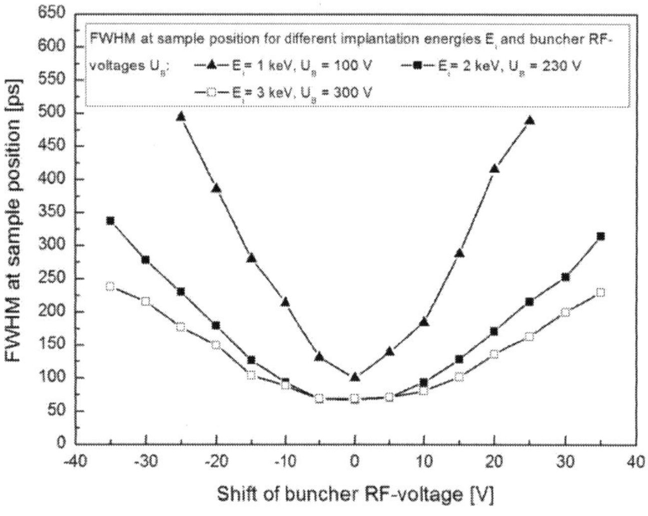

Figure 4. Influence of a shift of the buncher RF-voltage on the FWHM at sample position.

4. Combination of a buncher with a variable drift path

The other timing method is the combination of the double-slit buncher with a variable drift path (see figure 2) which is operated by a DC voltage. Therefore, the buncher RF-voltage is set to the value that is necessary for the implantation energy of 30 keV (the maximum of the positron energy). Figure 5 presents the relation between FWHM at sample position (and optimum drift path voltage) as a function of the positron energy.

The result is a substantially better FWHM of the positron pulse for lower positron energies (from 3 to 10 keV) compared to the usage of the buncher with varying RF voltage alone. The FWHM in this range is much lower than 90 ps. The minimum FWHM is 30 ps (compare with figure 3, the minimum there was 60 ps).

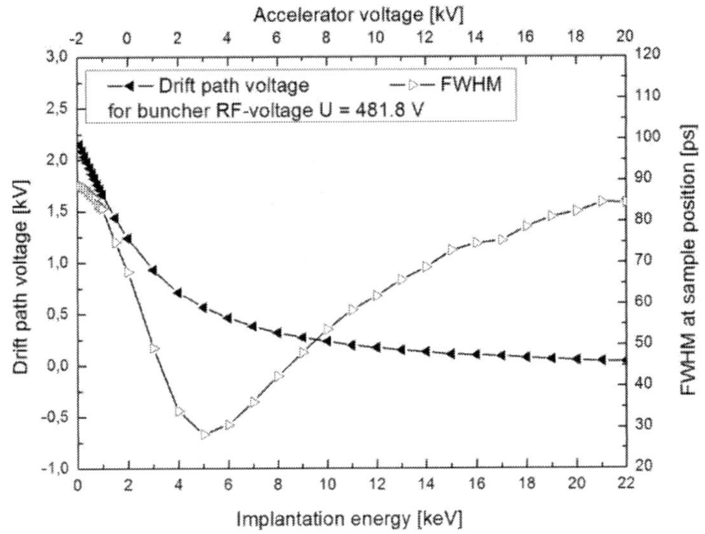

Figure 5. Optimum drift path voltage and FWHM at sample position for different positron energies.

5. Combination of a buncher with a drift path with individual voltages for each positron implantation energy

To simulate the results of a variation of both voltages (buncher RF amplitude and drift path´s DC voltage) it is necessary to fix the positron energy for each simulation. For each drift path voltage the buncher RF amplitude was varied with the aim to get the time focus at the sample position. Figure 6 presents the influence on the FWHM and gives a comparison between all three methods (see the dashed lines). One can see that in a certain area the FWHM is better as compared with the other methods. The FWHM has a parabolic dependence on the drift path voltage with a clear minimum.

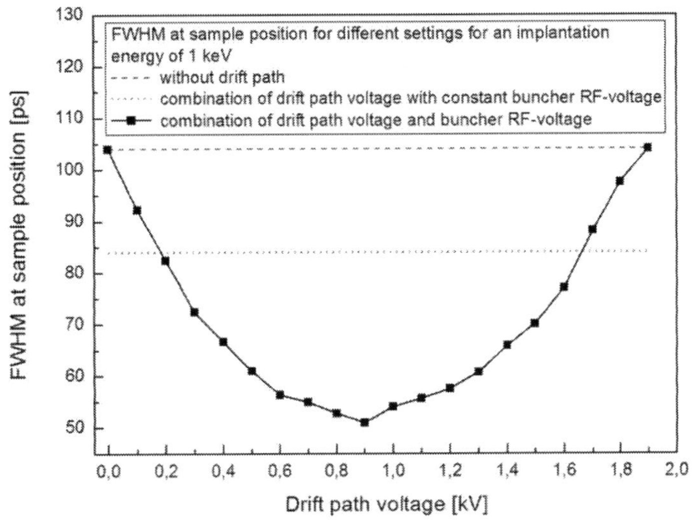

Figure 6. FWHM at sample position for the combination of drift path voltage and buncher RF-voltage for a positron energy of 1 keV.

A look at the relation between drift path voltage and buncher RF-voltage indicates a similar behavior for different positron implantation energies.

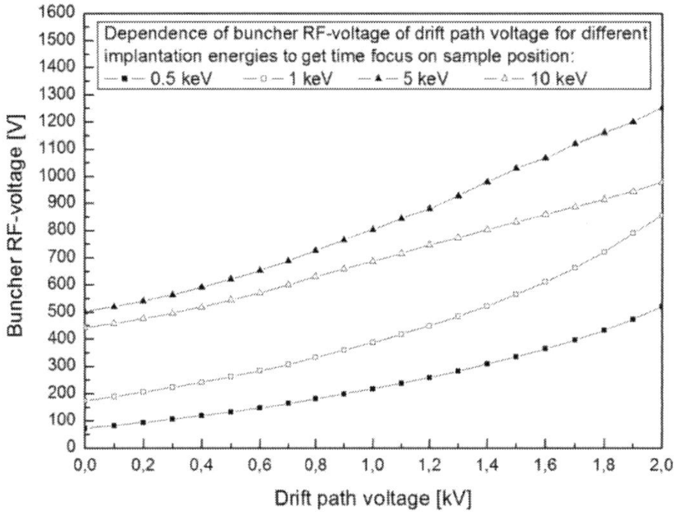

Figure 7. Relation between optimum buncher RF amplitude and drift path voltage for certain positron energies.

These simulations (like for figure 6) were done for different positron implantation energies to find the minimum FWHM (the minimum of the parabola in figure 6). Figure 8 compares all three methods (variation of the buncher RF-voltage, combination of a fixed buncher with a variable drift path, combination of a variable buncher with a drift path with individual voltages for each positron energy). It is obvious that the combination gives the sharpest positron pulses for all positron implantation energies. One can improve the FWHM of the positron pulses by more than a factor of 2 for some energies.

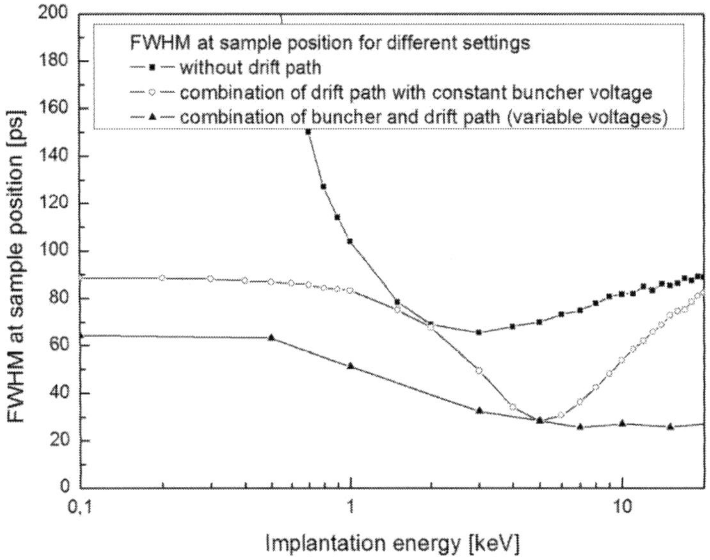

Figure 8. Comparison between all three methods of bunching.

References

[1] Krause-Rehberg R, Sachert S, Brauer G, Rogov A and Noack K 2006 EPOS - An intense positron beam project at the ELBE radiation source in Rossendorf *Applied Surface Science* **252** 3106–3110

International Workshop on Positron Studies of Defects (PSD 08) IOP Publishing
Journal of Physics: Conference Series **265** (2011) 012028 doi:10.1088/1742-6596/265/1/012028

Thermal analysis of EPOS components

M Werner[2], E Altstadt[2], M Jungmann[1], G Brauer[2], K Noack[2], A Rogov[2] and R Krause-Rehberg[1]

[1]Department of Physics, Martin-Luther-University Halle-Wittenberg, 06099 Halle, Germany
[2]Institute of Ion Beam Physics, Research Center Dresden-Rossendorf, 01314 Dresden, Germany

E-mail: marcojungmann@gmx.de

Abstract. We present a simulation study of the thermal behaviour of the moderator and the electrostatic lens of the positron source EPOS at the Research Center Dresden-Rossendorf. The positron moderator and the upper part of the electrostatic extraction einzellens are directly exposed to the primary electron beam with an energy of 40 MeV and a current of 1 mA. Thus, it was necessary to investigate whether the construction can stand the high temperatures. As a result we found that thin moderator foils with a thickness less than 40 μm will not reach too high temperatures. The thickness can be varied up to 0.2 mm, so we adjust the temperature in order to anneal lattice defects. These defects are caused by radiation. The wall of the extraction lens which is made from a stainless steel tube must be distinctly thinned to 0.05 mm to avoid damage temperatures. The time-dependent simulation shows, that the final temperatures are reached after less than a minute.

1. Introduction

The positron beam technique is a unique tool to study crystal lattice defects and open-volume cavities of nanometer scale [1]. The Forschungszentrum Dresden-Rossendorf provides an intense pulsed 40 MeV electron beam with high brilliance and low emittance (ELBE). It gives the unique possibility to construct an intense, pulsed positron beam for materials research. The EPOS System (**E**LBE **Po**sitron **S**ource) generates an intense pulsed beam of mono-energetic positrons by pair production on a tungsten converter and a moderator with a repetition time of 77 ns and a bunch length of 5 ps. A detailed description to realize the EPOS project has already been published [2]. After the positrons are generated in the converter, the positrons will be moderated and an electrostatic lens focuses the positrons on a point of 1.5 mm diameter to transport the positrons by means of a magnetic guidance field into the positron laboratory. Because the moderator and a part of the lens will be hit by the intense primary electron beam, it is necessary to investigate the temperature behaviour of both parts in the electron beam. The results can be used for an efficient choice of construction material and the geometry of the moderator and the electrostatic lens [3].

2. Evaluation of the heat generation rates

The heat generation rates in the moderator and the upper part of the lens were calculated with the Monte Carlo code MCNP-5, release 1.40 [4]. MCNP is a Monte Carlo N-Particle code that can be used for stochastically simulating neutron, photon or electron transports. The code

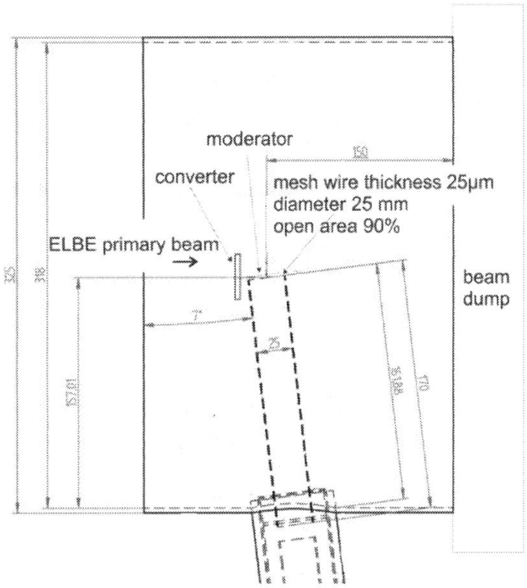

Figure 1. Part of the EPOS configuration. The ELBE electron beam hits the electron-gamma converter and is broadened from 5mm diameter to an angle so that the whole beam dump is exposed. The positron moderator is placed in front of the converter and parallel to the entrance plane of the upper tube of the positron focusing lens (see figure 2).

is able to treat an arbitrary three-dimensional configuration of materials. For neutrons and photons energy point wise cross-section data given in Evaluated Nuclear Data Libraries (such as ENDF/B-VI) are used to account for the various interactions with the materials. In case of electrons they are treated by means of a continuous-slowing-down model. The code allows the estimation of a broad range of integrals of the space-, energy- and direction-dependent particle fields together with their statistical errors. For the study presented in this report the mean values of the heat generation by photons and electrons averaged over user-defined volumes were calculated in a coupled electron/photon/neutron transport simulation. Figure 2 shows the arrangement and the definition of the estimation volumes (cells). The moderator foil was chosen to have a homogeneous thickness of 20 μm (cell 202). The electrostatic lens consists of 3 tubes. Only the upper lens tube is hit by the electron beam. Thus, only this tube is taken into the consideration and is shown in figures 1 and 2. The lens tube having a wall thickness of 1 mm (cell 241) is in the upper part, which is exposed to the high-energetic electron beam electrochemically thinned to a wall thickness of 50 μm (cell 210, 220 and 230 to 236). On top of this tube an electrostatic extraction grid of stainless steel is welded (cell 243). The results of the calculation for the different cells are listed in table 1. The energy input values are given as integral values in the volumes V_i for one electron with an energy of 40 MeV. The volumetric heat generation is obtained by:

$$Q_i^{vol} = \frac{E_i I}{V_i e}$$

where I is the current of the electron beam and I/e equals the number of electrons per seconds.

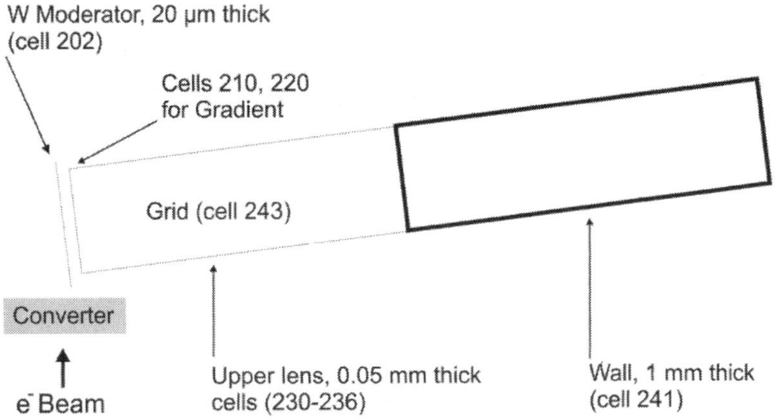

Figure 2. Arrangement of the upper part of the electrostatic lens and the moderator. The definition of the MCNP cells is indicated.

Table 1. Results of the energy calculation. The energy input values are given as integral values in the volume V_i for one electron with an energy of 40 MeV.

Cell	Volume V_i $[cm^3]$	Energy input [MeV]	Comment
202	1.414e-2	1.18e-2	W-moderator
210	3.000e-2	5.77e-5	additional cells to account for the
220	3.000e-2	2.42e-4	gradient in beam direction
230	3.919e-2	9.10e-3	ring1, hight 1 cm
331	3.919e-2	4.26e-3	ring2, hight 1 cm
332	3.919e-2	2.28e-3	ring3, hight 1 cm
233	3.919e-2	1.37e-3	ring4, hight 1 cm
234	3.919e-2	8.84e-4	ring5, hight 1 cm
235	3.919e-2	6.09e-4	ring6, hight 1 cm
236	3.919e-2	4.28e-4	ring7, hight 1 cm
241	7.389e+0	1.88e-2	lower part of the lens
243	1.461e-3	5.32e-4	grid at the top of the lens

3. Thermal calculations

3.1. Model

Based on the code ANSYS a finite element model is used to calculate the transient temperature distribution in the moderator, the grid and in the positron lens. Heat conduction within the materials and radiation heat transfer at the surfaces are considered. The heat generation density is obtained from table 1. An axisymmetric model was chosen assuming that the temperature gradient in beam direction is negligible.

3.2. Material properties

For the positron moderator two material options are considered: Pt and W. The lens and the grid are made of stainless steel. The extraction grid has an open area of 90%. To take this fact into account in the model reduced (or effective) material constants are used. The material properties from table 2 were used [5].

Table 2. Material properties for the thermal analysis.

Material	Tungsten	Platinum	Steel	Steel (grid)
Density [kg/m^3]	19259	21450	7800	780
Heat conduction coef. [W/m/k]	174	71.6	15	1.5
Specific heat [J/kg/k]	130	133	510	51
Emission coef.	$\epsilon(T)$	$\epsilon(T)$	0.24	0.24

3.3. Load and boundary conditions

The volumetric heat generation density is obtained by converting the value of table 1 using equation 1 assuming a current of $I = 1$ mA, thus a total electron power of 40 kW. One obtains the values given in table 3. The volumetric heat sources are "switched on" at $t = 0$ and remain constant afterwards. The heat generation rates in table 3 are valid for the full power mode. During the diagnostic mode 1% of these values are used. The initial temperature of the whole structure at t = 0 is $T_0 = 308$ K, the ambient temperature is assumed to be constant $T_a = 353$ K. For the volumetric heat generation rates in Pt we assumed the same values as in W.

Table 3. Heat generation rates.

Cell	Heat generation rate [W/m^3]	Comment
202	8.345e+8	moderator
220	8.067e+8	additional range for gradient (added to 230)
230	2.322e+8	
231	1.087e+8	
232	5.818e+7	
233	3.496e+7	upper part of lens (thickness 0.05 mm)
234	2.256e+7	
235	1.554e+7	
236	1.092e+7	
241	2.554e+6	lower part of lens (thickness 1 mm)
243	3.641e+6	grid

3.4. Results

Transient calculations were performed for different scenarios. The following parameters were varied:

- the power of the beam (1% - diagnostic mode and 100% - full power mode)
- the moderator material (W or Pt)
- the thickness of the moderator
- the heat generation in the grid (100% - upper bound; 10% - realistic case)

The calculations were calculated as a function of time after switching on the electron beam. They were stopped when a steady state temperature field was reached.

Temperature in the Moderator The heat-up of the moderator was calculated for Pt and W with a nominal thickness of 0.04 mm at 100% beam power. Figure 3 shows the maximum temperatures vs. time for W and Pt respectively.

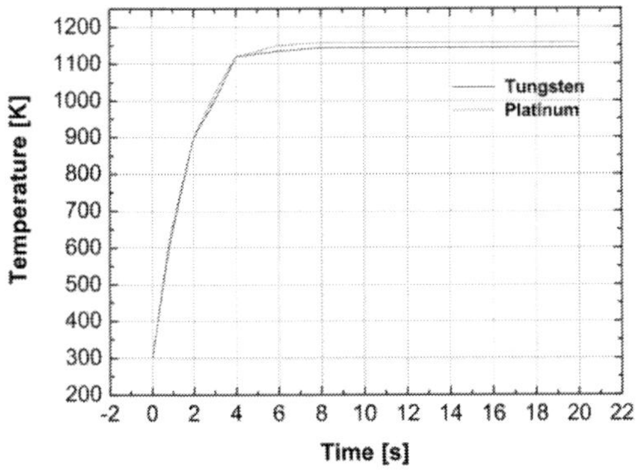

Figure 3. Temperature [K] vs. time [s] in the Pt and W moderator 100 % beam power.

It can be seen that the steady state temperature is reached at about $t = 8$ s in both cases. The final temperature in W is 1144 K and in Pt 1158 K. The somewhat smaller value in W correlates to the slightly larger emission coefficient at 1000 K (Pt: 0.12, W: 0.14). In the following we investigate the influence of the moderator thickness on the maximum temperature. It is obvious that the maximum temperature will increase with increasing thickness, since the volume to surface ratio is given by:

$$\frac{V}{A_{surf}} = \frac{s}{2 + 4s/D}$$

where s and D are thickness and diameter respectively ($s << D$). The thickness should be chosen in such a way that:

- the maximum temperature is sufficiently far from the melting temperature of the material (W: 3683 K; Pt: 2045 K) and
- the maximum temperature is high enough to enable the healing of the microstructural point defects caused by the irradiation.

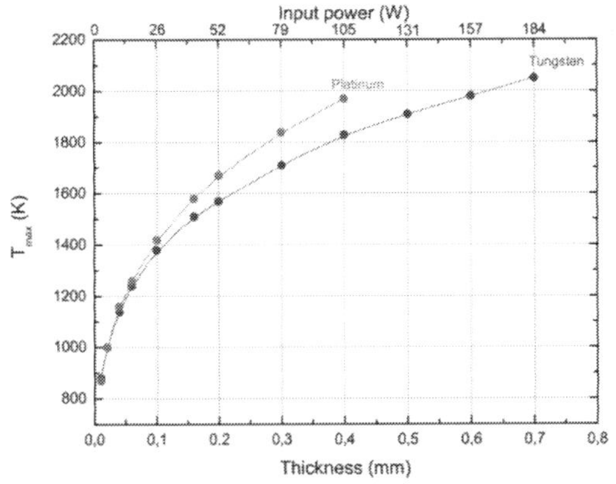

Figure 4. Maximum moderator temperature in dependence on thickness.

The thickness of the moderator was varied from 0.01 mm to 0.7 mm for W and from 0.01 mm to 0.48 mm for Pt. The dependence of T_{max} on the thickness is shown in figure 4. The total heat input, which is proportional to the thickness, is indicated in the upper horizontal axis. The temperature gradient in the moderator is rather small (5 K) and can be neglected.

Temperatures in the lens The temperature distributions in the lens (grid and tube) are discussed for the case of 100% heat generation. For the upper border case the influence of the emission coefficient of steel is investigated in the range $0.03 \leq \epsilon \leq 1$. The emission coefficient of steel strongly depends on the surface finishing; therefore the "nominal" value given in table 2 might be inaccurate. Figure 5 shows the temperature vs. time for 100% beam power at different locations. The steady state is reached approximately after 60 s. Figure 6 shows the steady state temperature distribution in the extraction tube.

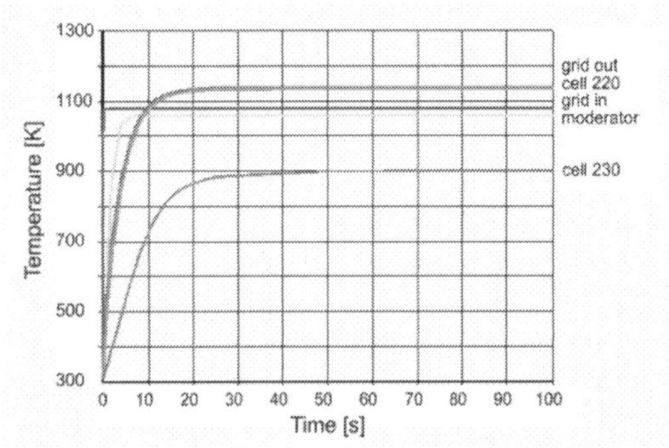

Figure 5. Temperature vs. time for different positions, 100 % beam power, emission coefficient $\epsilon = 0.24$.

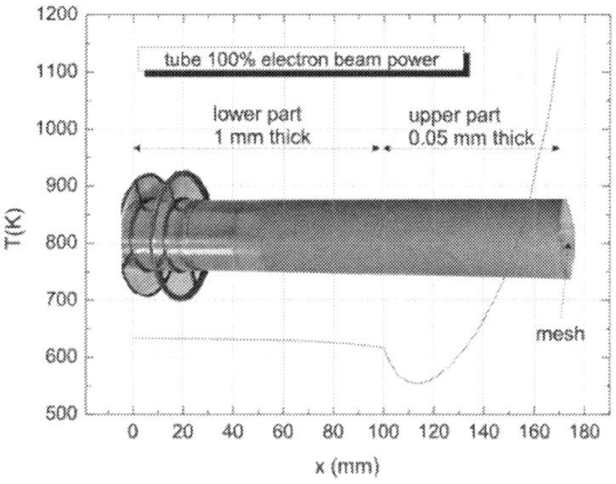

Figure 6. Temperature distribution [K] in the upper part of the extraction lens at $t = 785$ s (100% beam power).

The influence of the emission coefficient on the maximum (steady state) temperature is shown in figure 7 and figure 8. It can be seen that the hottest point is the end part where the grid

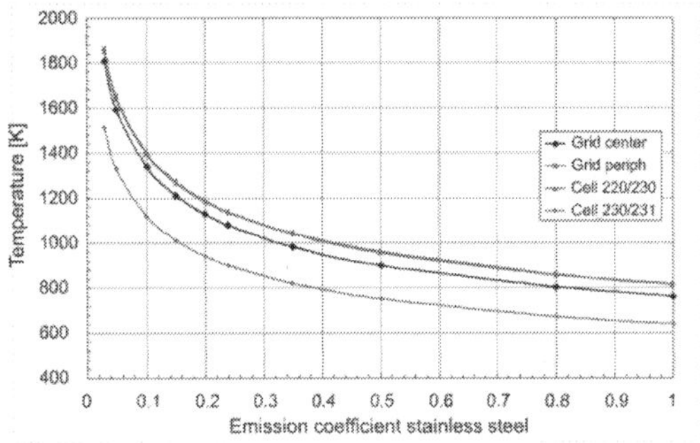

Figure 7. Maximum temperatures vs. emission coefficient at different positions in the upper part of the tube; 100% beam power.

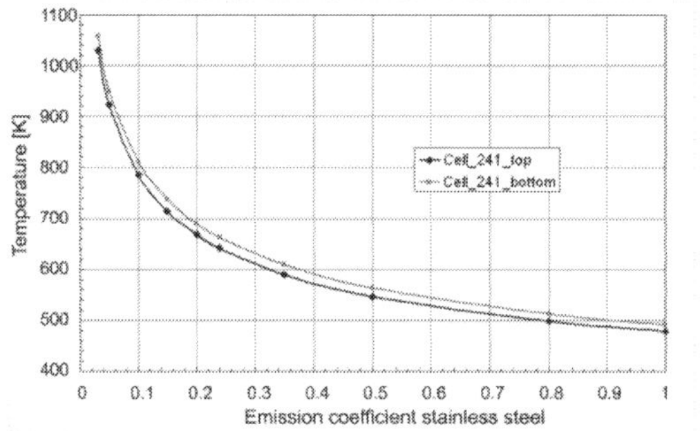

Figure 8. Maximum temperatures vs. emission coefficient at different positions in the lower part of the tube; 100% beam power.

is connected to the tube ("grid periph"). Here, the electron beam has the highest intensity. In the lower part of the tube (cell 241) the temperatures are significantly lower than in the upper part. It can be concluded that for $\epsilon \geq 0.05$ the maximum temperature in the tube is clearly lower than the melting temperature of steel (≈ 1760 K).

4. Conclusions

We performed a time-dependent thermal analysis of metallic parts (positron moderator and electrostatic extraction lens) of the positron source EPOS being directly exposed to the high-energetic, high-flux electron beam of the radiation source ELBE at the Research Center Dresden-Rossendorf. The temperature of the moderator can be adjusted by varying the thickness of the foil. It is possible to obtain high enough temperatures to ensure defect annealing under operation conditions. This will lead to a larger positron diffusion length, and thus to a higher efficiency of the electron-positron conversion. The ideal thickness of the moderator foil by the electron beam energy of 40 MeV is in the range of 0.5 to 2 mm, but in this region the temperature is too high. Because the annealing temperature of platinum is lower then tungsten, the moderator should be made of platinum with a thickness of 0.2 mm. In this case, the temperature on the moderator is 1550 K and the point defects can anneal.

The extraction lens tube is made of stainless steel. The upper part which is hit by the high-intense electron beam must be thinned electrochemically in order to reduce the heat load. A version having a wall thickness of only 50 μm has been realized and will be tested soon. We expect a temperature not higher than 1150 K. This is far from the melting temperature of stainless steel (1760 K). The mesh grid on top of the extraction tube which is also made from stainless steel will have a temperature of about 1100 K at full beam power. All this parts will be at a stable operation temperature after less than 60 s.

References

[1] Krause-Rehberg R and Leipner H 1999 *Positron Annihilation in Semiconductors* (Berlin: Springer-Verlag)
[2] Krause-Rehberg R, Sachert S, Brauer G, Rogov A and Noack K 2006 *Appl. Surf. Sci.* **252** 3106
[3] Werner M, Altstadt E, Jungmann M, Brauer G, Noack K, Rogov A and Krause-Rehberg R 2008 *Thermal Analysis of EPOS components* (Research Center Dresden-Rossendorf) report
[4] *MCNP-A General Monte Carlo N-Particle Transport Code, Version 5, X-5 Monte Carlo Team, LA-CP-03-0284 2003*
[5] 2007 Table of emissivity of various surfaces, Mikron Instrument Company, Inc.

Journal of Physics: Conference Series **265** (2011) 012029 doi:10.1088/1742-6596/265/1/012029

Application possibilities of Acqiris digital card DP240 for positron lifetime measurement

M Petriska[1], V Slugeň[1], V Kršjak[2] and A Zeman[3]

[1] Department of Nuclear Physics and Technology, Slovak University of Technology, Ilkovičova 3, 81219 Bratislava, Slovakia
[2] Institute for Energy, Joint Research Centre of the European Commission, P.O.Box 2, 1755ZG Petten, The Netherlands
[3] International Atomic Energy Agency, Department of Nuclear Sciences and Applications, Wagramer Strasse 5, A-1400, Vienna, Austria

E-mail: martin.petriska@stuba.sk

Abstract. Based on existing knowledge, a software for the digital positron lifetime (LT) spectrometer has been designed at the Slovak University of Technology. As a digitizer, Acqiris DP240 card was used. Start and stop timing signals have been digitised separately with sampling rate 1GS/s in 8 bit resolution. This sampling rate was not sufficient; therefore a joined delayed channel mode with 2GS/s sampling rate was used. Next applications for the study of advanced materials for nuclear industry are foreseen.

1. Introduction

Digital positron lifetime measurement systems start to be a standard part of positron annihilation laboratories [1,2]. It is due to evolution in fast ADC digitizer cards and significant lowering of their prices. Using digitizer cards in positron annihilation spectroscopy (PAS) LT systems, one gets new possibilities in measured data processing, automatic measurement setup, offline data correcting depended on other measurement parameters like temperature or magnetic field influence. From analogue setups in digital systems new task in software for detector data processing development is needed. The first role of the new software is to replace the task of analogue modules like SA (single channel analyzer), CFD (constant fraction discriminator), TAC (time to amplitude converter) and MCA (multi channel analyzer). The second role is to add new possibilities like automatic measurement setup and offline data filtering. The significant advantage of digital measurement is possibility to use different approximation methods without need to repeat measurement like in analogue system. For example, a small change in the window or in constant fraction (CF) value in an analogue system requested the repetition of measurement. In a digital system it is possible by software change using the same acquired data. This approach seems to be much more efficient.

2. Digitised dynode timing signal

The shape of two dynode signals obtained by digitizing with 1GS/s is in figure 1. While this type of voltage divider was used in [1], the same signal shape is expected. For lower sampling rate provided by Acqiris DP240, the second signal minimum is not adequately visible in our case. In spite of lower sampling rate we developed two setups, one with separated start and stop signal and second with joined delayed channel mode, with 2GS/s sampling rate.

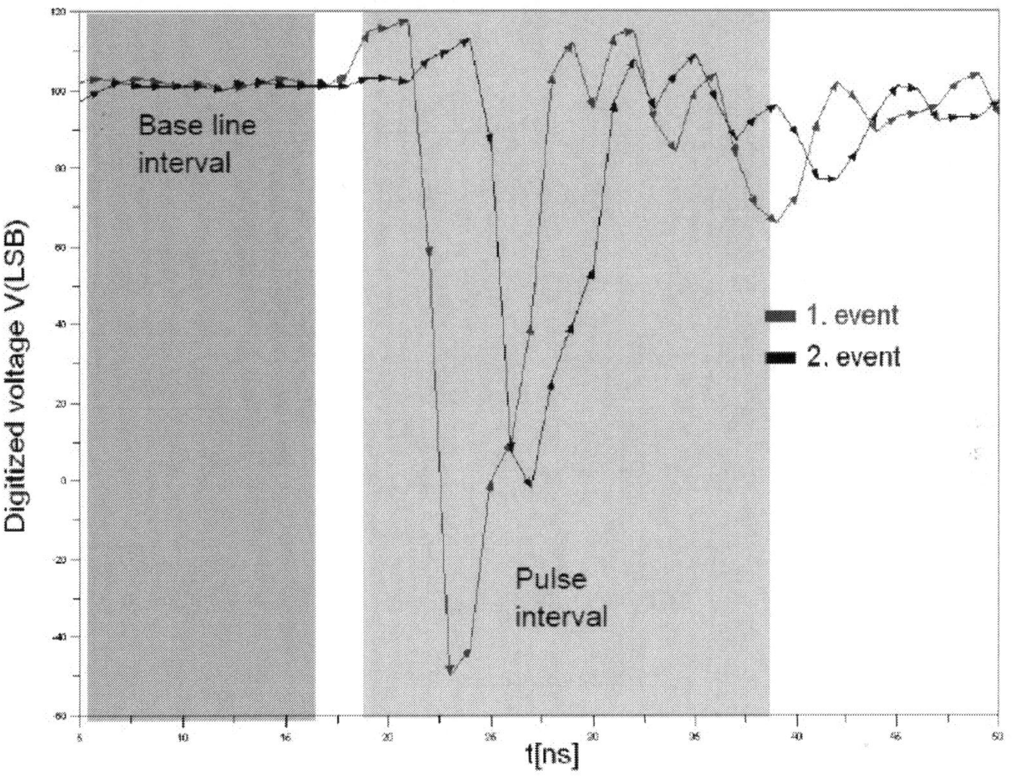

Figure 1. Two dynode pulses from two stop events sampled at 1GS/s with base line and pulse interval definitions.

3. Hardware

Hardware is based on existing analogue Fast-Fast setup [3] (see figure 2A). It consists of BaF_2 scintillators, XP2020Q photomultiplier tubes, voltage dividers with dynode outputs [4]. Two Ortec 583 constant fraction discriminators and Ortec 414A fast coincidence unit were used as trigger source for the digitizer card. In this case 2 channel mode with 1GS/s sampling frequency is done. As digitizer, Acqiris DP240 card was used.

4. Software for digital system with Acqiris DP240

It was found that off-line analysis method is reasonable for ability to develop various methods for time extraction without repeating measurement every time. Therefore the developed software was divided into two parts: data acquisition part and data analysis part.

The size of measured data (400 MByte for 1 million events and 2GS/s sampling rate) with external signal triggering is acceptable.

4.1. Data acquisition part

Data acquisition program was based on Acqiris C example source GetStartedVC distributed with digitiser card. Program was developed as command line tool. Needed parameters for the card setting were obtained from graphical tool – AcqirisLive distributed with acquisition card. All input parameters were stored in configuration file.

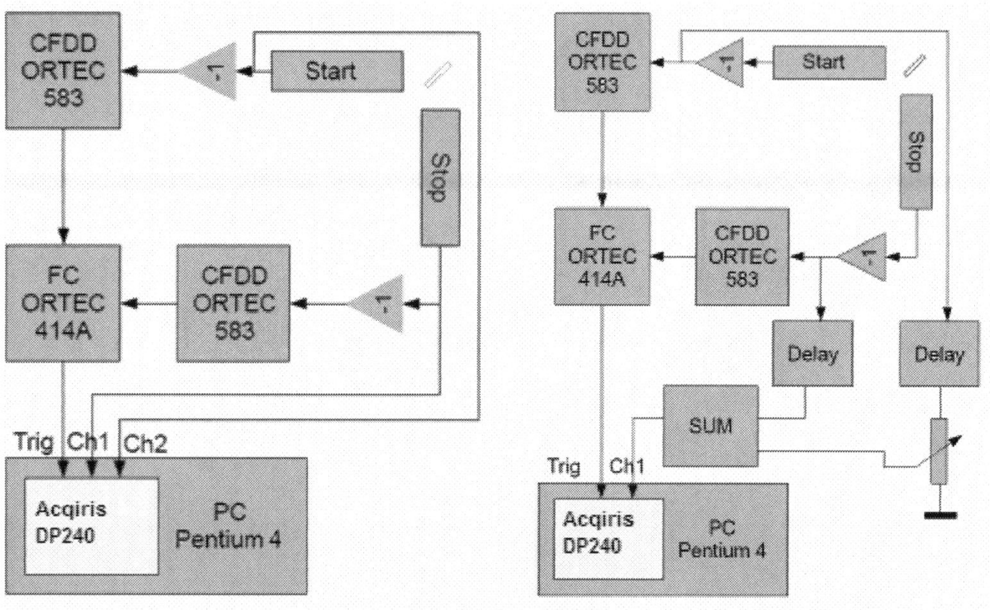

A) B)

Figure 2. Hardware for the digital PAS LT with 1GS/s sampling rate and separated start and stop signals digitising. A) Two channels mode with 1GS/s B) Join channel mode with 2GS/s.

4.2. Data analysis part

Two channel mode with separated Start and Stop pulses digitized with sampling rate 1GS/s were first preferred for measurement, but after number of experiments it was found that this mode is not suitable for the combination of BaF_2 detectors with dynode outputs with non typical dynode signal shape. From figure 1 rising edge in sampled pulse can be seen. Base line of signal was counted as average from first ten points. Timing information was extracted using digital constant fraction method [6]. Signal minimum was found after parabolic approximation in 3 points. First point of this approximation is one sample before minimum, second is minimum sample and third is one sample after minimum. Timing information was counted as constant fraction 20% in parabolic approximation from 4 points in rising edge. For higher sample rates more approximation points can be used. Resulting spectrum from separated start and stop digitised signals with 1GS/s sampling rate is shown in figure 3. This

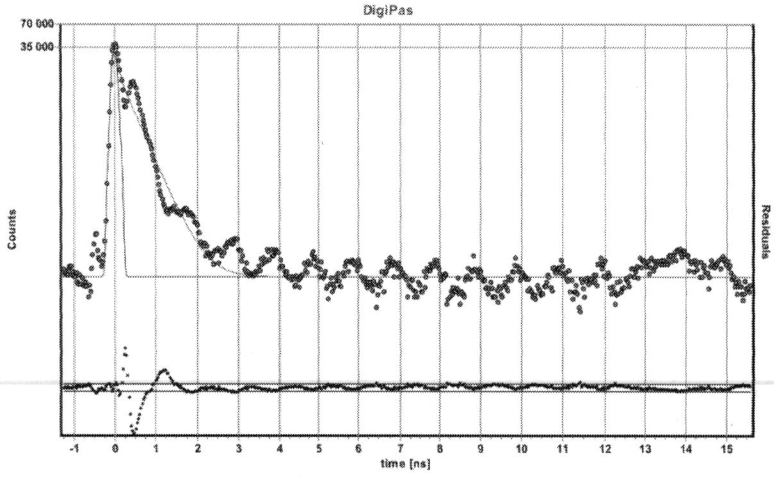

Figure 3. PAS LT spectra, Ni Sample, 2 channel mode, 1GS/s sampling rate – sinusoidal noise in spectrum.

spectrum was not usable for fitting in LT program [5]. Experiments in using different approximation methods with 1GS/s sampling rate and dynode outputs did not satisfy our requirements. Therefore the new setup with joined channel mode was used (figure 2B). Higher sampling rate was suitable to make precise approximation from more points in joined mode, but joining delayed two pulses together had disadvantages in noise increasing. Our dynode pulse has long periodic response; therefore at least 100ns delay between pulses was used. Regardless of long delay part of periodic response from start pulse was added to stop pulse and this brings the above mentioned noise to sampled stop pulse. The resulted spectrum obtained from joined channel mode can be seen in figure 4.

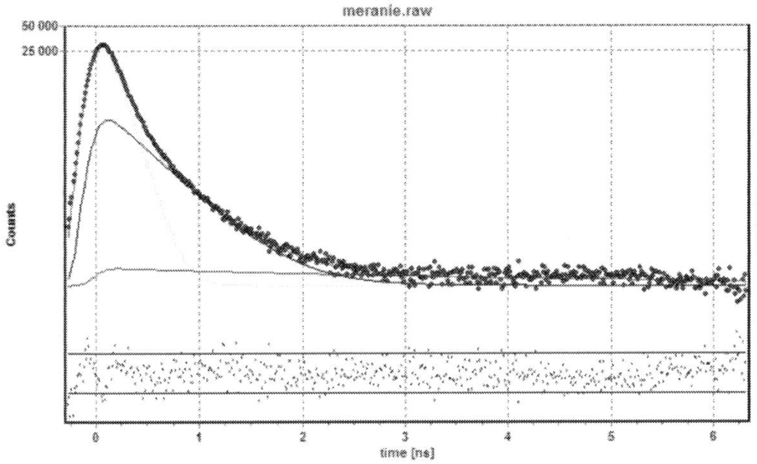

Figure 4. Ni spectra obtained with digital PALS and 2GS/s joined channels mode.

5. Results and discussion

Digital system with 2GS/s sampling rate in channel joined mode was tested in simultaneous measurement of two materials: Si and Ni with analogue and digital system. Results in table 1 were obtained by two components' fitting in program LT 9 [5]. A higher noise in digital system is the reason for joining two channels together.

Table 1. Results from simultaneous measurement on analogue and digital PAS LT system.

Setup	Material	τ_1 [ps]	τ_2 [ps]	I_1 [%]	I_2 [%]	FWHM [ps]	Bckg / signal
Digital	**Si**	**222.2** ± 2.0	1690±150	97.34±0.12	2.63±0.12	230.4	2.09%
	Ni	**110.6** ± 0.2	3910±370	98.81±0.05	1.19±0.05	214.9	2.89%
Analogue	**Si**	**222.9** ± 1.7	1609±43	97.40±0.08	2.60±0.08	244.4	0.05%
	Ni	**118.3** ± 0.5	3310±83	96.06±0.07	3.94±0.07	210.6	0.04%

6. Conclusion

The new data acquisition and analysis software was developed. It was confirmed, that the parameters of the Acqiris DP240 card and measurement detector system with BaF_2 and dynode outputs were not suitable to obtain significantly better results in comparison to the previous analogue measurement system. At least 2GS/s sampling rate was needed to have enough data to make pulse shape approximation without disturbing sinusoidal addition in output spectra. Results are comparable to others digital systems like [1] and [2]. Based on our current study, only acquisition cards with higher sampling rate (at least (4GS/s)) can be considered as suitable for better values of the time resolution and decreasing the spectrum background noise.

References

[1] Bečvář F, Čížek J, Procházka I and Janotová J 2005 The asset of ultra-fast digitizers for positron-lifetime spectroscopy *Nuclear Instr. Meth.* A **539** 372–385

[2] Nissilä J, Rytsolä K, Aavikko R, Laakso A, Saarinen K and Hatojärvi P 2005 Performance analysis of a digital positron lifetime spectrometer *Nuclear Instr. Meth.* A **538** 778–789

[3] Petriska M, Slugeň V, Zeman A and Kršjak V 2005 Three Detector PALA Set-Up-for Measurement of Irradiated Steels, presented at the *10th International Workshop on Slow Positron Beam Techniques for Solids and Surfaces (SLOPOS-10)* Doha, Qatar, March 19-25, 2005 (Book of Abstracts, p 75)

[4] Čížek J, Bečvář F and Procházka I 2000 Three-detector setup for positron-lifetime spectroscopy of solids containing ^{60}Co radionuclide *Nuclear Instr. Meth.* A **450** 325–337

[5] Kansy J 1996 Microcomputer program for analysis of positron annihilation lifetime spectra *Nuclear Instr. Meth.* A **374** 235–244

[6] Bardelli L, Poggi G, Bini M, Pasquali G and Taccetti N 2004 Time measurements by means of digital sampling techniques: a study case of 100 ps FWHM time resolution with a 100 MSample/s, 12 bit digitizer *Nuclear Instr. Meth.* A **521** 480–492

Journal of Physics: Conference Series **265** (2011) 012030 doi:10.1088/1742-6596/265/1/012030

Study of defect structure with new software for numerical analysis of PAL spectra

J Kansy and D Giebel

Institute of Material Science, Silesian University, Bankowa 12, 40-008 Katowice, Poland

E-mail: kansy@us.edu.pl

Abstract. The positron trapping model is implemented into the code of LT 9-2 version of LT software. This allows to search directly for the positron trapping parameters of lifetime spectra relating to samples containing defects of vacancy type. The method of analysis, together with the possibility of simultaneous analysis of many lifetime spectra, enables one to reduce, in comparison with the conventional way of analysis, the number of free parameters used by the fitting procedure. LT 9-2 is employed to analyze artificial spectra simulated according to 2- and 3-state trapping model and the results are discussed with regard to the spectrum statistics and the way in which the calculation process is conducted.

1. Introduction

The LT program [1], and especially its newest version (LT 9), is one of the most popular software for positron lifetime spectra analysis. Usually, it is its simplest option, i.e. deconvolution of the lifetime spectrum into a few exponential components, which is chosen by most users. However, the program offers more options that can be used when dealing with some particular problems.

In the present paper we describe a new version (LT 9-2) of LT program which is especially useful for defect structure analysis by positron annihilation lifetime spectroscopy. To its code, the program has implemented the two- and three-state trapping models, which in connection with a possibility of simultaneous analysis of many lifetime spectra allows to reduce the effective number of free parameters used by the fitting procedure. This version is well tested for both artificial (simulated) lifetime spectra as well as the experimental ones.

The aim of this paper is to show the advantages of LT 9-2 and encourage the users to employ it in their investigations of defects in materials by PALS.

2. Defect analysis, theoretical models

The LT 9-2 program offers two theoretical models that can be used in analyses of positron lifetime spectra originated from samples containing defects of vacancy types:

1. *The Conventional Model (CM)* defined as a linear combination of exponential lifetime components

$$S_{CM}(t) = \sum_i \frac{I_i}{\tau_i} \exp\left(-\frac{t}{\tau_i}\right) . \quad (1)$$

2. *The Simple Trapping Model (STM).* LT 9-2 implements well-known [2,3] two- and three-state trapping models. As the two-state trapping model is a particular case of the three-state one it is the latter that will be discussed below. The model takes into account positron trapping (without the possibility of de-trapping) for two different kinds of defects. The trapping and positron annihilation processes (figure 1) are described by the system of differential equations

$$\frac{dc_f}{dt} = (-\lambda_f - \kappa_1 - \kappa_2)c_f$$
$$\frac{dc_1}{dt} = -\lambda_1 c_1 + \kappa_1 c_f$$
$$\frac{dc_2}{dt} = -\lambda_2 c_2 + \kappa_2 c_f \quad (2)$$

with the initial conditions $c_f(0) = 1$ and $c_{1,2}(0) = 0$.

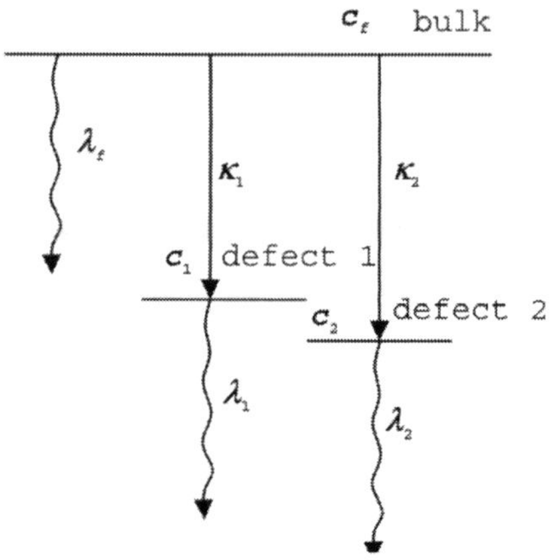

Figure 1. The positron trapping and annihilation processes assumed in the STP model.

κ_1 and κ_2 are the trapping rates of the two types of defects, λ_f, λ_1, λ_2 are the annihilation rates (reciprocal of lifetimes τ_f, τ_1, τ_2) in the bulk and in the defects and c_f, c_1, c_2 denote the probabilities of finding the positron (at instant t) in the bulk, in the first or in the second type of defects. So the shape of the lifetime spectrum is

$$S^{STM}(t) = \lambda_f c_f + \lambda_1 c_1 + \lambda_2 c_2 \quad (3)$$

where, according to the solution of equations (2)

$$c_f = \exp(-\lambda t),$$
$$c_i = \frac{\kappa_i}{\lambda - \lambda_i}\left[\exp(-\lambda_i t) - \exp(-\lambda t)\right] \quad i = 1,2,$$

and $\lambda = \lambda_f + \kappa_1 + \kappa_2$. (4)

It can be easily noticed that the three-state trapping model can be transformed into a sum of three exponential components

$$S^{STM} = S^{CM} = I_0 \lambda \exp(-\lambda t) + \sum_{i=1}^{2} I_i \lambda_i \exp(-\lambda_i t) \quad (5)$$

where λ is given by equation (4),

$$I_i = \kappa_i / (\lambda - \lambda_i) \text{ and } I_0 = 1 - I_1 - I_2, \quad (6)$$

so formally there is no difference between the description of lifetime spectrum by STM or CM. The advantage of using the STM can be gained if one simultaneously analyzes more than one spectrum measured for the same material after its different treatment (more detailed explanation will be given in section 5).

3. Resolution

The experimental lifetime spectrum $S_{ex}(t)$ is a convolution of the theoretical function $S(t)$ (given by equation (1) or equation (3)) with the apparatus resolution function $R(t)$

$$S_{ex}(t) = \int_0^\infty R(t - t_0) S(t_0) dt_0 . \quad (7)$$

The R function is approximated by a sum of a few Gaussians

$$R(t)=\sum_{i=1}^{n} I_i \frac{1}{\sqrt{2\pi}\sigma_i}\exp\left[-\frac{(t-t_{oi})}{2\sigma_i^2}\right] \tag{8}$$

Each Gaussian is described by its intensity I_i, the width $(FWHM)_i = 2\sqrt{\ln 2}\sigma_i$ and the center position t_{oi}.

4. Analysis of series of spectra

LT 9-2, similarly to a few previous versions of this program, can process a series of lifetime spectra simultaneously. Usually, many spectra in a series are described by the same parameters (e.g. the parameters relating to the resolution curve, to the source contribution, etc.). Such parameters are defined as *common* while the parameters relating to a single spectrum are called *local.* The introduction of common parameters reduces the amount of possible degrees of freedom in the mathematical problem, which makes the solution more unequivocal and reliable.

5. Defect structure analysis: Conventional model or STM?

As it was mentioned above, the descriptions of a single spectrum by CM (equation (1)) or by STM (equation (3)) are equivalent so an analysis of a single spectrum can be performed with any of the models with identical effectiveness. However, STM is more effective when we simultaneously analyze a series of lifetime spectra originating from samples of the same material after a thermal or mechanical treatment if the treatment does not change the structure of the bulk. In such a case the lifetime τ_f relating to positron annihilation in the bulk can be defined as a *common* parameter. CM does not depend directly on the bulk lifetime as the first lifetime in CM (equation (5)) depends not only on the positron lifetime in bulk but also on the trapping rates of defects (equation (4)), therefore it cannot be chosen as a common parameter. This means that for a series consisting of n spectra, STM requires by n-1 free parameters less than CM— that makes the analysis with STM advantageous.

6. Testing calculations for simulated spectra

Below we describe main features concerning the input data, typical program operation and the results.

6.1. 2-state-trapping model

The spectra used in the test were simulated according to the 2-state trapping model, (i.e. according to equations (3)-(4) at the assumption that $\kappa_2 = 0$). Each spectrum contained an additional component

Table 1. Values of the parameters used in simulations according to 2-state-trapping model and the respective values found by the fitting for both series of spectra of statistics 30×10^6 counts and 3×10^6 counts.

	sample		source		resolution function				background
parameter	τ_f [ps]	τ_1 [ps]	C_s [%]	τ_s [ps]	$(FWHM)_1$ [ps]	I_2 [%]	t_{01} [channels]	$(FWHM)_2$ [ps]	B [counts]
assumed values	120.0	200.0	10.0	386	270	40.0	5.00	350	600(30×10^6) 60 (3×10^6)
found values									
(30M)	120.2	200.0	9.9	387	272	38.1	5.14	348	599.8
(3M)	122.3	201.7	9.2	393	272	37.4	5.18	342	59.9

τ_f - bulk lifetime, τ_1 – lifetime in defect, c_s –source contribution, $t_{01,2}$- shifts of the centers of Gaussians, $(FWHM)_{1,2}$- widths of the Gaussians, I_2- contribution of the second Gaussian, B –background level.

imitating a "source contribution" and was convoluted with a "resolution function" (equation (7)) consisting of two shifted Gaussians. The spectra contained 2000 channels of width of 28 ps. Each spectrum was multiplied by 30 million (first series) or 3 million (second series) and a constant background (4% of total counts) was added. Eventually, the points in the smooth spectra were scattered by a Poisson noise. The values of the model parameters taken to simulations were common for all the spectra in a series except for the trapping rate (κ_1) which changed from spectrum to spectrum. The assumed values of the common parameters are listed in table 1, whereas figure 2 shows the various values of κ_1.

The analysis of the simulated spectra was carried out in the following steps: At the beginning only the first spectrum of the series was fitted with initial values of $\tau_f = 100$ ps, $\tau_1 = 180$ ps, $\kappa_1 = 1$ ns^{-1}, $c_s = 9\%$ and $\tau_s = 380$ ps. The values of other parameters (relating to background and the resolution function) were estimated by the program automatically. After the fitting the program displayed preliminary values of the searched parameters (figure 3). These values were used as the initial values in the common analysis of the whole series of spectra.

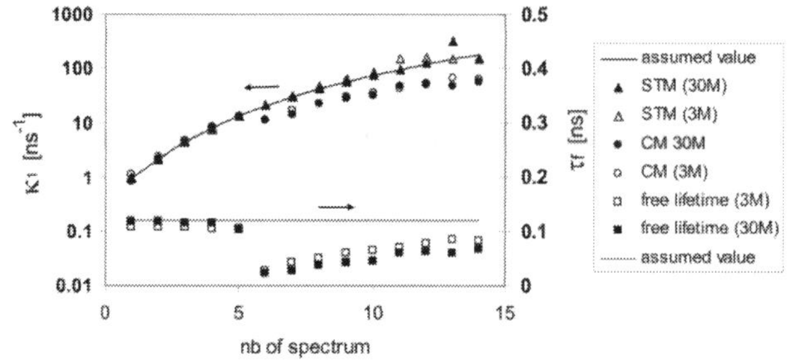

Figure 2. *Left axis*: The assumed (*lines*) and determined values (*points*) of κ_1 for the spectra of statistics 30 and 3 $\times 10^6$ counts (denoted by 30M and 3M), simulated according to the 2-state-trapping model. Triangles represent results obtained with STM and circles – with CM. *Right axis*: The assumed values of τ_f (*line*) and the values determined with CM.

	Value	Fixed	Global
LIFETIME of sample comp. no. 1	0.205280		yes
SOURCE CONTRIBUTION	9.000000		yes
LIFETIME of SOURCE comp. no. 1	0.380000		yes
Trapping const. KAPPA 1	1.013502		
FREE positron LIFETIME	0.126040		yes
FWHM	0.267400		yes
FWHM no. 2	0.348519		yes
SHIFT of ESG no. 2	4.693976		yes
FRACTION of ESG no. 2	41.14294		yes
Shift of "ZERO"	0.000000		
BACKGROUND	602.31381		

Figure 3. The parameters determined from analysis of first spectrum of series spectra with 30×10^6 counts) and their statuses before the final calculations (values of lifetimes - in ns).

Note: The best practice, which allows to avoid unnecessary arbitrariness in the analysis of positron lifetime data is to determine the instrument resolution function by a measurement of a reference sample with known lifetime(s) and kept it fixed in the subsequent analysis of lifetime spectra. It seems, however, that for a long series of spectra, measured in identical conditions, there is an alternative way of determining of the resolution function. It consists in choosing the parameters of the resolution function as *common* for the all analyzed spectra. One of the aims of the test described was to determine whether such a procedure proves itself successful in obtaining good results.

Before the common analysis started the status of each of the listed parameters was defined. All of the parameters were chosen as *common* except κ_1, t_{01} and B (figure 3) which were defined as *local.*

Table 1 contains the final values of *common* parameters obtained from the fitting and the assumed values of these parameters. The found values are very close to the assumed ones both for the series with 30×10^6 counts as well as for the series with 3×10^6 counts in a spectrum.

The found and assumed values of κ_1 are compared in figure 2. Some discrepancies between them are seen for $\kappa_1 > 80$ ns^{-1}. As expected, the discrepancies are much higher for spectra of lower statistics.

6.2. Conventional model

For comparison the same series of spectra (which were generated according to 2-state-trapping model) were analyzed with the help of CM. In this case the *local* parameters were $\tau = \lambda^{-1}$ (where λ is defined by equation (4)), I_0 and I_1 (equation (6)). The status of *common* had $\tau_1 = \lambda_1^{-1}$ as well as the source and resolution parameters. The found values of τ, τ_1, I_0 and I_1 were recalculated into τ_f and κ_1 on the basis of formulas which result from equations (4) and (6)

$$\tau_f = \frac{\tau\tau_1}{I_0\tau_1 + I_1\tau} \tag{9}$$

and

$$\kappa_1 = I_1 / I_0 \left(\tau_f^{-1} - \tau_1^{-1}\right) \tag{10}$$

The calculated values of τ_f are shown in figure 2 (right axis). Only a few initial values of τ_f (relating to the spectra with low trapping parameters) are close to their correct values. When κ_1 is higher than 10 the discrepancies between the found and correct values of τ_f are large. Therefore in calculations of κ_1 from equation (10) the best value of τ_f, i.e. the first one, was used. Such determined values of κ_1 are shown also in figure 2 (left axis). They are much lower than the respective correct values when κ_1 excesses ~10 ns^{-1}. The comparison of the resulting values of κ_1 presented in figure 2 depicts evidently the advantage of direct using the STM in analysis of lifetime spectra.

6.3. 3-state trapping model

This time a series of lifetime spectra was simulated according to the 3-state-trapping model (equations (3)-(4)). The simulation procedure was similar to those described in the previous example. Each spectrum statistics was 30×10^6 counts (first series) or 3×10^6 counts (second series). The source, resolution and background parameters were identical with those chosen in the previous case. The common parameters used in simulations are shown in table 2 and values of the trapping rates (κ_1 and κ_2) in figure 4.

We attempted fitting the data in several ways. It turned out that satisfactory results can be obtained

Table 2. Values of the common parameters used in simulations according to the 3-state-trapping model and the respective values found by the fitting.

	sample			source	resolution function				background
parameter	τ_f [ps]	τ_1 [ps]	τ_1 [ps]	c_s [%]	(FWHM)$_1$ [ps]	I_2 [%]	t_{01} [channels]	(FWHM)$_2$ [ps]	B [counts]
assumed values	120.0	200.0	300.0	10.0	270	40.0	5.00	350	600(30×10^6) 60 (3×10^6)
found values									
(30M)	120.8	200.6	304.3	10.0	270	40.2	4.95	350.8	599.8
(3M)	120.2	200.1	300.4	10.0	269.4	41.0	4.89	352.4	60.1

τ_f - bulk lifetime, τ_1 – lifetime in first type of defects, τ_2 – lifetime in second type of defects c_s –source contribution, $t_{01,2}$- shifts of the centers of Gaussians, (FWHM)$_{1,2}$- widths of the Gaussians, I_2- contribution of the second Gaussian, B –background level.

when the initial values of the model parameters are not too far from their right values and when the fitting procedure consists of several stages. The starting values of *source contribution* and *source lifetimes* are also crucial. Below we described a successful attempt:

- *Stage 1.* First spectrum (described by low trapping rates) is fitted with starting parameters τ_f = 100 ps, $\kappa_1 = \kappa_2$ = 0.1 ns^{-1} (free parameters), τ_1 = 190 ps, τ_2 = 320 ps, c_s = 12% and the correct value of τ_s = 386 ps (a fixed parameter). Starting values of the resolution and background parameters are estimated automatically by the program and treated as free ones.
- *Stage 2.* The whole series is included into common fitting with the previously obtained values of parameters as the starting values. The parameters τ_f, τ_1, τ_2, c_s and the resolution and background parameters have the *common* status, κ_1 and κ_2 are *local* while τ_s is still fixed at its correct value.

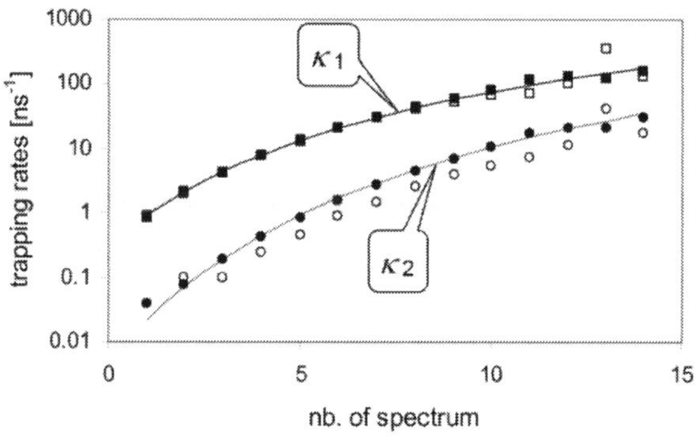

Figure 4. The assumed (*lines*) and determined values of κ_1 and κ_2 (*points*) for the first series of spectra (30×10^6 counts – *solid symbols*) and the second series of spectra (3×10^6 counts – *opened symbols*) simulated according to the 3-state-trapping model.

- *Stage 3.* The last spectrum of the series (described by the highest trapping rates) is analyzed separately. The fitting procedure starts from the current values of parameters obtained on stage 2. Only τ_1, τ_2, κ_1, κ_2 are free. All of the other parameters are fixed.
- *Stage 4.* The series is analyzed from the last spectrum to the first one with previously determined values of parameters. The statuses of parameters are chosen like in the stage 2.

Table 2 contains the final values of common parameters whereas the determined values of κ_1 and κ_2, compared to their correct values, are shown in figure 4. It is observed that the found values shown in table 2 differ only by about 1% of the correct values. Contrary to the expectations the lifetimes determined for the series with statistics 3×10^6 counts are even better than those obtained for the series with statistics 30×10^6 counts, although values of the resolution parameters are a bit worse. It seems that the small deviations of the resolution parameters are the reasons for the systematic shifts of κ_2 from its correct values observed for the series with statistics of 3×10^6 counts (figure 4). Similarly to the 2-state trapping model, some bigger discrepancies between expected and found values of κ_1 are observed for κ_1>80 ns^{-1}.

7. Conclusions

The testing calculations with the help of simulated spectra show that direct applying of the trapping model to the lifetime spectra analysis allows to obtain much better values of parameters characterizing the positron trapping process than in the case when the spectra are decomposed in the conventional way.

For the 3-state-trapping model the convergence of the fitting procedure is slow and its success strongly depends on the choice of values of initial parameters as well as on the details of the fitting process. Usually, the satisfactory results are achieved after many attempts.

Acknowledgement
Partial financial support of the Ministry of Science and Higher Education of Poland under the grant number 581/T/2006 is acknowledged.

References

[1] Kansy J 1996 *Nucl. Instr. Meth. A* **374** 235
[2] Bergersen B and Stott M J 1969 *Solid State Commun.* **7** 1203
[3] Connors D C and West R N 1969 *Phys. Lett. A* **30** 24

AUTHOR INDEX

Institute of Physics
Dirac House, Temple Back
Bristol BS1 6BE UK